SEMICONDUCTOR PHOTOELECTROCHEMISTRY

SEMICONDUCTOR PHOTOELECTROCHEMISTRY

Yu. V. Pleskov and
Yu. Ya. Gurevich

A. N. Frumkin Institute of Electrochemistry
Academy of Sciences of the USSR
Moscow, USSR

Translation Editor
P. N. Bartlett

University of Warwick
Warwick, England

CONSULTANTS BUREAU • NEW YORK AND LONDON

Library of Congress Cataloging in Publication Data

Gurevich, IU. IA. (IUrii IAkovlevich)
  Semiconductor photoelectrochemistry.

  Translation of: Fotoelektrokhimiia poluprovodnikov.
  The names of the authors on the 1983 Russian edition appear in reverse order.
  Bibliography: p.
  Includes index.
  1. Photochemistry. 2. Photoelecticity. 3. Semiconductors. I. Pleskov, IU. V. (IUrii
Viktorovich) II. Title.
  QD715.G8613   1985   541.3′5   85-17411
  ISBN 0-306-10983-2

The Russian text underlying this translation was prepared by the authors
especially for this edition. This translation is published under an agreement
with the Copyright Agency of the USSR (VAAP).

©1986 Consultants Bureau, New York
A Division of Plenum Publishing Corporation
233 Spring Street, New York, N.Y. 10013

Printed in the United States of America

Editor's Foreword

Interest in semiconductor electrochemistry has expanded rapidly in the past few years spurred on by the search for economic methods for solar energy conversion. Semiconductor electrochemistry and photoelectrochemistry is not, however, restricted to this single area of application. Over the same period there have been many interesting and exciting developments in other areas of semiconductor electrochemistry including the use of laser etching, electroreflectance studies and radiation electrochemistry. All of these areas are among those covered by Pleskov and Gurevich in this book. Their text is a comprehensive study of the electrochemistry and photoelectrochemistry of semiconductors and as such should prove a worthy successor to 'Electrochemistry of Semiconductors' by Myamlin and Pleskov which, since its publication in 1967, has established itself as a standard reference text for workers in this field.

In editing the English translation of the present text I have at all times attempted to maintain a clear, concise style while at the same time remaining true to the arguments and ideas of the authors. I hope that I have succeeded in this endeavor and I hope that this book will be welcomed by experts and beginners alike in the expanding area of semiconductor electrochemistry.

I would like to express my thanks to Christine Mahoney for typing the manuscript in its finished form and for uncomplainingly dealing both with my handwriting

and my various corrections. Finally I would like to
thank Ken Derham of Plenum Publishing for his help and
advice in the preparation of the manuscript.

P.N.B.

Preface

The task we set before ourselves is to present, from a unified standpoint, the whole diverse range of photoelectrochemical processes which occur at semiconductors in contact with electrolyte solutions. In recent years these processes have attracted added interest because of their great potential especially for energy conversion.

Our basic aim is to describe the advances in the theory and practice of the photoelectrochemistry of semiconductors, on the one hand, and to review the problems and prospects of this rapidly developing area, on the other. Nevertheless, we felt it would not be justifiable to confine ourselves strictly to photo-electrochemistry. Thus a second task arises, namely, to acquaint the reader with the state-of-the-art in the electrochemistry of semiconductors. As such, this monograph may be looked upon as a successor to the book Electrochemistry of Semiconductors by V.A. Myamlin and Yu.V. Pleskov published by Plenum Press in 1967. Our objective is thus twofold: to make the book self-contained as well as comprehensible even to a reader not knowledgeable in the field.

This dual objective has naturally molded in its own way the manner of presentation and the choice of contents. Part I (Chapters 1-4) is general: it gives a bird's eye view of the modern electrochemistry of semiconductors, or more precisely, a protracted introduction to Part II which deals with the photoelectrochemistry

proper (Chapters 5-11). Experimental data presented in
Part I primarily illustrates the fundamental concepts
of the electrochemistry of semiconductors. Therefore
the list of references is rather selective. It does,
however, include certain reviews which give a deeper
insight into the problems. In contrast, Part II outlines
in more detail the available experimental data. Accor-
dingly, the list of references is much longer, though
not complete (it largely lists the works published by
the end of 1980, while recent additions can be found
at the end).

We have given wide coverage to researchers working
in different countries on various aspects of the photoelectrochemical behavior of semiconductors. Special
mention should be made of the ideas of Professor H.
Gerischer that have had such a stimulating impact on
the advancement of this field as a whole. We would
also like to emphasize the contribution made by the late
J.F. Dewald and V. A. Tyagai to the development of the
photoelectrochemistry of semiconductors at different
stages.

We express our sincere gratitude to Professor L.I.
Krishtalik, Dr. Z.A. Rotenberg and Dr. S.F. Timashov for
the invaluable remarks they made while reading the
manuscript.

We hope that our efforts will contribute to the
general progress in the electrochemistry of semiconductors as well as to the solution of the urgent
practical problems facing science, in particular, and
mankind, in general.

                                        The Authors

Introduction

Photoelectrochemistry in general studies the processes occurring under the influence of illumination at all types of electrodes. The photoelectrochemistry of semiconductors is concerned with those photoelectrochemical processes which are determined by the specific features of the semiconductor electrode.

The reactions of photoexcited species in the solution at metal electrodes have long been traditional objects of research for photoelectrochemists. During the past two decades, however, the range of problems under examination has increased considerably. The large-scale invasion of new ideas, approaches, and experimental methods from many diverse spheres of modern physics, above all solid-state physics, has appreciably accelerated the rate of development of physical chemistry in general and of photoelectrochemistry in particular.

In recent decades the photoelectrochemistry of semiconductors was essentially developing under the influence of concepts from the physics of semiconductors and of surfaces. The increasing interest in the photoelectrochemistry of semiconductors in recent years has increased both the breadth and depth of this newest aspect of physical chemistry. Today the photoelectrochemistry of semiconductors concerns itself with the following processes:

(i) the electrode reactions on semiconductors determined by the photoexcitation of reagents in solution and in the adsorption layer; as mentioned above, this is a traditional sphere of photoelectrochemistry. However, the discharge processes of photoexcited reagents at semiconductor electrodes are distinguished by a number of specific qualitative features discussed below.

(ii) Electrode reactions determined by the photoexcitation of the electron ensemble of the solid; such reactions are highly specific to semiconductor electrodes on which they were analyzed for the first time, although the concept of the photoexcitation of electrons in a metal as the cause of photoelectrochemical effects had been mentioned in the literature.

Other features of the photoelectrochemistry of semiconductors are closely associated with the successes attained in the study of metal electrodes. Among these are the photoemission of electrons from semiconductors into solution and the electroreflectance of the semiconductor/electrolyte interface. However, these effects in the semiconductor/electrolyte system differ substantially from those in the metal/electrolyte system. Therefore mechanical transfer of results from one sphere to the other is out of the question.

Photoelectrochemistry also encompasses the study of processes which are the reverse of those described above. In particular, among such processes are the electrochemical reactions on semiconductors accompanied by emission of light in the optical and adjacent ranges of the spectrum. At present this sphere of research is in its infancy.

The progress in the study of the photoprocesses in electrochemical systems is indissolubly linked with the improvements in available light sources, in particular the laser which is finding its first uses in electrochemistry. It is precisely in the photoelectrochemistry of semiconductors that the most impressive achievements have been made. They are connected with the efficient use of lasers to obtain fine reliefs on the surface of semiconductors immersed in electrolyte solutions. With further development, the laser electrochemistry of semiconductors may become a source of valuable information of both a fundamental and an applied nature.

The applications of the photoelectrochemistry of semiconductors have become increasingly important in recent years. This is primarily true in the field of solar energy conversion. Although this research is

still at the development stage, great hopes are placed
on the construction of cheap and efficient photoelectrochemical cells. The research into the photoanodic etching
of semiconductors also seems quite promising. For example, photoelectrochemical methods can already compete
with traditional methods in recording holograms. Thus,
the photoelectrochemistry of semiconductors embraces an
ever increasing wide range of problems.

The history of the photoelectrochemistry of semiconductors is rooted in the 19th century. It is
difficult to cite the "very first" work on the photoelectrochemistry of semiconductors but this is apparently
the work by Becquerel (1) written almost 150 years ago.
He discovered the generation of an electric current when
one of the two electrodes submerged into a dilute solution of acid was exposed to light. Based on present-day
scientific knowledge it seems quite probable that the
"Becquerel effect" was, at least in some cases, caused
by the internal photoeffect in the semiconducting films
on the surface of the metal electrodes, although on
metals with a clean surface it was almost undoubtedly due
to photoemission.

Some notions which serve as the basis of the photoelectrochemistry of semiconductors date back to the
works dealing with the photoelectrochemical behavior of
adsorbed layers and oxidized metals. This work was
carried out in the 1940s (2). Semiconductor photoelectrochemistry as a branch of science began, as did the electrochemistry of semiconductors as a whole, in the mid-1950s
when Brattain and Garrett (3) succeeded in establishing
links between the electrochemical, and, in particular,
the photoelectrochemical properties of single crystal
semiconductors, and specific features of their electronic
structure.

At the same time, the photoelectrochemistry of semiconductors was based on the fundamental principles of
classical theoretical electrochemistry; the electrochemical school of A.N. Frumkin played a key role.

The photoelectrochemistry of semiconductors finally
became an independent branch of electrochemical physics*
late in the 1950s. This was largely due to works by

---

*This term is now officially adopted by the Interna-
tional Society of Electrochemistry to denote the fields
in electrochemistry which are most closely related with physics.

Dewald who brought to light the detailed mechanism of the
generation of a photopotential at a semiconductor electrode (4). These early stages in the development of the
electrochemistry and photoelectrochemistry of semiconductors are dealt with in (5).

In the mid-1960s there was large-scale interest in
optical and photoelectrical methods for the study of
electrochemical reactions and interfaces. This also
stimulated further development of the photoelectrochemistry of semiconductors, in as much as some of the non-
traditional methods required the use of semiconductor
electrodes. For example, the method of attenuated total
reflection requires the electrode to be transparent to
the light. At that time, the semiconductor/electrolyte
interface was first used to study electrooptical phenomena.
Subsequently, largely due to the works by Tyagai (6) it
was possible to obtain information not only about the
band structure of semiconductors, but also about the
structure of the double layer and the reactions at semiconductor electrodes. A fundamental step forward in the
development of the photoelectrochemistry of semiconductors
is connected with the works of Gerischer who proposed a
theory of semiconductor electrode photodecomposition (7).

Finally, the photoelectrochemistry of semiconductors
received a fresh and powerful impetus in the mid-1970s
soon after Fujishima and Honda (8) demonstrated the
photodecomposition of water into hydrogen and oxygen, and
thus the conversion of light into the chemical energy of
the products of electrode reactions, using a cell consisting of a semiconductor (titanium dioxide) and a metal
electrode immersed in an aqueous electrolyte. The impact
of this work was increased by the impending energy crisis
and highlighted the use of photoelectrochemical methods
for solar energy conversion. In 1975, Archer, the author
of the first review on the electrochemical aspects of
solar energy conversion, wrote: "There is very little
research work carried out on photoelectrochemical cells
today" (9). During the past 4-5 years, however, the
developments in this field can be compared to an av-
alanche. The characteristics of semiconductor photoelectrodes are now being studied in dozens of lab-
oratories in many countries, and many hundreds of papers
have been published. The progress in photoelectrochemistry stimulated further developments in electrochemistry of semiconductors in general. (It should be
pointed out that at the turn of the 1970s much less
interest was displayed in the electrochemistry of semiconductors because applications proved more modest than

had been expected.)   As a result, during the past few
years many more semiconductor materials have become
objects of electrochemical examination than during the
previous twenty years.   Present trends show that the
electrochemistry of semiconductors is now becoming, to
a considerable extent, the photoelectrochemistry of
semiconductors.

Thus, the photoelectrochemistry of semiconductors
and the electrochemistry of semiconductor materials are
going through their second spring.   A number of new
concepts have emerged, some previous concepts have been
filled with new, more profound content, new methods have
been elaborated and practical applications extended.

Contents

Notation[†]

$$a_{n,p}$$ — capture coefficient for electrons (n), holes (p).

c — velocity of light (3×10^8 ms^{-1}).

c_o — electrolyte concentration.

c_{ox}; c_{red} — concentration of oxidizing (reducing) agent in solution.

$c_{ox,s}$; $c_{red,s}$ — concentration of oxidizing (reducing) agent at the plane of discharge.

C — capacity.

C_{el} — capacity of the diffuse ionic layer in the electrolyte.

C_H — capacity of the Helmholtz layer.

C_{sc} — capacity of the space charge layer in the semiconductor.

d — spatial period of a holographic relief.

$D_{n,p}$ — diffusion coefficient for electrons (n), holes (p).

$\mathcal{D}_{ox}$ — density of unoccupied electron states in solution.

$\mathcal{D}_{red}$ — density of occupied electron states in solution.

$\mathcal{D}_{sc}^{occup}$ — density of occupied electron states in the semiconductor.

$\mathcal{D}_{sc}^{vacant}$ — density of unoccupied electron states in the semiconductor.

[†]To avoid repetition (for example, in denoting energies and electrode potentials) and also excessive use of indices, there are some slight departures from the IUPAC electrochemical nomenclature.

e — absolute value of the electronic charge (1.6×10^{-19} C).

E — electron energy.

$E^{\ddagger}$ — activation energy (in particular cases $E^{\ddagger} = E^a_{n,p}$; $E^c_{n,p}$).

$E_{A,D}$ — equilibrium energy level of an electron on an acceptor, donor.

$E_{A*,D*}$ — quasi-equilibrium energy level for an electron on a photoexcited acceptor (donor) in solution.

E_c — energy of the conduction band edge in the semiconductor bulk.

$E_{c,s}$ — energy of the conduction band edge at the surface.

E_F — Fermi energy of an electron in a metal and in a degenerate semiconductor.

E^o_{ox} — equilibrium energy level of an electron on an oxidizing agent in solution.

E^o_{red} — equilibrium energy level of an electron on a reducing agent in solution.

E_R — solvent reorganization energy.

E_{ss} — energy level of a surface state.

E_t — energy of a localized level in the semiconductor bulk.

E_v — energy of the valence band edge in the semiconductor bulk.

$E_{v,s}$ — energy of the valence band edge at the surface.

E_{vac} — potential energy of an electron in vacuum.

E — electrode potential.

E^o — equilibrium electrode potential.

E_{corr} — corrosion (mixed) potential.

E^o_{dec} — equilibrium potential for semiconductor decomposition reaction.

$E^o_{dec,n}$; $E^o_{dec,p}$ — equilibrium potential for semiconductor decomposition with the participation of electrons (n), holes (p).

E_{fb} — flat band potential.

E_{ph} — photopotential.

E^{oc}_{ph} — open-circuit photopotential.

E^o_{redox} — equilibrium potential of a redox reaction.

E^*_{redox} — quasi-equilibrium electrode potential under steady state illumination.

f — Fermi-Dirac distribution function.

f — fill factor.

F — electrochemical potential (Fermi level) of electrons in the semiconductor.

F_{dec}^{o} — electrochemical potential of electrons in solution in equilibrium with an electrode at the semiconductor decomposition potential.

$F_{dec,n}^{o}$; $F_{dec,p}^{o}$ — electrochemical potential for semiconductor decomposition reactions with the participation of electrons (n), holes (p).

F_{met} — electrochemical potential (Fermi level) of electrons in a metal.

$F_{n,p}$ — quasi-Fermi level of electrons (n), holes (p).

F_{redox} — electrochemical potential of electrons in a solution containing a redox couple.

F — Faraday constant (96480 $Cmol^{-1}$).

$g_{n,p}$ — rate of generation of electrons (n), holes (p).

G — free energy.

h — half depth of a holographic relief.

$\hbar$ — Planck's constant divided by 2π. (6.58 x 10^{-16} eVs).

i — $\sqrt{-1}$.

i — current density.

i^{*} — current density associated with excitation of reagents in solution.

i_{corr} — corrosion current density.

i_{dec}^{o} — exchange current density for semiconductor decomposition reaction.

i_{n} — electron current density.

$i_{n,p}^{o}$ — exchange current density through the conduction band (n), valence band (p).

$i_{n,p}^{a}$ — anodic current density through the conduction band (n), valence band (p).

$i_{n,p}^{c}$ — cathodic current density through the conduction band (n), valence band (p).

$i_{n,p}^{lim}$ — limiting (saturation) current density for electrons (n), holes (p).

i_{p} — hole current density.

i_{ph} — photocurrent density.

i_{redox}^{o} — exchange current density for a redox reaction.

$i_{sh.c}$ — short circuit current density of a photocell.

I_{p} — photoemission current density.

I_{T} — thermoemission current density.

J_{inc} — intensity of the incident light flux.

J_{o} — intensity of the light flux entering the semiconductor.

k — Boltzmann constant (8.6 x 10^{-5} eVC^{-1}).

k — light absorption coefficient.

$k^a_{n,p}$; $k^c_{n,p}$ — rate constants for reactions corresponding to the currents $i^a_{n,p}$, $i^c_{n,p}$.

L_D — Debye length in the semiconductor.

L_{el} — thickness of the space charge layer in the electrolyte solution.

L_H — thickness of the Helmholtz layer.

$L_{n,p}$ — diffusion length for electrons (n), holes (p).

L_{sc} — thickness of the space charge layer in the semiconductor.

m_o — mass of the electron (9.11×10^{-31} kg).

m_c — effective mass of the electron in the conduction band.

m_v — effective mass of the hole in the valence band.

M — current multiplication factor.

n — concentration of electrons in the conduction band.

n — refractive index.

n_o — equilibrium concentration of electrons in the semiconductor bulk.

n^* — quasi-equilibrium concentration of electrons in the illuminated semiconductor.

n_s — surface concentration of electrons.

n^o_s — equilibrium surface concentration of electrons.

n — complex refractive index.

$N(E)$ — number of photons of energy E incident on unit surface area of the semiconductor.

$N_{c,v}$ — density of states in the conduction band (c), valence band (v).

$N_{D,A}$ — concentration of donor (D) or acceptor (A) impurities in the semiconductor bulk.

N_{ss} — concentration of surface states.

N_t — concentration of recombination centers in the semiconductor bulk.

p — concentration of holes in the valence band.

P_o — equilibrium concentration of holes in the valence band.

p^* — quasi-equilibrium concentration of holes in the illuminated semiconductor.

p_s — surface concentration of holes.

p^o_s — equilibrium surface concentration of holes.

P — light flux power entering the semiconductor.

q — coefficient of modulation of a holographic relief.

$q_{i,f}$ — solvent coordinates corresponding to the initial (i) or final (f) states.

Q_{sc} — space charge in the semiconductor per unit area.

R — resistance.

R — light reflection coefficient.

$R_{n,p}$ — rate of generation of electrons (n), holes (p).

s — surface recombination velocity.

s_{el} — electrochemical recombination velocity.

t — time.

T — absolute temperature.

$u_{n,p}$ — mobility of electrons (n), holes (p).

$U_{i,f}$ — terms for initial (i) or final (f) states.

v — velocity.

w_p — photoelectric work function.

w_T — thermodynamic work function.

x — coordinate normal to the interface.

Y — quantum yield.

Y — dimensionless potential ($Y=e\phi_{sc}/kT$).

z — charge number.

α — linear light absorption coefficient ($\alpha=2k\omega/c$).

$\alpha_{n,p}$ — transfer coefficient for a cathodic reaction occurring through the conduction band (n), the valence band (p).

δ — thickness of the reaction layer.

δ_D — thickness of the diffusion layer in solution.

$\Delta n, \Delta p$ — deviation of electron (n) or hole (p) concentration from the equilibrium value.

ΔE_{ext} — voltage externally applied to a photoelectrochemical cell.

$\Delta\phi_{Demb}$ — Dember potential.

$\hat{\varepsilon}=\varepsilon_1+i\varepsilon_2$ — complex dielectric constant.

ε_o — permittivity of free space (8.86×10^{-14} Fcm^{-1}).

ε_{el} — static dielectric constant of the solution.

ε_{sc} — static dielectric constant of the semiconductor.

ζ — light flux modulation coefficient.

η — efficiency of light energy conversion.

η — overvoltage.

$\eta^{a,c}$ — anodic (a) or cathodic (c) overvoltage.

η_{el} — overvoltage in the Gouy-Chapman layer.

η_H — overvoltage in the Helmholtz layer.

η_{sc} — overvoltage in the space charge region.

θ — angle of incidence of light.

κ — wave number.

λ — wavelength of light in vacuum.

λ_B — de Broglie wavelength.

λ_n — wavelength of light in a medium of refractive index n.

μ_i — chemical potential of species i.

$\tilde{\mu}_i$ — electrochemical potential of species i.

ξ — electric field.

ρ — charge density.

$\rho(E)$ — density of electron states.

σ — conductivity of a semiconductor.

σ — dispersion of a statistical distribution.

$\tau_{c,v}$ — time taken to establish thermodynamic equilibrium between carriers in the conduction band (c), in the valence band (v).

τ_{cv} — time taken to establish thermodynamic equilibrium between the bands.

$\tau_{n,p}$ — lifetime of electrons (n), holes (p).

$\tau_{n,s}$; $\tau_{p,s}$ — reaction time for surface processes with the participation of conduction band electrons (n); holes (p).

ϕ — electric potential.

$\phi(el)$ — electric potential in the solution bulk.

$\phi(el/sc)$ — Galvani potential at the semiconductor/electrolyte interface.

$\phi(sc)$ — electric potential in the semiconductor bulk.

ϕ_{el} — potential drop in the diffuse ionic layer of the electrolyte.

ϕ_H — potential drop in the Helmholtz layer.

ϕ_{sc} — potential drop in the space charge region of the semiconductor.

χ — electron affinity of the semiconductor.

ψ — Volta potential.

ψ' — psi-prime potential.

ψ — electronic wave function.

ω — angular radiation frequency.

ω_0 — threshold frequency ("red boundary").

ω_1^0 — librational frequency characterizing the solvent.

ω_p — plasma frequency.

Indices

Superscripts: o — equilibrium
                    * — excited
                    a — anodic
                    c — cathodic
                lim — limiting
                max — maximum

Subscripts:
- A — acceptor
- c — conduction band
- corr — corrosion
- D — donor
- dec — decomposition of semiconductor
- el — electrolyte
- H — Helmholtz layer
- met — metal
- n — electron
- ox — oxidizing agent
- p — hole
- ph — photo
- red — reducing agent
- redox — redox process
- s — surface
- sc — semiconductor
- t — local levels in the semiconductor bulk
- v — valence band

Chapter 1

The Fundamentals of Semiconductor Physics

The passage of electric current in physicochemical and, in particular, in electrochemical systems under the influence of, for example, an external electric field, illumination, changes in the composition of the solution or other such processes is accompanied by the movement of charge carriers. In metal electrodes these carriers are electrons, while in semiconductor electrodes this role is played by electrons and positively charged electron holes. These two classes of materials with electronic (in the broad sense of the word) conductivity are called conductors of the first type. Electrolyte solutions (electrolytes for short) in which the charge is carried by ions are called conductors of the second type.

Thus, the semiconductor/electrolyte interface is a contact between conductors of the first and the second types. Unlike the metal/electrolyte interface the current may be carried by the mobile charges of both signs on each side of the contact.

In examining the processes occurring at the semiconductor/electrolyte interface effective use is made of the qualitative notions and mathematical formalism which are well-developed in the physics of semiconductors in general and in the physics of the semiconductor surfaces in particular. For this reason, it is expedient to discuss some of the electrical and optical characteristics of semiconductors, as well as their surface properties, before considering the photoelectrochemical phenomena at the interface.

§ 1.1. General Characteristics of Semiconductor Materials

Semiconductors are substances with electronic conductivity which has an intermediate value between the conductivity of metals (10^6 - 10^4 ohm^{-1}·cm^{-1}) and dielectrics (less than 10^{-10} ohm^{-1}·cm^{-1}). It should be borne in mind, however, that the value of the conductivity in itself is not sufficient to testify to the semiconductor nature of the material. For example, it is necessary to make a distinction between semiconductors and semimetals. The latter also exhibit electronic conductivity within the limits mentioned above. Without going into a detailed discussion of the differences in the physical nature of the materials in question (10 - 13), we will only emphasise the main feature of semiconductors. That is that in the energy spectrum of those electrons which determine the most important physical and physicochemical properties there is a gap. It is this bandgap which determines the main features peculiar to semiconductors (10-17) and, in particular, to their electrical, optical and electrochemical properties (5, 18-21).

It should also be pointed out that while the electronic structure of metals is radically different from the electronic structure of semiconductors, the difference between semiconductors and dielectrics (insulators) is of quantitative rather than qualitative character. Like semiconductors, dielectrics have a bandgap. That is why a number of their optical, thermal and photoelectrochemical properties prove to be rather similar. In view of this, one could speak of non-metallic substances as a whole without singling out semiconductors as a special class. Indeed as the temperature is increased the electronic conductivity of dielectrics can become comparable with that of semiconductors. However, the term "semiconductor" is often interpreted in a more limited sense as a member of a group of typical materials whose semiconducting properties are clearly seen even at room temperature (300 K).

Germanium (Ge) and silicon (Si), elements of group IV of the periodic table, are typical semiconductors. Atoms of these elements possess four valence electrons and form crystalline lattices of the diamond type with covalent bonding between atoms. The elementary cell of the diamond type lattice is shown in Fig 1.1.

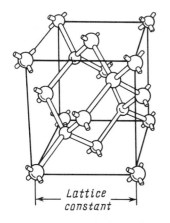

Lattice constant

Fig. 1.1. Crystalline structure of diamond.

Semiconducting compounds formed by the elements of group III of the periodic system (Ga, In) with the elements of group V (P, As, Sb) are referred to as $A^{III}B^V$ semiconductors (GaAs, InSb, GaP, InP, etc.). They also form crystalline lattices of the diamond type. The only difference is that the nearest neighbors of any A^{III} atom are all B^V atoms and vice versa. Due to partial redistribution of electrons the A^{III} and B^V atoms are oppositely charged in this structure. This is why the bonds in $A^{III}B^V$ crystals are partially ionic rather than completely covalent. The binary compounds of the elements of groups II and VI of the periodic table, $A^{II}B^{VI}$ (ZnTe, ZnSe, ZnO, CdS, CdTe), are also semiconductors with a diamond-like structure, although the ionic character of their bonds is even more pronounced. More complicated semiconductor compounds $A^{II}B^{IV}C_2^V$ (for example, $ZnSeP_2$, $CdGeAs_2$) have a similar structure. Many of these semiconductors form alloys which are themselves semiconductors: for example, Ge—Si, GaAs—GaP and so on.

Some elements of group VI, selenium (Se) and tellurium (Te) and also compounds of the type $A^{IV}B^{VI}$ (for example, PbS, PbTe, SnTe) which on average have five valence electrons per atom, also display semiconducting properties. Many compounds of the elements of group VI, with the elements of groups I-V (Cu_2O, Bi_2Te_3 being the most thoroughly studied examples), are semiconductors, as are the compounds of the elements of group VI with transition and rare-earth metals (TiO_2, WS_2, NiS, $MoSe_2$ and others). A separate group, which will not be dealt with in this monograph, is made up by organic and liquid semiconductors (see, for example, (22-24)).

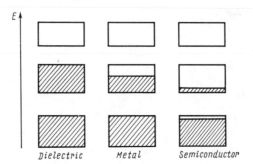

Fig. 1.2. The scheme of filling by
electrons of the allowed energy bands
in a dielectric (insulator), a metal
and a semiconductor. The rectangles
represent schematically the allowed
bands. The hatching shows the energy
regions in the allowed bands filled
with electrons.

The quantum theory of solids presents a complete
and rigorous description of the nature of current carriers in semiconductors and of the laws governing their
motion. This theory is based on the concept of elementary excitations of the many body quantum system. Its
important results can be summed up as follows:

(i)   The energy spectrum of electrons in an ideal
crystal consists (see Fig. 1.2) of intervals of energy
filled with energy levels (allowed bands) separated by
intervals in which there are no electronic energy levels
(forbidden bands, or bandgaps)[*].

(ii)   At absolute zero the electrons fill the lowest
energy levels. Depending on the number of electrons,
they fill several of the lowest allowed bands leaving
the higher bands empty. A crystal in which the lower
bands are completely filled at T=0, whilst the higher
bands are completely empty, is a dielectric (Fig. 1.2).
The metallic electronic structure occurs when at least
one allowed band is only partially filled with electrons

---

[*]According to the principles of quantum chemistry, the allowed
bands emerge as a result of the splitting of the electronic orbitals
of individual atoms (ions) due to the formation of the crystal.
This is why the corresponding allowed bands, like the initial orbitals, are sometimes called s-, p-, d-, etc. bands.

at T=0. The upper of the filled bands is called a valence band, while the lower of the unfilled bands is
called a conduction band. It is the distribution of
electrons in these two bands which exerts a decisive influence on the properties of the crystal. (Note that in
a broader sense all filled bands may be called valence
bands, and all unfilled ones conduction bands.)

(iii) Within each band different states of the
electron are characterized both by their energy E and
their quasi-momentum **p**. Quasi-momentum is an analogue
of the usual momentum associated with free particles and
can assume any value within certain limits in the three-
dimensional momentum space. The value of E is an unambiguous function of **p**. The dependence E(**p**) is called
the dispersion law and is specific to each crystal and
to each of its energy bands.

The electronic conductivity of semiconductors as
expected from the band structure can be caused by the
electrons of atoms of the basic substance (intrinsic
conductivity) and by the electrons of impurity atoms
(extrinsic conductivity). Apart from doping with impurities, the shortage or excess of atoms of one of the
components in a semiconductor compound (deviation from
stoichiometry) or the presence of other defects in the
crystal lattice, for example, vacancies, interstitials
and so on, may be sources of current carriers.

In intrinsic semiconductors at T > 0 the generation
of current carriers occurs as a result of the thermal excitation of some of the electrons from the valence (filled)
band to the conduction band with the corresponding thermal
rupture of some of the chemical bonds. Simultaneously,
an equal number of positively charged holes are created
in the valence band (see Fig. 1.3). These holes, like
electrons in the conduction band, may participate in the
flow of current. In the electric field the holes behave

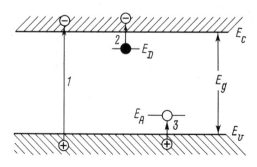

Fig. 1.3. The types of thermal
electron transitions giving rise
to electrical conductivity in
semiconductors; 1 - the emergence of an electron-hole pair;
2 - donor ionization (n-type
semiconductor); 3 - capture of
a valence electron by an acceptor
(p-type semiconductor).

like particles possessing a positive charge equal in absolute value to the charge of the electron. They are
also characterized by the definite values of their quasi-
momentum, and thus, like electrons in the solid, they may
be regarded as charged quasi-particles.

The mechanism underlying extrinsic impurity conductivity is somewhat different. Impurities and defects
are subdivided into donors and acceptors. Donors give
up excess electrons to the conduction band of a semiconductor, thereby creating electronic conductivity, in the
narrow sense of the word (n-type semiconductors). Acceptors capture valence electrons from atoms of the basic
substance (into which they have been introduced) producing hole conductivity (p-type semiconductors) (see
Fig. 1.3).

Atoms of the group V elements (P, As, Sb) in Ge and
Si are typical examples of donors. When such an atom is
introduced into the crystal lattice, it substitutes for
the host atom of Ge or of Si on one of the lattice sites.
At the same time, four out of its five valence electrons
form covalent bonds with the neighboring atoms of the
host substance. The remaining electron is then in "excess" of the bonding requirements for this lattice.
Without localizing itself in any elementary cell, it becomes a conduction electron leaving the impurity atom
(donor) with a single positive charge.

Atoms of the group III elements (B, Al, Ga, In) are
typical acceptors in Ge and Si. Capturing one of the
valence electrons of Ge or Si in addition to their three
valence electrons, they form four covalent bonds with
their close neighbors in the lattice and are converted
into negatively charged ions. At the spot where the
electron was captured a hole is left, which, like the
electron produced by the donor impurity, becomes a charge
carrier but now of positive sign and in the valence band
of the semiconductor. A similar explanation applies in
the case of the $A^{III}B^{V}$ compounds; thus impurities of some
group VI elements (S, Se, Te) have a donor effect replacing a B^{V} atom whilst the group II elements (Zn, Cd)
have an acceptor effect by replacing the A^{III} atom.
Furthermore, donor and acceptor impurities may also be
multilevel. For example, Zn in Ge is a double acceptor
because in addition to its two valence electrons it can
capture two more, thereby creating two holes. Cu and Au
may exist in Ge in a singly, doubly or triply charged
state, forming one, two or three holes respectively.

All of the above examples refer to substitutional impurities. Li in Si serves as an example of an interstitial impurity. While being located on interstitial sites, it readily gives up its valence electron behaving as a typical donor. In many $A^{IV}B^{VI}$ semiconductors the vacancies of A^{IV} atoms are sources of holes, while the vacancies of B^{VI} atoms are sources of conductivity electrons. It is clear from what has been said that doping is an efficient method of obtaining semiconductors with required electrical properties. It is precisely these doped semiconductor materials (i.e., semiconductors with extrinsic conductivity) that are of immense interest for photoelectrochemical studies.

Let us note that at a sufficiently high (more than 10^{18} - 10^{19} cm^{-3}) doping level (i.e., in heavily doped materials), qualitative changes in the properties of semiconductors occur. These are connected in particular with the formation of narrow bands; furthermore at such high values of carrier concentration the effects of degeneracy (metallization) (11, 15-16) start to become important.

Concluding the discussion of the links between the electronic structure and physicochemical characteristics of semiconductor materials we now examine the dependence of the width of the bandgap on the directly measured properties of a semiconductor. The width (thermal) of the bandgap $E_g = E_V - E_C$ which, as we have seen, is the most important characteristic of the crystal, directly depends on the strength of the chemical bonds. The heat of formation of the crystal (per bond), the difference in electronegativities of the A and B components of a compound, the shortest distance between atoms and certain other quantities have been chosen to characterize the bond strength (see, for example, (25)). Experimental data demonstrates that there is a definite correlation between E_g and these quantities: the width of the bandgap tends to become larger as the heat of atomization and the difference in electronegativities increases, and to decrease as the interatomic distance increases (see Fig. 1.4). Although these trends are not universal and are of semi-empirical character, they may be useful in certain cases, for instance when the values of E_g of new materials need to be estimated.

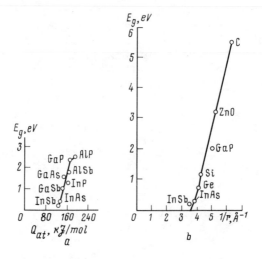

Fig. 1.4. The dependence of the
width of the bandgap, E_g, on (a)
the heat of atomization per bond
and on (b) the inverse interatomic
distance.

Ξ 1.2.   E q u i l i b r i u m   C o n c e n t r a t i o n s
         o f   C u r r e n t   C a r r i e r s   i n
         S e m i c o n d u c t o r s

In the absence of external effects (illumination,
electric field, etc.) the equilibrium concentrations of
electrons and holes in a semiconductor are determined
by temperature, the width of the bandgap, the concentrations of defects and impurities and, finally, the
dispersion laws, $E(p)$, for the conduction and the valence bands. Let us note that the current carriers in
semiconductors are concentrated, in general, in rather
narrow energy regions: electrons -- near the lower
edge (bottom) of the conduction band, E_c, in an energy
range of the order of kT - the energy of thermal movement where T is the absolute temperature and k the
Boltzmann constant (at room temperature, T = 300 K and
kT = 0.025 eV); holes -- within the same range $\sim$ kT,
but near the upper edge (top) of the valence band E_v.
Within these narrow energy ranges the complex dependences
of $E(p)$ usually becomes simpler.

In the simplest case they can be presented for the
conduction band and for the valence band as follows:

$$E = E_c + p^2/2m_c \quad ; \quad E = E_v - p^2/2m_v \qquad (1.1)$$

where m_c and m_v are the so-called effective masses of
electrons in the conduction band and of holes in the
valence band respectively. These are not, in general,
the same as m_0, the mass of a free electron, and depending
upon the semiconductor the values of the effective
masses can vary from one hundredth of m_0 to several
m_0. Using Eq. (1.1), it is possible to calculate the
equilibrium distribution of the charge carriers in the
bands. Electrons as particles (quasi-particles) with
half-integral spin are described by Fermi-Dirac statistics. The probability of the state with energy E being
occupied by an electron is given by the Fermi-Dirac function:

$$f(E,T) = \left[1 + exp \left(\frac{E - F}{kT} \right) \right]^{-1} \qquad (1.2)$$

The value of F, the electrochemical potential or Fermi
level for the electrons, is in general also dependent
on temperature. From general thermodynamic principles
it follows that the electrochemical potential can be defined as an increment in the free energy of the system
when one particle, an electron in this particular case,
is added under conditions where the pressure and temperature are kept constant (26-28). Without discussing
in detail the thermodynamic meaning of F, we shall regard this value as a certain characteristic energy depending on the nature of the semiconductor, its
composition, temperature, and so on. It follows from
Eq. (1.2) that the Fermi level may be defined formally
as the energy of that quantum state for which the probability of occupation is one half.

Taking into account Eq. (1.2), the thermodynamic
equilibrium concentration, n_0, of electrons in the conduction band is

$$n_0 = \int_{E_c}^{\infty} \rho_c (E) f(E,T) dE \qquad (1.3)$$

where $\rho_c(E)$ is the density of states in the conduction
band. The value of $\rho_c(E) dE$ represents the number of
quantum states which corresponds to the interval of

energies from E to E + dE per unit volume of the crystal. It can be shown that (26-27) the number of quantum states in a unit volume of a gas of quasi-particles which are characterized by a quasi-momentum p is $d^3p/(2\pi\hbar)^3$ where $2\pi\hbar$ = h, Planck's constant. Here the phase volume d^3p is equal to the volume of a spherical layer $d^3p = 4\pi p^2 dp$. Using the relation in equation (1.1) for p = $|p|$ we find

$$p = \sqrt{2m_c(E-E_c)} \; ; \; dp = (m_c/2)^{1/2} (E-E_c)^{-1/2} \, dE \quad (1.4)$$

Introducing a factor of two to take into account the two possible spin orientations, we obtain

$$\rho_c(E) = \frac{1}{2\pi^2\hbar^3} (2m_c)^{3/2} (E-E_c)^{1/2} \quad (1.5)$$

It should be noted that in Eq. (1.3) the energy of the upper edge of the conduction band should be taken as the upper integration limit. However, since the function f (E,T) diminishes rather rapidly with increasing E for values of E > F, taking the upper limit to infinity does not significantly change the value of the integral.

Substituting Eqs. (1.2) and (1.5) into Eq. (1.3), using a new integration variable z = $(E-E_c)/kT$ and introducing the dimensionless quantity ζ = $(F-E_c)/kT$, we can represent Eq. (1.3) as

$$n_o = N_c \, \Phi_{1/2}(\zeta) \quad (1.6)$$

where

$$N_c = 2[2\pi m_c kT/(2\pi\hbar)^2]^{3/2}$$

is the effective density-of-states in the conduction band, and the dimensionless function

$$\Phi_{1/2}(\zeta) = \frac{2}{\sqrt{\pi}} \int_0^\infty \frac{z^{1/2}}{1 + exp(z-\zeta)} \, dz \quad (1.7)$$

is the Fermi-Dirac integral with index 1/2. For large negative values of ζ the integral (1.7) can be evaluated analytically

$$\Phi_{1/2}(\zeta) = \begin{cases} e^{\zeta} & \zeta < 0, \ |\zeta| \gg 1 \\ \dfrac{4}{3\sqrt{\pi}} \zeta^{3/2} & \zeta \gg 1 \end{cases} \tag{1.7'}$$

Interpolation formulae and tables of values have been published for $\Phi_{1/2}(\zeta)$ (29-31).

Similarly the expression for the thermodynamic equilibrium concentration of positive holes in the valence band, p_o, may be readily obtained. Taking into account Eq. (1.2), the probability of a given quantum state with energy E not being occupied by an electron is given by the formula

$$(1-f) = \left[1 + exp\left(\frac{F-E}{kT} \right) \right]^{-1} \tag{1.8}$$

whilst for the density of states $\rho_v(E)$ near the edge of the valence band using Eq. (1.1), instead of Eq. (1.5), we obtain

$$\rho_v(E) = \frac{1}{(2\pi\hbar)^3} (2m_v)^{3/2} (E_v - E)^{1/2} \tag{1.9}$$

Thus, the equilibrium concentration p_o of the holes in the valence band

$$p_o = \int_{-\infty}^{E_v} f(E,T)\rho_v(E)\,dE \tag{1.10}$$

may be written as

$$p_o = N_v \Phi_{1/2}(\zeta') \tag{1.11}$$

where

$$N_v \equiv 2[2\pi m_v kT/(2\pi\hbar)^2]^{3/2}$$

is the effective density-of-states in the valence band and $\zeta' \equiv (E_v - F)/kT$.

The expressions for the concentrations (1.6) and (1.11) are considerably simplified for the most important case of a non-degenerate semiconductor. Under these circumstances the Fermi-Dirac distribution (1.2) turns into the Boltzmann distribution

$$f(E,T) = exp\left(\frac{F-E}{kT}\right) \tag{1.12}$$

It follows from the comparison of Eqs. (1.2) and (1.12) that a semiconductor is not degenerate if the Fermi level F lies within the band gap and is several kT from the band edges E_C and E_V. In this case, when n_0 and p_0 are calculated it is only the "tail" of the Fermi-Dirac distribution that is important, and this "tail" may be approximated by the Boltzmann distribution.

For the non-degenerate semiconductor, in formula (1.6) we may assume that, for all z, $exp\ (z - \zeta) \gg 1$. Then, the Fermi-Dirac integral will be as follows (see also Eq. (1.7'))

$$\Phi_{1/2}(\zeta) = e^\zeta \frac{2}{\sqrt{\pi}} \int_0^\infty e^{-z} z^{1/2}\ dz = e^\zeta \tag{1.13}$$

and, consequently,

$$n_0 = N_C\ exp\left(\frac{F-E_C}{kT}\right) \tag{1.14}$$

Similarly, for the holes we obtain

$$p_0 = N_V\ exp\left(\frac{E_V-F}{kT}\right) \tag{1.15}$$

The expressions (1.14) and (1.15) explain the meaning of the term "effective density of states." The exponential factor in Eq. (1.14) formally describes the probability of filling a quantum state that has a single energy E_C. Equation (1.14) shows that for a non-degenerate semiconductor the concentration of mobile electrons from a continuous distribution of states in the band is just the same as that given by a model in which each unit volume contains N_C states all of energy E_C. Similarly, Eq. (1.15) shows that when the concentration of holes is calculated, the valence band can be replaced with a set of states of energy E_V, and with the number of states per unit volume N_V.

The substitution of numerical values for the constants in the expressions for N_c and N_v gives

$$N_{c,v} = 2.5 \times 10^{19} \left(\frac{m_{c,v}}{m_o} \right)^{3/2} \left(\frac{T}{300} \right)^{3/2} cm^{-3}$$

It is possible to show (see, for example, (15,16)) that the formulae (1.14) and (1.15) do not essentially change when the dispersion law for the bands is more complicated than (1.1), and with an appropriate choice of the density of states they are widely applied.

The product of the concentrations of electrons and holes for a non-degenerate semiconductor does not depend on the position of the Fermi level or the conductivity type. In accordance with Eqs. (1.14) and (1.15)

$$n_o p_o = n_i^2 = N_c N_v \, exp \, (-E_g/kT) \tag{1.16}$$

where n_i is the concentration of electrons (or holes) in the intrinsic semiconductor, i.e., when $n_o = p_o$. For example, at T = 300 K for silicon $n_i = 1.5 \times 10^{10} cm^{-3}$, and for germanium $n_i = 2.5 \times 10^{13} cm^{-3}$.

Equation (1.16) is also valid for non-degenerate semiconductors irrespective of the concentrations of donors or acceptors, although the values of n_o and p_o naturally vary considerably with the doping density. Furthermore Eq. (1.16) is similar to the law of mass action and is directly analogous to the equation relating for example, to the equilibrium concentrations of H^+ and OH^- ions in water since these concentrations are small when compared with the concentration of non-dissociated molecules. In this case the value of n_i^2 plays the role of a chemical equilibrium constant.

Unlike the product $n_o p_o = n_i^2$, the ratio n_o/p_o, as is seen from Eqs. (1.14) and (1.15), is mainly determined by the position of the Fermi level F with respect to the band edges. This, in turn, depends on the temperature and doping. In the bulk of the spatially homogeneous semiconductor the position of F can be deduced by application of the conservation principle to the total number of electrons. In other words, from the condition of electroneutrality

$$n_O + N_A^- = p_O + N_D^+ \tag{1.17}$$

where N_D^+ is the concentration of the positively charged ionized donors and N_A^- is the concentration of negatively charged acceptors which have captured electrons. In the general case the position of F is determined by the transcendental equation which results from Eq. (1.17) after substitution from Eqs. (1.6) and (1.11) for n_O and p_O and with

$$N_D^+ = N_D (1-f_D) \quad ; \quad N_A^- = N_A (1-f_A) \tag{1.18}$$

In (1.18) $N_{D,A}$ are the concentrations of donors and acceptors, while $f_{D,A}$ are the probabilities of their filling with electrons and thus also depend on F. In the simplest model the functions $f_{D,A} = f_{D,A}(E_{D,A}, T)$ coincide with the Fermi distribution (1.2) where $E_{D,A}$ are the energies of the donor and acceptor electron levels.

Without dealing at length with the general analysis of Eq. (1.17), let us examine the most important particular cases.

For an intrinsic semiconductor, $N_A^- = N_D^+ = 0$, it follows from Eqs. (1.16) and (1.17) that

$$n_O = p_O = n_i \tag{1.19}$$

and for the position of Fermi level we use Eqs. (1.14) and (1.15) to get

$$F = E_c - \frac{1}{2} E_g - \frac{1}{2} kT \ln \frac{N_c}{N_V} \tag{1.20}$$

At T=0 the Fermi level is located precisely in the center of the bandgap. As the temperature goes up, it shifts towards the band which has the lower density of states. Let us also note that the width of the bandgap E_g also changes (usually decreases) with increasing temperature and this increases the resulting shift of F towards the band edge.

Similar relations also apply to the so-called fully compensated semiconductors where $N_D^+ \neq 0$ and $N_A^- \neq 0$ but where $N_D^+ = N_A^-$. In such materials the donor electrons are captured by the acceptors, and the behavior is that of an intrinsic semiconductor.

In doped n-type semiconductors if the donors are fully ionized ($N_D^+ = N_D$), the condition of electroneutrality (1.17) leads to the relation

$$n_o = p_o + N_D \qquad (1.21)$$

It follows from Eqs. (1.16) and (1.21) that if $N_D \gg n_i$, then $n_o \gg n_i \gg p_o$ and $n_o \simeq N_D$. In other words, the concentration of electrons in the conduction band under these circumstances considerably exceeds the concentrations of holes in the valence band. Under these conditions, electric current in the bulk of the semiconductor is carried completely by electrons. That is why the electrons are called "majority carriers" and the holes "minority carriers".

The value of F is obtained directly from Eq. (1.14)

$$F = E_c - kT \ln \frac{N_c}{n_o} \quad , \quad n_o \simeq N_D \qquad (1.22a)$$

Similarly, in a p-type semiconductor, provided the acceptors are completely ionized ($N_A^- = N_A$), and the condition $N_A \gg n_i$ is fulfilled, the holes are the majority carriers. In this case

$$F = E_v + kT \ln \frac{N_v}{p_o} \quad , \quad p_o \simeq N_A \qquad (1.22b)$$

Thus, in n-type semiconductors the Fermi level, F, is located in the upper half of the forbidden band, while in p-type semiconductors it is in the lower half (see Fig. 1.15). Narrow-band semiconductors with highly differing values of m_c and m_v may be exceptions to this rule.

Now let us examine the problem of degenerate carriers in semiconductors. Degeneracy is a manifestation of quantum effects which are known to become important if the de Broglie wavelength of the particle, $\lambda_B = \hbar/p$ (where p is absolute value of its momentum) is comparable with or greater than the characteristic linear dimensions of the problem. For an n-type semiconductor and using Eq. (1.1) the absolute value of the quasi-

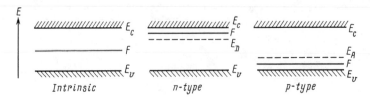

Fig. 1.5. The position of the Fermi level, F,
with respect to the band edges, E_c and E_v, in
semiconductors of different conductivity types.
The dotted lines denote the positions of the
energy levels of the donor (E_D) and acceptor
(E_A) impurities.

momentum of a thermalized electron in the conduction
band is $p \simeq \sqrt{2kTm_c}$. Taking this into account, we obtain
an approximate value for λ_B of $\lambda_B \sim \hbar/\sqrt{kTm_c}$. In this
case, a characteristic dimension of the problem is the
mean distance between electrons in the conduction band
(or between the impurity atoms in the crystal) and is
of the order of $n_0^{-1/3}$. Thus, the condition for the degeneracy of carriers in the bulk of an n-type semiconductor is $\hbar/\sqrt{kTm_c} \gtrsim n_0^{-1/3}$ or

$$T \lesssim \frac{\hbar}{km_c} n_0^{2/3} \tag{1.23}$$

The temperature at which the equality in Eq. (1.23) holds
is called the temperature of degeneracy, T_{deg}. When
$T < T_{deg}$ Boltzmann statistics are no longer applicable
to the electrons, and it is necessary to use the more
general Fermi-Dirac statistics. At constant temperature
the value of N_0 at which degeneracy occurs is given by
the inequality in Eq. (1.23). Assuming that $T = 300$ K
and $m_c = m_0$, we find $n_0 \simeq 10^{19}$ cm^{-3}.

    The condition for degeneracy may also be formulated
in a somewhat different way: the concentration of carriers in the bands, according to Eqs. (1.6) and (1.10),
depends on the differences $(E_c - F)$ and $(E_v - F)$. That is
why degeneracy occurs when the Fermi level is close
enough to the edge of, or located within, the pertinent
band. It follows from this and Eq. (1.23) that the
semiconductor is not degenerate if the Fermi level is
within the interval

$$E_v + 3kT \lesssim F \lesssim E_c - 3kT \tag{1.24}$$

When one of these inequalities (1.24) is broken, degeneracy of either electrons or holes will be found.

   If the electrons are strongly degenerate (taking
an n-type semiconductor as an example) so that
$exp\{(E_C - F)/kT\} << 1$, the Fermi level, as in metals, is
located within the conduction band. These degeneracy
effects can be important even at room temperature for
some narrow bandgap compounds (HgSe, HgTe, grey tin) and
particularly for heavily doped semiconductor materials.
Taking into account the location of the Fermi level,
such strongly degenerate semiconductors should formally
be referred to as low-grade metals. Indeed, they possess
a number of metallic properties.

   Using the asymptotic value for the Fermi integral
at $\zeta >> 1$ (see Eq. (1.7')), we get from Eq. (1.6)

$$n_o = \frac{4}{3\sqrt{\pi}} N_C \left(\frac{F - E_c}{kT} \right)^{3/2} \tag{1.25}$$

in accordance with which, unlike Eq. (1.14), $n_o > N_C$.

   At absolute zero, according to Eq. (1.2), in a
degenerate semiconductor all states in the conduction
band, such that $E > F$, are unoccupied, while all states
with $E < F$ are occupied. So the value $E_F = F - E_c$,
($E_F > 0$) is the maximum energy of electrons in the conduction band at T=0. This value which plays an important role in the physics of metals is often called the
Fermi energy.

   Substituting for N_C (see Eq. (1.6)) in Eq. (1.25)
we find

$$E_F = \frac{(3\pi^2)^{2/3} \hbar^2 n_o^{3/2}}{2m_c} \tag{1.26}$$

which coincides with the well known expression in the
theory of metals (11,13,26,27).

   It should be emphasized, in connection with what
has been said above, that the electrochemical potential,
F, of electrons does not generally characterize the
energy of any particular particle (quasi-particle) in
the system, although it has the dimensions of energy.

Metals and degenerate semiconductors at low temperatures
are an exception.   In these materials the value of the
electrochemical potential may be taken equal to E_F,
that is to say, the energy of electrons on the Fermi surface measured from the bottom of the conduction band.
It is precisely for this reason that the electrochemical
potential (Fermi level) of electrons in metals is often
called the Fermi energy.   The use of the term "Fermi
energy" and the designation of E_F for semiconductor materials is in general unjustified and leads to a confusion in terminology.

§ 1.3. Light Absorption by Semiconductors

     Let us examine a light flux of a moderate inten-
sity such that the electromagnetic wave does not cause
a significant change in the energy structure of the
solid.   Moreover, we shall assume that the optical characteristics of the medium are independent of the light
intensity (linear approximation).   Let us also assume
that the wavelength of the electromagnetic wave considerably exceeds the lattice constant.   Under these conditions the interaction between the radiation and the
material is described by the introduction of two dimensionless positive parameters, the refractive index n
and the absorption (extinction) coefficient k.

     The electric field, ξ, of a monochromatic electro-
magnetic wave propagating in a vacuum along the x direction is described by the expression

$$\xi = \xi_o e^{i(\kappa x - \omega t)}$$

where ξ_o is the amplitude of oscillation of the wave's
electric field, $\kappa \equiv \omega/c$ is the wave number (ω is the
angular radiation frequency, c is the velocity of light),
t is the time and $i = \sqrt{-1}$.   When the wave is propagated
in a medium, the wave number is generally a complex
quantity

$$\kappa = \frac{\omega}{c} (n + ik) \qquad\qquad (1.27)$$

where $n(\omega)$ is the refractive index and $k(\omega)$ is the absorption coefficient.   The value $n = n + ik$ is sometimes
called the complex refractive index.

     The magnitudes of $n(\omega)$ and $k(\omega)$ are not completely
independent and are connected by two integral relations

called the Kramers-Kronig relations (see, for example, (16,18)).

It follows from Maxwell's equations that $n = (\hat{\varepsilon} + i\sigma/\omega)^{1/2}$, where $\hat{\varepsilon}(\omega)$ and $\sigma(\omega)$ are the dielectric permeability and electric conductivity of the medium respectively and are, in general, complex numbers (for the sake of simplicity it is assumed that the magnetic susceptibility is equal to 1). For an insulator where $\sigma = 0$ we get

$$n + ik = [\hat{\varepsilon}(\omega)]^{1/2} = (\varepsilon_1 + i\varepsilon_2)^{1/2} \qquad (1.28)$$

Equation (1.28) is sometimes formally used to describe conducting media, in particular semiconductors, assuming that $\hat{\varepsilon}$ includes the two components in the above expression for n. Numerical values of n for a number of semiconductors are given in Appendix 1.

The measurable macroscopic properties of the substance can be expressed through n and k. In particular for the field in the medium we have

$$\xi = \xi_o \, e^{\,i\omega(nx/c \, - \, t)} e^{\, - \, k\omega x/c} \qquad (1.29)$$

According to Eq. (1.29), v, the phase velocity of an electromagnetic wave in the medium, decreases by a factor of n (v = c/n), and the wave gradually attenuates; its amplitude decreases exponentially with distance. The density of the wave energy and therefore the intensity of the light J is proportional to the square of the modulus of the electric field ξ. According to Eq. (1.29), this means that J (and also the number of photons per unit volume) decreases with increasing distance, x, in accordance with the law

$$J = J_o \, e^{-\alpha x} \qquad (1.30)$$

The quantity $\alpha = 2k\omega/c$ in Eq. (1.30) has the dimensions of reciprocal length and is called the linear light absorption coefficient. The coefficient α through $k(\omega)$ depends, in a complex manner, on the frequency ω and the detailed microscopic characteristics of the medium. The quantity J_o is the density of the luminous flux which entered the sample

$$J_o = J_{inc} \ (1-R) \qquad (1.31)$$

where J_{inc} is the incident light flux density and R the coefficient of light reflection at the interface. The reflection coefficient $R(\omega)$ is also expressed as a function of n and k. For normally incident light

$$R = \frac{(n - 1)^2 + k^2}{(n + 1)^2 + k^2} \qquad (1.32)$$

The medium is non-absorbing in the frequency region over which $\varepsilon_2 = 0$, so that, from Eq. (1.28), k = 0 and $n = \varepsilon_1^{1/2}$. The medium is weakly absorbing if $\varepsilon_1^2 \gg \varepsilon_2^2$. Then, from Eq. (1.28), we can make the following approximation ($\varepsilon_1 > o$)

$$n = \varepsilon_1^{1/2} \ ; \quad k = \varepsilon_2 / 2\varepsilon_1^{1/2} = \varepsilon_2/2n \qquad (1.33)$$

Bearing in mind that the length of the electromagnetic wave in the medium, λ_n, is equal to $\lambda_n = 2\pi c/n\omega$ and using Eq. (1.33), we find that the condition for absorption corresponds to $\alpha^{-1} \gg \lambda_n/2\pi$; in other words the distance over which the wave in the medium is noticeably absorbed is large when compared with its wavelength. Finally, if the inequality $\varepsilon_1^2 \ll \varepsilon_2^2$ (or $\alpha^{-1} \ll \lambda_n/2\pi$) holds the medium is strongly absorbing.

In describing the optical properties of semiconductors, it is possible to single out schematically several different regions and types of optical absorption*. The lowest frequency (long-wave) region relates to the far infrared part of the spectrum. The absorption of radiation here is relatively small and occurs mainly through the interaction of light with the oscillations of the lattice (phonons). The next region corresponds to the visible and the adjacent parts of the spectrum on both sides and is of paramount interest in the study

---

*Formally, the optical range of the wavelengths of electromagnetic radiation is from 0.01 to 2,000 μ and is subdivided into ultraviolet (from 0.01 to 0.4 μ), visible (from 0.4 to 0.8 μ) and infrared (from 0.8 to 2,000 μ) Note that a wavelength, $\lambda = 2\pi c/\omega$, of 1.24 μ and a wave number of 8,060 cm^{-1} correspond to a quantum of energy equal to 1 eV.

Fig. 1.6. The different types of
phototransition in a semiconductor:
1 - intraband transition; 2,4 -
transitions with the participation
of impurity levels (i.e., of donor
or acceptor levels which are un-
ionized in the dark); 6 - interband
transition. The wavy arrows (2,4
and 6) denote those which occur under
the influence of light quanta; the
straight arrows (3 and 5) those
which occur as a result of thermal
excitation.

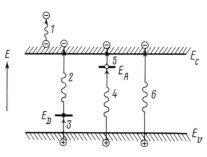

of photoelectrochemical phenomena. This region is characterized by a sharp spectral structure resulting from
absorption caused by electronic transitions. The three
different types of transition are shown in Fig. 1.6.
Firstly, there is absorption by free carriers within the
bands (intraband transitions). The energy of light is
converted into a high (optical) frequency current and,
eventually, it converts into Joule heat. Secondly, there
is impurity absorption. Energy is absorbed by charge
carriers localized on impurity atoms or structural imperfections in the lattice. This leads to either transitions of carriers from the ground state of the impurity
center to the excited state or to the ionization of the
impurity. Thirdly, in this frequency region an extremely significant phenomenon is observed. This is a sharp
growth of absorption caused by the interband optical
transitions in which the energy of the photon is used
to create electron-hole pairs. This is called fundamental absorption.

With increasing radiation frequency, we switch over
to a spectral region characterized by a drop in the
semiconductor's absorption of luminous energy and also
by a rapid decrease in reflection. At even higher frequencies when the energies reach the order of dozens of
electron-volts, a substantial absorption of electromagnetic waves is again observed. This is caused, in the
main, by photoexcitation of electrons from low-lying
filled bands (core electrons) into the free states in
the conduction band.

Let us now deal in greater detail with the mechanism of the photogeneration of electron-hole pairs, produced by the transition of electrons from the valence
band to the conduction band. Interband electron transitions are subdivided into direct and indirect (see

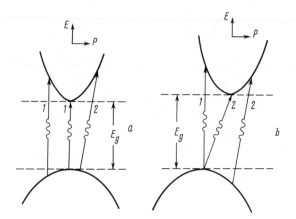

Fig. 1.7. The scheme for direct (1)
and indirect (2) optical transitions
from the valence band to the conduction band : a) maxima and minima of
the bands coincident; b) maxima and
minima of the bands non-coincident.

Appendix 1). The meaning of these terms is illustrated
in Fig. 1.7 where the dependences of the electron energy on quasi-momentum in the conduction band $E_C(p)$ (upper
curves) and in the valence band $E_V(p)$ (lower curves)
along one of the directions in p-space are presented
schematically. In the simplest situation the dependences $E_C(p)$ and $E_V(p)$ are described analytically by Eqs.
(1.1), and the minima and maxima of the curves $E_C(p)$ and
$E_V(p)$ lie at $p = 0$.

One electron from the valence band and one photon
of energy $\hbar\omega$ take part in a direct transition. The
momentum of the photon is of the order of $\hbar\omega/c$ and is
usually negligible when compared with the quasi-momentum
of the electron. Thus the quasi-momentum of the electron
does not change during a direct transition in an ideal
crystal lattice, and such transitions appear as "vertical."

Along with the law of the conservation of quasi-
momentum, the law of energy conservation must also apply
and it is in no way permissible, unlike the photon's momentum,to ignore its energy, $\hbar\omega$, and so

$$E_C(\mathbf{p}) - E_V(\mathbf{p}) = \hbar\omega \qquad (1.34)$$

When the maxima and minima of the bands coincide,
as shown in Fig. 1.7a, the minimum radiation frequency

ω_{min} (red boundary), required to carry out a direct (vertical) interband transition, is given by the following relation

$$\hbar\omega_{min} = E_g$$

In the case where the maxima and minima of the bands do not coincide, Fig. 1.7b, the law of conservation of quasi-momentum excludes the possibility of the absorption of photons with energies close to the bandgap, thus direct transitions are only possible for photon energies greater than $\hbar\omega_{min}$. A fundamental absorption near the edge of the band does, however, become possible if lattice non-idealities are taken into account. Such interband transitions are called indirect (or non-vertical). As shown in Fig. 1.7b, when such transitions take place, the quasi-momentum of the electron changes substantially. Physically it stems from the fact that a "third body," which ensures the necessary changes of quasi-momentum, takes part in indirect transitions. The role of such a "third body" may be played by impurities, vacancies, dislocations, etc., and also by phonons (quanta of the lattice oscillations). In the latter case the major part of the energy in the interband transition is transmitted to the electron by the photon (quantum of light), whereas the phonon supplies the missing quasi-momentum or takes away the excess.

There are semiconductor materials in which the bottom of the conduction band and the top of the valence band are located at the same or almost the same point in momentum space. Gallium arsenide and indium antimonide are examples of such materials. In semiconductors like germanium and silicon, on the other hand, the maxima and minima of the bands do not coincide. Materials of the two different types are sometimes referred to as "direct" and "indirect" respectively.

In sufficiently pure crystals indirect transitions are usually less probable than direct ones, in as much as an additional small parameter, connected with electron-phonon interaction, is involved. In contrast, in relatively heavily doped semiconductors, the probability of indirect transitions is larger, due to the interaction between the electrons and the impurities.

The quantum energy dependences of the probability for these two types of transitions are also different.

In particular, near the fundamental absorption edge the optical absorption coefficient α_g for direct interband transitions in non-degenerate semiconductors is given by the expression

$$\alpha_g = A_d (\omega - \omega_{min})^{1/2} \qquad (1.35)$$

where the value of A_d is independent of $(\omega - \omega_{min})$.

For an indirect optical interband transition occurring with the participation of one phonon

$$\alpha_g = A_i (\omega - \omega_{min})^2 \qquad (1.36)$$

and the magnitude of A_i, which like A_d is independent of $(\omega - \omega_{min})$, is determined by the phonon density and may vary substantially with temperature. Some optical characteristics of a number of semiconductor materials are given in Appendix 1.

Such an interpretation of interband optical transitions is based on a single particle model. In the final state, however, there are two particles, an electron and a hole, which, as a result of interactions with one another, may form a two-particle state referred to as an exciton (7, 12, 15, 18, 32). An exciton is an electrically neutral quasi-particle. It represents a bound state formed by an electron and a hole as a result of their Coulomb attraction. Having formed an exciton, the electron and the hole move through the crystal as a single entity. Being mobile, the exciton does not form spatially-localized states. However, due to the internal energy of the exciton's bond, the full energy of the semiconductor, plus exciton system, is less than that of the semiconductor plus electron in the conduction band and hole in the valence band. Within the framework of the band model the energy levels of the exciton are within the forbidden band (see Fig. 1.8).

The energy of the exciton's bond and the distribution of the aforementioned levels depends on the properties of the semiconductor, and in the simplest model (Wannier-Mott exciton) they are described by formulae which are similar to those for the hydrogen-like atom. The energy of the levels with respect to the bottom of the conduction band is $E_{exc} = E^0_{exc}/s^2$ ($s = 1,2...$) where

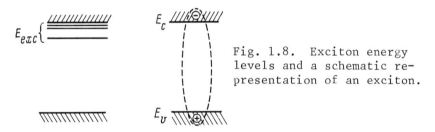

Fig. 1.8. Exciton energy levels and a schematic representation of an exciton.

E_{exc}^{O} is the energy of the deepest exciton level, which is equal (in CGSE system) to

$$E_{exc}^{O} = - m_{cv} e^4 / (2 \hbar^2 \varepsilon_{sc}^2) \qquad (1.37)$$

where m_{cv} is the reduced effective mass of the electron and the hole, $m_{cv}^{-1} = m_v^{-1} + m_c^{-1}$, and ε_{sc} is the dielectric permeability of the semiconductor. As s increases the value of $|E_{exc}|$ rapidly decreases, "pressing" itself against the bottom of the conduction band.

The lifetime of the exciton is limited. Firstly, annihilation of the electron and the hole is possible; triplet excitons exist for a longer time when compared with singlet excitons[*]. Secondly, the decomposition of the exciton is possible. In particular, local electric fields in the semiconductor give rise to forces which act independently upon the electron and the hole in opposite directions. Under the influence of these fields, dissociation of the exciton into a "free" electron and a "free" hole is possible.

The photogeneration of exciton states may play an essential role in the photoelectrochemical processes at semiconductor surfaces. Taking into account the effects caused by excitons is of particular importance when the frequencies of irradiation are less than the fundamental absorption edge ($\hbar \omega < E_g$) since the free exciton is formed by the absorption of a photon with an energy of $E_g - |E_{exc}|$. In addition, because excitons are electrically neutral, they can cross the space charge region and bring "reagents" (electrons or holes) to the

---

[*]Note that the state of two particles with spin 1/2 is called a triplet if the spins are oriented in the same direction (resulting spin unity), and a singlet if the spins are oriented in opposite directions (resulting spin zero).

interface at semiconductor surface potentials where the
"independent" approach of these particles by themselves
is highly improbable.

Let us now deal with the absorption of light by
electrons and holes through intraband transitions.
This absorption in the normal frequency range, in
the semiconductor bulk, is relatively small. However intraband transitions may play an essential role in
certain phenomena occurring close to the interface. The
interaction between electrons and the electromagnetic
field in the conduction band (taking an n-type semiconductor as an example) is approximately described within
the framework of Drude's theory. This leads to the following expression for the complex refractive index (see,
for example, (11)):

$$(n + ik)^2 = 1 + \frac{\omega_p^2}{\omega^2} \frac{i\omega\tau}{1 - i\omega\tau} \tag{1.38}$$

where $\omega_p^2 = e^2 n_0/\varepsilon_0 m_c$ is the so-called electronic plasma
frequency and τ is the phenomenologically introduced relaxation time for electrons in the conduction band. In
the radiation frequency range, which interests us, usually $\omega\tau \gg 1$; moreover, in accordance with Eq. (1.38),

$$n^2 - k^2 = 1 - (\omega_p/\omega)^2 \quad ; \quad 2nk = \omega_p^2/(\omega^3\tau) \tag{1.39}$$

Equations (1.38) and (1.39), after the corresponding
changes have been introduced, can also be applied to p-
type semiconductors.

The existence of free carriers in the semiconductor
may produce an additional effect. Until now we have
dealt with non-degenerate semiconductors where almost
all the states in the conduction band are unoccupied.
Under the conditions of strong degeneracy when the Fermi level lies within the conduction band, the situation
changes. Since the electronic states located below F
are already filled, fundamental absorption arising from
transitions into these states is impossible. As a result, the absorption edge is shifted towards higher energies by an amount E_F (given by Eq. (1.26)) for indirect transitions, and by even a greater amount for direct
transitions (see Fig. 1.9). The frequency ω_{min} in Eqs.
(1.35) and (1.36) should, under these conditions, be replaced by $\omega_{min} + E_F/\hbar$. Such a shift of the red boundary
is called the Burstein-Moss shift.

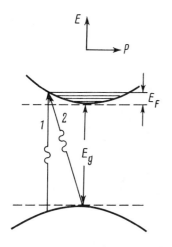

Fig. 1.9. Optical interband transitions in an n-type semiconductor when some of the electron levels in the conduction band are occupied: 1 - a direct transition; 2 - an indirect transition.

§ 1.4. Non-Equilibrium Electrons and Holes; Recombination Processes

Under the conditions of thermodynamic equilibrium, the thermal generation of carriers is accompanied by the simultaneous reverse process of recombination. The resultant concentrations of electrons and holes are therefore equal to their equilibrium values calculated above. Under the influence of an external perturbation such as illumination, non-thermal generation must also be considered and the semiconductor is no longer in thermodynamic equilibrium. In this case the concentrations of electrons (n) and holes (p) are no longer equal to their thermodynamic equilibrium values (n_o and p_o); the non-equilibrium carrier concentrations in the bands are

$$n = n_o + \delta n , \quad p = p_o + \delta p \qquad (1.40)$$

In general, under these circumstances, changes also occur in the concentrations of bound charges on the local electronic levels of the impurities, defects, etc. This is also true of metals but since the concentration of free electrons in metals is much higher, the relative change is usually negligible.

Let us assume that, under an external influence, in a unit volume of the semiconductor, in unit time, g_n electrons in the conduction band and g_p holes in the valence band are generated, where in the general case

$g_n \neq g_p$. The quantities g_n and g_p are called the generation rates. Let us assume that R_n is the rate of the reverse process, that is, the number of electrons in the conduction band vanishing in unit time as a result of recombination with both free holes and those bound on local energy levels. R_p is the corresponding rate of removal of free holes. In the absence of any current the change in time t of the non-equilibrium concentrations n and p in the bands is determined by the equations

$$\frac{d(\delta n)}{dt} = g_n - R_n \ , \quad \frac{d(\delta p)}{dt} = g_p - R_p \qquad (1.41)$$

It should be emphasized that the values of g_n and g_p signify only the "excess" generation due to the influence of an external perturbation and do not include the "equilibrium" transitions (thermal generation). Under the conditions of thermodynamic equilibrium $\delta n = \delta p = 0$ and $g_n = g_p = 0$, and therefore, according to Eq. (1.41), the values of R_n and R_p should also in turn be zero. A phenomenological approach is widely used for the quantitative description of the kinetics of non-equilibrium electron processes, in which the quantities above are written as follows

$$R_n = \frac{n - n_0}{\tau_n} \ , \quad R_p = \frac{p - p_0}{\tau_p} \qquad (1.42)$$

Equations (1.42) are definitions of the mean lifetime of non-equilibrium electrons in the conduction band, τ_n, and of holes in the valence band, τ_p. In other words, τ_n^{-1} is the probability that a particular excess electron vanishes from the conduction band in unit time as a result of its recombination with either free or bound holes. Similarly, τ_p^{-1} is the probability of recombination per unit time for a particular excess hole.

In the general case, the lifetimes τ_n and τ_p depend not only on the physics of the elementary act of recombination, but also on the values of n and p. Therefore they are not characteristic of the given semiconductor but, rather, are also dependent on the experimental conditions. In a number of most of the important cases, however, when the deviations from the equilibrium concentrations are relatively small (see below), the phenomenological quantities τ_n and τ_p in Eq. (1.42) can be

regarded as constants. Under these conditions, it follows from Eqs. (1.41) and (1.42) that the transient behavior of the excess concentrations of electrons and
holes, when the external perturbation is switched on or
off, is described by an exponential law. In this case
the mean lifetime τ_n (or τ_p) is the time during which
the concentration of non-equilibrium electrons (or holes)
falls to $1/e$ of its original value. In the general case
when τ_n and τ_p change together with n and p, the relations
(1.42) determine the instantaneous lifetimes. The values
of τ_n and τ_p have been shown experimentally to cover a
wide range of timescales, from many hours to 10^{-8} s or
even less.

Physically, the recombination processes in semiconductors, resulting in the restoration of equilibrium, can
be subdivided into the three main classes:

(i)   Direct band-to-band recombination.

(ii)   Recombination with participation of impurities
        and defects.

(iii) Surface recombination.

The processes of surface recombination will be examined when the properties of the semiconductor/electro-
lyte interface are described. Here we shall deal with
processes of the first two types (see Fig. 1.10).

Direct band-to-band recombination (Fig. 1.10a) may
occur with the emission of light quanta (photons), that
is, radiative recombination, or the excess energy and quasi-
momentum may be transmitted to the oscillations of the
lattice as phonons. There are also processes where the
energy and quasi-momentum are transmitted to a third
particle, either an electron or a hole, and this type
of process is called an impact recombination (Auger recombination). Other, more complicated types of elementary recombination acts are also possible (see, for
example (15), Chapter 17).

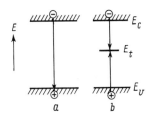

Fig. 1.10. The different
types of electron-hole recombination; a) Direct
band-to-band recombination;
b) recombination through a
localized energy level, E_t.

Irrespective of the detailed mechanism of direct recombination the full number of recombining pairs in a unit volume is $c_d np$ where the coefficient c_d is determined by the probability of the elementary act of recombination, the density of states in the bands and the temperature. At equilibrium the number of recombining pairs (which is equal to the rate of thermal generation) is $c_d n_o p_o$. Hence, defining the quantity R_d so that $R_d = R_n = R_p$ (under the conditions of direct recombination, it is obvious that $R_n = R_p$), we obtain

$$R_d = c_d(np - n_o p_o) \qquad (1.43)$$

Using Eqs. (1.40) and (1.42) and assuming that $\delta n = \delta p$, we find that the lifetime, $\tau_d = \tau_n = \tau_p$, of non-equilibrium pairs is

$$\tau_d = \frac{\delta p}{R_d} = \frac{1}{c_d(n_o + p_o + \delta p)} \qquad (1.44)$$

Thus, the lifetime τ_d, as has already been mentioned, depends on δp. However, if the value of δp is smaller than the equilibrium concentration of majority carriers (but not necessarily smaller than the equilibrium concentration of minority carriers), then τ_d does not depend on δp. In particular, for an n-type semiconductor when $n_o \gg p_o$,

$$\tau_d = (c_d n_o)^{-1} \qquad (1.45)$$

Thus, the lifetime τ_d is inversely proportional to the equilibrium concentration of the majority carriers.

The rate of the radiative band-to-band recombination (the constant c_d, in Eqs. (1.43), (1.44) and (1.45)) is large in relatively narrow-gap semiconductors. This type of recombination may also predominate in wide-gap direct transition semiconductors (for example, GaAs, CdS). It is more often the case, however, that the dominant processes are recombination via intermediate local energy levels caused by impurities and/or structural defects, which serve as "recombination centers" and which we shall now examine.

Recombination with the participation of recombination centers is a two-step process. It consists of the capture of an electron from the conduction band by the empty recombination center and the capture of a hole from the valence band by the occupied recombination center (Fig. 1.10b). It should be pointed out that the recombination centers are not themselves changed and thus play the role of a catalyst in the process of recombination.

Let us examine the simplest recombination centers, which are able to capture or donate only one electron and consequently have only two states. Such centers can be characterized by a single local energy level for electrons E_t in the forbidden band. If N_t is the total concentration of centers, then at equilibrium the concentration of filled centers is $N_t f_t^O$, and that of empty centers is $N_t(1-f_t^O)$ where $f_t^O(E_t, T)$ is the Fermi function (Eq. (1.2)). When equilibrium is perturbed, the concentration of filled and empty centers are equal to $N_t f_t$ and $N_t(1-f_t)$ respectively, where f_t is the non-equilibrium probability of the center being filled and is no longer expressed by the function $f_t^O(E_t, T)$.

The rate of capture of electrons from the conduction band by the recombination centers (see Fig. 1.11) is equal to $R_C = a_n N_t n(1-f_t)$, where a_n is the capture coefficient for electrons. The rate of the reverse process of re-emission, the ejection of electrons from filled centers to the conduction band, is $R_C' = a_n' f_t N_t$, where a_n' is the so-called coefficient of re-emission. At thermodynamic equilibrium $R_C = R_C'$, and $n = n_o$, $f_t = f_t^O$, which means that $a_n' f_t^O = a_n n_o(1-f_t^O)$. Making use of the expression ensuing from Eq. (1.2)

$$\frac{1 - f_t^O}{f_t^O} = exp\left(\frac{E_t - F}{kT}\right) \tag{1.46}$$

and with Eq. (1.14) for n_o, we obtain

$$a_n' = a_n n_1 \tag{1.47}$$

where $n_1 = N_c exp\{-(E_c-E_t)/kT\}$ is the characteristic parameter of the model. It is evident that $E_c - E_t$ is nothing but the ionization energy of the center. Let us note that the magnitude of n_1 formally coincides with

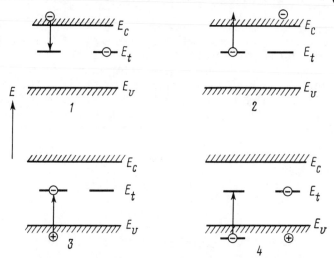

Fig. 1.11. Four kinds of elementary processes involved in recombination via
local levels within the bandgap (initial
and final states are shown); 1 - capture
of an electron by a trap; 2 - thermal
ejection of an electron from a trap to
the conduction band; 3 - capture of a
hole by a trap; 4 - thermal ejection of
a hole from a trap to the valence band.

the concentration of electrons in the conduction band
when the Fermi level coincides with the energy level of
the center ($F = E_t$). Using Eq. (1.47), the number of
electron captures occurring in unit volume and unit time,
$R_n = R_c - R_c'$, can be written as follows

$$R_n = a_n N_t [n(1 - f_t) - n_1 f_t] \qquad (1.48)$$

The number of hole captures is calculated in a similar
way, and may be written as

$$R_p = a_p N_t [p f_t - p_1 (1 - f_t)] \qquad (1.49)$$

where

$$p_1 \equiv N_v exp\{-(E_t - E_v)/kT\}$$

Using Eqs. (1.41), (1.48) and (1.49), we get

$$\frac{dn}{dt} = g_n - a_n N_t [n(1 - f_t) - n_1 f_t] \qquad (1.50)$$

$$\frac{dp}{dt} = g_p - a_p N_t [pf_t - p_1(1 - f_t)] \qquad (1.51)$$

In addition, the condition of electroneutrality must be taken into consideration

$$p - p_o = n - n_o + N_t(f_t - f_t^o) \qquad (1.52)$$

so that the total charge is conserved. The system of three equations (1.50) - (1.52) completely describes the time dependence of the three quantities: the concentrations of non-equilibrium electrons $n(t)$ and holes $p(t)$, and of the degree of filling of the traps $f_t(t)$ in the course of recombination.

As Eqs. (1.50) and (1.51) are non-linear, in the general case, the relaxation of concentrations $n(t)$ and $p(t)$ is not exponential. In particular, the kinetics of recombination for non-steady-state photoexcitation is, as a rule, non-exponential. Moreover, even when the problem is linearized, the decay of the non-equilibrium concentrations is described by a sum of two exponentials, and there are two relaxation times τ_1 and τ_2. These times are the same for electrons and holes and, in rather a complicated manner, they depend on the parameters a_n, a_p, n_1, p_1.

Let us examine in greater detail the stationary (but not equilibrium) state which is established at sufficiently long times after the external constant generation of electron-hole pairs is switched on. Under these conditions $dn/dt = dp/dt = 0$ and $R_n = R_p = R_t$. Equating the expressions (1.48) and (1.49), it is easy to find f_t, the steady state degree of filling of recombination centers (while $f_t \neq f_t^o$) and then, substituting this value f_t in Eq. (1.48) or (1.49), the magnitude of the stationary recombination rate

$$R_t = \frac{N_t a_n a_p (np - n_o p_o)}{a_n(n + n_1) + a_p(p + p_1)} \qquad (1.53)$$

Let us examine the case when the excess concentrations may be regarded as equal, $\delta p = \delta n$. From Eq. (1.52) this will be a valid approximation when the concentration of recombination centers is relatively low. Besides, if the excitation intensity is small enough and

$\delta p \ll (n_0 + n_1)$ and $(p_0 + p_1)$, then $\tau_t = \delta p / R_t$ is given by (33)

$$\tau_t = \frac{(N_t a_p)^{-1}(n_0 + n_1) + (N_t a_n)^{-1}(p_0 + p_1)}{n_0 + p_0} \qquad (1.54)$$

Under the above conditions the lifetime of electrons and holes is the same (as it was for direct recombination), so that $\tau_n = \tau_p = \tau_t$.

If the concentration of recombination centers is such that $\delta n \neq \delta p$, the lifetimes of electrons and holes

$$\tau_n = \frac{\delta n}{R_t} \quad \text{and} \quad \tau_p = \frac{\delta p}{R_t}$$

are different even at the lowest excitation intensities.

For an n-type semiconductor, when $n_0 \gg p_0$, n_1, it follows from Eq. (1.54) that

$$\tau_t = (a_n N_t)^{-1} \qquad (1.55)$$

Similarly, for a p-type semiconductor $\tau_t = (a_p N_t)^{-1}$. Note that the lifetime τ_t (Eq. (1.55)), unlike τ_d (Eq. (1.45)), does not depend on n_0.

From this we may conclude that the role of local levels in the processes of recombination becomes less significant (all other conditions being equal) the closer the energy level is to the edge of one or other of the bands. In particular, from Eq. (1.54), as $E_c - E_t$ (or $E_t - E_v$) is decreased, n_1 (or p_1) increases, thus resulting in the growth of τ_t. Physically this means that most of the charge carriers captured from the nearest band are immediately lost back into that band. Such "shallow" levels are good suppliers of current carriers, electrons or holes, and considerably change the electrical conductivity (doping impurities). On the other hand, deep-lying levels insignificantly influence the equilibrium electrical conductivity but are capable of significantly changing the lifetime. Such impurities with "deep" levels are usually recombination centers.

In conclusion let us note that the parameters a_n and a_p are approximately equal to the product of the thermal velocity of electrons or holes (v_T) and the cross-section of their capture by a center (σ^t) so that $a_{n,p} = v_T \sigma^t_{n,p}$. Depending on the number of charges on the recombination center, due to the change in $\sigma^t_{n,p}$, the values of $a_{n,p}$ can vary over a broad range,

$$a_{n,p} = 10^{-5} \text{ to } 10^{-15} \text{ cm}^3/\text{s}$$

The magnitudes of $a_{n,p}$ and E_t are the key characteristics of the impurity in the semiconductor and fully determine its influence on the equilibrium and kinetic properties. Being aware of these characteristics, it is often possible, with the help of the equations above, to select the required type of dopant and to calculate the concentration needed to obtain the required recombination characteristics for the semiconductor material.

Ξ 1.5. T r a n s p o r t   P h e n o m e n a

When an electric current is flowing in a semiconductor, the change in concentration of carriers in unit volume is determined not only by generation and recombination, but also by their transport. Under these conditions, the following equations serve as direct generalization of Eq. (1.42) with due account of Eq. (1.43):

$$\frac{\partial n}{\partial t} = g_n + \frac{1}{e} \frac{\partial i_n}{\partial x} - \frac{n - n_o}{\tau_n} \qquad (1.56)$$

$$\frac{\partial p}{\partial t} = g_p - \frac{1}{e} \frac{\partial i_p}{\partial x} - \frac{p - p_o}{\tau_p} \qquad (1.57)$$

These are continuity equations for the electrons and the holes where i_n and i_p are the current densities determined by the motions of the electrons and holes (we shall assume for the sake of simplicity that the quantities in Eqs. (1.56) and (1.59) depend only on one coordinate).

The current density i_n and i_p within the framework of the usually applied phenomenological approach are presented as sums of diffusion and migration terms:

$$i_n = eD_n \left(\frac{\partial n}{\partial x} - \frac{e}{kT} n \frac{\partial \phi}{\partial x} \right) \qquad (1.58)$$

$$i_p = - eD_p \left(\frac{\partial p}{\partial x} + \frac{e}{kT} p \frac{\partial \phi}{\partial x} \right) \qquad (1.59)$$

where D_n and D_p are the diffusion coefficients of electrons and holes, and are independent of n and p. The first terms in Eqs. (1.58) and (1.59) are proportional to the concentration gradients, while the second terms are proportional to the gradient of the electric potential ϕ in the semiconductor.

In writing Eqs. (1.58) and (1.59) it is assumed that the carriers in the semiconductor are nondegenerate, and the Nernst-Einstein equation relating the diffusion coefficients $D_{n,p}$ and the mobilities $u_{n,p}$ of both sorts of carrier is used

$$u_{n,p} = eD_{n,p}/kT$$

(if the semiconductor is degenerate the generalized relations

$$u_n = eD_n \frac{d(\ln n)}{dF} \text{ and } u_p = eD_p \frac{d(\ln p)}{dF}$$

are used).

We should note that the expressions for the currents and the notion of a diffusion coefficient itself are only sensible if changes in the concentrations n and p have insignificant effects on the free-path lengths $\ell_{n,p}$. In particular, for holes it is necessary that $|dp/dx| \ell_p \ll p$.

If $i_n = i_p = 0$, it follows from Eqs. (1.58) and (1.59) that

$$n(x) = n_o e^{\phi(x)/kT} , \quad p(x) = p_o e^{-e\phi(x)/kT} \qquad (1.60)$$

The potential ϕ is defined so that the equilibrium magnitudes of the concentrations n_o and p_o are obtained when $\phi = 0$. Equations (1.60) show that in the absence of any current the concentrations of electrons n(x) and holes p(x) follows the Boltzmann distribution.

The potential ϕ in Eqs. (1.58) and (1.59) is determined by Poisson's equation

$$\frac{\partial^2 \phi}{\partial x^2} = - \frac{1}{\varepsilon_{sc}\varepsilon_o} \rho \qquad (1.61)$$

where ε_o is the so-called dielectric permeability of the vacuum*, ε_{sc} is the relative dielectric permeability of the semiconductor (which in the cases examined below may be regarded as equal to the static value), and ρ is the volume charge density

$$\rho = e(p + N^+ - n - N^-) \qquad (1.62)$$

In Eq. (1.62) N^+ and N^- are the concentrations of the positively and negatively charged impurities.

The magnitudes of N^+ and N^- generally depend on n and p (and therefore on t) and are linked to these by recombination kinetic equations of the type examined above. In the overwhelming number of cases, however, the effect of this dependence on the magnitude of the volume charge density ρ can be neglected. It is necessary, firstly, that the temperature of the semiconductor is not too low; "shallow" donor and acceptor levels are then completely ionized. Secondly, the concentration of "deep" levels, which play the role of the recombination centers, should not be too big. If this is the case their contribution to the magnitudes N^+ and N^- is small (although it is precisely these levels that may exert a decisive influence on the lifetime of the carriers τ_n and τ_p). Within the framework of these assumptions

$$N^+ = N_D \text{ and } N^- = N_A.$$

Substituting from Eqs. (1.58) and (1.59) in (1.56) and (1.57), and Eq. (1.62) in (1.61), we get the following system of equations

$$\frac{\partial n}{\partial t} = g_n + D_n \left[\frac{\partial^2 n}{\partial x^2} - \frac{e}{kT} \frac{\partial}{\partial x}\left(n \frac{\partial \phi}{\partial x} \right) \right] - \frac{n-n_o}{\tau_n} \qquad (1.63)$$

$$\frac{\partial p}{\partial t} = g_p + D_p \left[\frac{\partial^2 p}{\partial x^2} + \frac{e}{kT} \frac{\partial}{\partial x}\left(p \frac{\partial \phi}{\partial x} \right) \right] - \frac{p-p_o}{\tau_p} \qquad (1.64)$$

---

* The transition from SI system used here to CGSE is achieved by replacing ε_o^{-1} by 4π.

$$\frac{\partial^2 \phi}{\partial x^2} = - \frac{e}{\varepsilon_o \varepsilon_{sc}} (p + N_D - n - N_A) \qquad (1.65)$$

Equations (1.63) - (1.65) describe at the preset g_n and g_p, together with the boundary and initial conditions, the spatial distribution of three quantities; the concentrations of electrons, n, and holes, p, and the potential distribution, ϕ.

A general analysis of the resulting self-consistent system of equations (1.63) - (1.65) encounters serious mathematical difficulties, in as much as it includes nonlinear equations, two of which are in partial derivatives. At the same time, taking into account relative magnitudes of the parameter for real physical systems, the mathematical interpretation in many crucial cases can be substantially simplified.

In the bulk of the semiconductor, away from the narrow surface region, where the space charge dominates (see Section 3.2), the use of Poisson's equation (1.65) is superfluous. It is assumed instead that $\rho = 0$, and taking into account Eqs. (1.65) and (1.17), we find that

$$p - p_o = n - n_o \qquad (1.66)$$

which is called the condition of quasi-neutrality[*].

With this condition it is possible to reduce the two equations (1.63) and (1.64) to a single equation in either p or n.

Further simplifications can be achieved if the material in question is an extrinsic semiconductor: n-type so that $n_o \gg p_o$, or p-type so that $p_o \gg n_o$. Let us divide the two parts of Eq. (1.58) by D_n and the two parts of Eq. (1.59) by D_p and sum them. Using the relation

---

[*]The transition from Eq. (1.65) to (1.66) is analogous to that used in the theory of transport processes in electrolytes. This is analyzed in detail in (34 - 36). Actually, the fact that $L_{sc}/L \ll 1$ where L_{sc} is the thickness of the space charge region ($10^{-6} - 10^{-4}$ cm, see Section 3.2), and L is the sample thickness, is used. In addition, in the semiconductor the condition $\tau_{n,p} \gg \varepsilon_{sc}\varepsilon_o/\sigma$ where σ is the static electrical conductivity should apply; the latter condition is always fulfilled.

$\partial n / \partial x = \partial p / \partial x$ following from Eq. (1.66) we get

$$\frac{i_n}{D_n} + \frac{i_p}{D_p} = - (n + p) \frac{e^2}{kT} \frac{\partial \phi}{\partial x} \qquad (1.67)$$

Using Eq. (1.67), the second term in the right hand part of Eq. (1.59) describing the migration of holes in the electric field can be written as follows

$$\frac{p}{n + p} (i_p + \frac{D_n}{D_p} i_n)$$

If $i_n \lesssim i_p$, so that the hole current accounts for a considerable portion of the total current, then for an n-type semiconductor the expression above turns out to be much smaller than i_p. This means that the migration current of minority carriers under these conditions can be neglected.

Note that for the majority carriers, as a result of similar transformations, the term $n/(n + p)$ occurs before the bracket in the corresponding expression; this is a significant quantity and hence the migration current of the majority carriers cannot be ignored.

Thus, it is possible to make use of the following approximate equation, derived from Eqs. (1.57) and (1.59), in order to find the concentrations of minority carriers (i.e. holes) in the quasi-neutral region

$$\frac{\partial p}{\partial t} = D_p \frac{\partial^2 p}{\partial x^2} + \frac{p_0 - p}{\tau_p} + g_p \qquad (1.68)$$

It is important that Eq. (1.68) is linear. From Eq. (1.68) it follows that the parameter $L_p = \sqrt{D_p \tau_p}$ (or $L_n = \sqrt{D_n \tau_n}$ in the case of a p-type semiconductor) which has dimensions of length and is called the diffusion length plays an important part in the description of the transport processes. This length characterizes the distance traversed by non-equilibrium minority carriers during their lifetime τ_p (or τ_n).

The concentration of electrons, when Eq. (1.68) has been solved, can be found from the condition of quasi-neutrality (1.66), although usually it is sufficient to assume that $n \simeq n_0$ (the corrections are small with

respect to the parameter p_0/n_0). Finally, proceeding
from the known concentration of electrons from Eq. (1.63)
it is possible to find the potential distribution. If we
assume that $n = n_0$, then when $g_n = 0$, this turns out to
be linear (ohmic). In the surface region of the semiconductor where, in general, there is a space charge, the
simplifications discussed above cannot be used (this
problem is dealt with in detail in Section 3.2).

Now let us examine the quantities g_n and g_p in Eqs.
(1.63), (1.64) and (1.68) describing the generation, and
in particular, photogeneration of non-equilibrium electrons and holes. When impurities are present, photogeneration is possible, even when $\hbar\omega < E_g$, by the excitation into the conduction band of electrons from the
impurity levels, or by excitation of electrons from the
valence band to those levels with the concomitant formation of holes in the valence band (see Fig. 1.6).
Optical generation of this kind is called extrinsic
(impurity). In this case $g_n \neq g_p$.

If the energy of photons $\hbar\omega > E_g$, then non-equilibrium electrons and holes are formed as a result of interband transitions. Such optical generation is called intrinsic, and here $g_n = g_p = g$. The intensity of intrinsic generation is usually of many orders of magnitude
greater than that of impurity generation.

Suppose that $J(x)$ is the monochromatic luminous
flux density of photons with a frequency ω in the bulk
of the semiconductor at a distance x from the illuminated surface. According to the definition of the optical absorption coefficient, the number of the absorbed
photons per unit time and unit volume is $J(x)\alpha$. Thus
the rate of optical generation, g, may be represented
as follows

$$g(x,\omega) = \alpha(\omega)\beta(\omega)J(x) \qquad (1.69)$$

where $\beta(\omega)$ is a dimensionless coefficient taking into
account the number of electrons (holes) generated as a
result of the absorption of one photon (the magnitude
$\beta(\omega)$ is often called the quantum yield of the internal
photoeffect). The functional dependence of $J(x)$ in Eq.
(1.69) is given in Eq. (1.30).

In the very important case of intrinsic optical
generation it can often be assumed that $\beta(\omega) = 1$. Then
taking into account Eqs. (1.69) and (1.30) we find that
for $g = g_n = g_p$

$$g = \alpha(\omega) J_o e^{-\alpha(\omega)x} \qquad (1.70)$$

As is seen from Eq. (1.70), the following condition is fulfilled

$$\int_o^\infty g(x)\,dx = J_o \int_o^\infty \alpha e^{-\alpha x}\,dx = J_o \qquad (1.71)$$

This means that in the entire semi-infinite bulk of the semiconductor $0 < x < \infty$ the luminous flux will be fully absorbed. Depending on the properties of the semiconductor, the magnitude $\alpha(\omega)$ in Eq. (1.70) is given by Eq. (1.35) or (1.36).

In conclusion, let us consider another point. According to Eq. (1.71), the full number of pairs generated by the light coincides with the number of absorbed photons. However, if the energy of the photons is large enough ($\hbar\omega \gtrsim 3E_g$), the number of generated pairs may turn out to be appreciably greater, and this is formally described by the condition $\beta > 1$ (see Eq. (1.69)). The physical reason for this effect is that the photoelectrons and photoholes have considerable kinetic energy which enables them to generate additional charge carriers by impact ionization of the atoms of the crystal lattice. For hard X-rays or γ-radiation the values of β may be very large and different for electrons and holes. This fact should be taken into consideration, in particular when describing the radiation-induced electrochemical processes examined in Section 11.4.

Chapter 2

THERMODYNAMIC PROPERTIES OF THE SEMICONDUCTOR/ ELECTROLYTE SOLUTION INTERFACE

The semiconductor/solution interface is a contact of two conducting media. Hence some of its properties are similar to those of the contacts between a semiconductor and a metal or between two semiconductors. At the same time, the interface in question is a contact of conductors with fundamentally different types of conductivity, one electronic, the other ionic, and with different states of aggregation of the contacting media: one solid, the other liquid. For these reasons, it has a number of unique features, including the problem of a thermodynamic description of electrons in the solution and of the equilibrium at the interface. A thermodynamic analysis, proceeding from purely general physical and chemical principles, brings to light some of the special features inherent to this interface.

2.1. The Work Function and Equilibrium Interphase Potentials

If contacting bodies are capable of exchanging particles, then at thermodynamic equilibrium the electrochemical potential of these particles must have the same value in all parts of the system. (This potential is reckoned from an energy level which is chosen arbitrarily but is the same for all parts of the system.) Conversely if the electrochemical potential changes in space, in particular when its gradient is nonzero, currents

necessarily appear in the system. Thus, the equilibrium
condition for the system merely signifies the absence of
any currents.

    If we denote the potential energy of an electron in
vacuum close to the surface yet outside the range of
purely surface forces as E_{vac}, then the electrochemical
potential (or Fermi level), F, in the solid relative to
E_{vac} is the thermodynamic (equilibrium) electronic work
function of the solid (see Fig. 2.1)

$$w_T = E_{vac} - F \qquad\qquad (2.1)$$

The magnitude w_T can be determined, for example, by
measuring the thermoemission current. We shall now carry
out a more detailed analysis to reveal the physical
meaning of the thermodynamic work function of a semiconductor.

    Let us imagine a closed cavity inside a solid (metal
or semiconductor) at a certain temperature T. At thermodynamic equilibrium the cavity will contain an electron
gas of thermoemitted electrons, with the current of
thermoemission from solid into the cavity being balanced
by the reverse current from the electron gas to the solid.
Hence, the magnitude of this current, which is called the
thermoemission saturation current, can be calculated irrespective of the electronic structure of the solid. With
the assumption that the electrons in the cavity behave
as an ideal gas the thermoemission saturation current
density (see, for example, (37)) is given by:

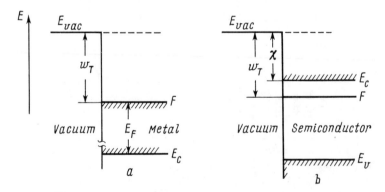

Fig. 2.1. Work function and characteristic
energy levels at the interfaces: a) metal/
vacuum; b) semiconductor/vacuum.

$$I_T = AT^2 exp\,(-w_T/kT) \qquad\qquad (2.2)$$

where $A \equiv 4\pi e k^2 m_0/(2\pi\hbar)^3$ is a universal constant equal
to 120 A cm^{-2}·grad^{-2}. It is essential that, irrespective
of the details of the model (an in-depth analysis can be
found, for example, in (38)), the quantity I_T should be
the same for metals and semiconductors when the appropriate value of the thermoemission work function is used.
It should be pointed out that in metals the Fermi level
coincides with the maximum energy of electrons in the
conduction band (ignoring the rather small "spread" of
the Fermi distribution in metals at $T \neq 0$). This is why
w_T for a metal is interpreted as the work necessary to
withdraw an electron with maximum energy (this energy,
when measured from the bottom of the conduction band, is the
Fermi energy E_F) from the metal to a vacuum. At the same
time, when applying Eq. (2.2) to non-degenerate
semiconductors, confusion may arise because in this case
the Fermi level, F, is located within the bandgap and,
consequently, w_T does not correspond to the work required
to remove an electron actually at that level. The reason
for this paradoxical conclusion is that the resultant
thermoemission current from a solid depends not only on
the energy of the electrons crossing the interphase
boundary, but also on their concentration in the conduction band. For non-degenerate semiconductors, unlike
metals, this concentration is proportional to
$exp\,(F - E_C/kT)$ (See Eq. (1.12)). Thus, the thermoemission current turns out to be proportional, firstly,
to the exponential function characterizing the density
of electrons in the conduction band and, secondly, to a
factor of

$$exp\left(-\frac{\chi}{kT}\right)$$

characterizing the probability of thermoexcitation of
electrons from the bottom of the conduction band where
$\chi \equiv E_{vac} - E_C$ and is the electron affinity (see Fig.
2.1). As a result, in Eq. (2.2) for the thermoemission current from semiconductors the exponential
contains (as was the case for a metal) only the work
function w_m, that is to say, the difference between
E_{vac} and the Fermi level, F (and not, for example, the
quantity χ).

The above example shows that even when the Fermi
level does not correspond to the energy of some particular particle, it is still the Fermi level that
determines the magnitude of the equilibrium current

across the interface. Note that the current I_T is
physically analogous to the exchange current in electrochemical systems (see section 4.1). It also follows
that the semiconductor work function may be largely dependent on the doping, provided it changes the position
of the Fermi level (see Fig. 1.5). When donor impurities
are introduced, the magnitude of w_T is supposed to diminish, and when acceptor impurities are introduced w_T
will increase.

The semiconductor work function in a vacuum depends
on the charging of the surface, therefore, it is highly
sensitive to surface contamination, the formation of imperfections, adsorption processes, and so on. The
charging of the surface of the semiconductor is accompanied by band bending. When the charge on the surface
is negative, that is, when a compensating positive charge
emerges in the semiconductor, the bands are bent upwards
(see Fig. 2.2), whilst if the surface charge is positive,
they are bent downwards. (The magnitude of $F - E_c$ in the
bulk of the semiconductor naturally remains constant.)
As a result, the change in the work function is

$$\Delta w_T = - e\phi_{sc} \qquad (2.3)$$

where ϕ_{sc} is the potential of the semiconductor surface
with respect to the bulk. The electron affinity,
$\chi = E_{vac} - E_c$, remains constant when the bands are bent
at the surface.

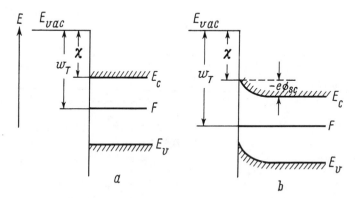

Fig. 2.2. The change in the work
function at the semiconductor/
vacuum interface due to charging
of the semiconductor surface: a)
no charge; b) positive charge on
the semiconductor.

The difference in Fermi levels between two solids or the difference in their thermodynamic work functions determines the contact potential difference $\Delta\psi$:

$$e\Delta\psi = F_2 - F_1 = w_{T,1} - w_{T,2} \qquad (2.4)$$

The contact potential difference is the difference in potentials between the points in the vacuum which are beyond the range of any surface forces but are close to the non-contiguous surfaces of the two different conductors which are assumed to be in electronic equilibrium. As seen from Fig. 2.3, the value of $\Delta\psi$ is the difference in potential between two points located in the same phase and, therefore, it can be measured experimentally.

A physical interpretation of the origin of the contact potential difference can be given as follows. When two materials are placed in contact, equilibrium is established and the levels F_1 and F_2 are equalized. (If, in particular, the two materials are electronic conductors, it is sufficient to connect them by a metal wire.) As a result of establishing equilibrium, the tops of the potential barriers, which restrict the surfaces of the two materials, will be at different heights while the potential energy of an electron, within the slit between the materials, will no longer be independent of position. The latter means that an electric field emerges between the two materials and charges appear on their surfaces.

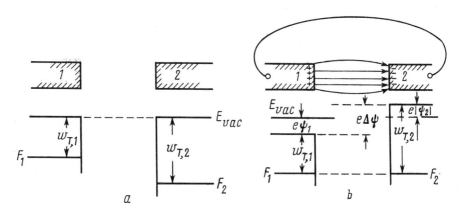

Fig. 2.3. The emergence of a contact potential difference between two electronic conductors: a) conductors electrically disconnected; b) conductors electrically connected and at equilibrium.

Note that the electronic work function of the charged conductor is the work required to transfer an electron from either conductor (1 or 2) to a point located in the vacuum near the surface of that conductor but beyond the range of purely surface forces. Thus, it does not include the change in the energy of the electron in the external electric field. When a non-charged conductor is considered, this work coincides with the work required for the transfer to an infinitely remote point. If the energy of the electron at this point, E_{vac}^{o}, is chosen to be zero then the position of the common Fermi level of the two contacting conductors is

$$F_{1,2} = w_{T,1} + e\psi_1 = w_{T,2} + e\psi_2 \qquad (2.5)$$

where $\psi_{1,2}$ are the so-called external (Volta) potentials of bodies 1 and 2: the difference in electric potential between the point just outside the solid and the infinitely remote point in the vacuum.

Thermodynamic equilibrium in the electrode/solution system is achieved by electron-ion exchange at the interface. Since, in this case, the establishment of equilibrium is controlled by charged particles, as in the contact between two electronic conductors, there will be some contact potential difference at equilibrium. At the same time, one can introduce the concept of the potential difference between two points located inside the different contacting bodies. This quantity is called the Galvani potential. In other words, the Galvani potential is the difference in electrostatic potentials between the two contacting phases, which compensates for the difference in their chemical potentials. It should be emphasized that since we mean the difference of potentials between points in different phases, the Galvani potential, unlike the contact potential difference, cannot be found directly by experiment.

In conclusion let us note that the concept of a contact potential difference (Volta potential) also applies to the electrode/electrolyte system. In this particular case it represents the difference in potential between two points in the same phase (the vacuum, or to be more precise, the vapor phase) which are near to the surfaces of the two contacting bodies, the electrode and the electrolyte solution (see Fig. 2.4). This is precisely analogous to the contact potential difference between two electronic conductors, and can also be measured directly.

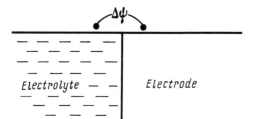

Fig. 2.4. The calculation of
the Volta-potential difference
for an electrode/electrolyte
system.

⊑ 2.2. The Equilibrium Electrode
Potential

The electrode potential E, which is the potential
difference between two identical metal terminals of an
electrochemical circuit, is the quantity usually measured
in electrochemical experiments. Such a circuit, with a
semiconductor electrode serving as one of its elements,
is represented schematically in Fig. 2.5. Apart from the
semiconductor electrode, it contains a reference electrode, whose potential is conventionally taken as zero
in the measurement of the electrode potential. An
essential feature specific to the materials used as
reference electrodes is that the Galvani potential at
the interface of such an electrode, with a solution of
specially chosen composition and concentration, has an
unknown but constant value which is not changed by
accidental influences (for details about reference electrodes see, for example, (39)).

The circuit in Fig. 2.5 is "properly disconnected,"
in as much as both its terminals are made of the same

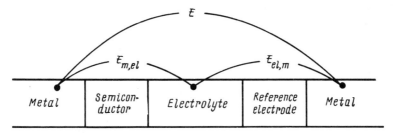

Fig. 2.5. Schematic representation
of a "properly disconnected" electrochemical circuit.

metal.   The electrode potential is made up of the potential drops at the various interfaces, that is, of
the Galvani potentials at the metal/semiconductor, semi-
conductor/electrolyte contacts, etc. (40,41), and also
of the ohmic voltage drops in the bulk phases when an
electric current is flowing.

In order to calculate the equilibrium electrode
potential, E^o, let us subdivide the potential difference
in the circuit into two parts, $E_{m,el}$ and $E_{el,m}$ (see
Fig. 2.5).   The quantity $E_{el,m}$, the potential difference
between the ends of the "half-circuit" electrolyte|
reference electrode|metal, is clearly independent of
both the properties of the semiconductor electrode and
the processes at the semiconductor/electrolyte interface.
These exert their most direct influence on the magnitude
of $E_{m,el}$ and, through this quantity, on the electrode
potential E^o.   In order to find the magnitude of $E_{m,el}$
let us calculate the Galvani potential at the semicon-
ductor/electrolyte interface.   In accordance with the
general thermodynamic principles, for a particle of the
i-th sort, which may be either in the solid, for example
in the semiconductor (sc), or in the electrolyte solution
(el), the condition for interphase equilibrium is given
by the relation

$$\tilde{\mu}_i^{sc} = \tilde{\mu}_i^{el} \tag{2.6}$$

where $\tilde{\mu}_i$ are the electrochemical potentials in the solid
and liquid phases; these are related to the chemical
potentials, μ_i, by

$$\tilde{\mu}_i = \mu_i + z_i F \phi \tag{2.7}$$

where z_i is the charge on the i-th particles (positive or
negative), F is the Faraday constant and ϕ is the electric potential in the phase.*

---

*It should be borne in mind that the magnitude $\tilde{\mu}_e$ in physical
chemistry is usually related to 1 mole of substance and is expressed in J/mol, whilst the magnitude of F in physics refers to
one particle and is expressed in electron-Volts; the transition
from F to $\tilde{\mu}_e$ is achieved by multiplying F by the Avogadro constant
$N_A = 6.02 \cdot 10^{23}$ mol^{-1}, taking into account the fact that 1 eV =
$1.61 \cdot 10^{-19}$ J; the Faraday constant is $F = eN_A$.

Let us assume that the electrolyte solution in the part of the cell where the semiconductor electrode is located contains a redox couple, and that equilibrium between the electrode and the solution is achieved by the reversible reaction of electron exchange

$$Ox + ne^- \; \underset{\longleftarrow}{\longrightarrow} \; Red \qquad (2.8)$$

where Ox and Red denote oxidized and reduced form of the solvated species and n is the number of electrons transferred in the reaction (the charge number). The condition for equilibrium for the reaction in (2.8) is

$$\tilde{\mu}_{red} = \tilde{\mu}_{ox} + n\tilde{\mu}_e^{sc} \qquad (2.9)$$

where $\tilde{\mu}_{red}$ and $\tilde{\mu}_{ox}$ are the electrochemical potentials of the components of the redox couple and $\tilde{\mu}_e^{sc}$ is the electrochemical potential of electrons in the semiconductor. Taking into account Eq. (2.7) and bearing in mind that $z_{ox} - z_{red} = n$, we get from Eq. (2.9)

$$nF[\phi(el) - \phi(sc)] = \mu_{red} - \mu_{ox} - n\mu_e^{sc} \qquad (2.10)$$

The difference $\phi(el) - \phi(sc)$, by definition, is the Galvani potential at the semiconductor/electrolyte interface. It follows from Eq. (2.10) that the Galvani potential, $\phi(el/sc)$, depends on the chemical potential, μ_e^{sc}, of electrons in the semiconductor.

Furthermore, at thermodynamic equilibrium

$$\tilde{\mu}_e^{met} = \tilde{\mu}_e^{sc} \qquad (2.11)$$

where $\tilde{\mu}_e^{met}$ is the electrochemical potential of electrons in the metal contacting the semiconductor. Hence, using Eq. (2.7), we get

$$F[\phi(met) - \phi(sc)] = \mu_e^{met} - \mu_e^{sc} \qquad (2.12)$$

where $\phi(met)$ is the electric potential in the metal.

Combining Eqs. (2.12) and (2.10), we find

$$nF[\phi(met) - \phi(el)] = n\mu_e^{met} - \mu_{red} + \mu_{ox} \qquad (2.13)$$

As a result, we get for the equilibrium electrode potential

$$E^O = \frac{1}{nF} (\mu_{ox} - \mu_{red}) + V \qquad (2.14)$$

where

$$V = E_{el,m} + \frac{1}{nF} \mu_e^{met}$$

and is independent of the properties of the semiconductor electrode.

Thus the electrode potential, with respect to a given reference, is, unlike the Galvani potential at the semiconductor/solution interface, determined only by the chemical potentials of the components of the redox couple in solution. In particular, it follows that impurities introduced into the semiconductor material, which determine its electric properties, have no influence on the equilibrium electrode potential. However these same impurities do influence, as has been stated before, the semiconductor work function and, consequently, the magnitude of the Galvani potential.

Since μ_{ox} and μ_{red} may be written in the form (26,28)

$$\mu_{ox,red} = \mu_{ox,red}^O + RT\ln c_{ox,red} \qquad (2.15)$$

where $R = N_A \cdot k$ is the universal gas constant, $c_{ox,red}$ is the concentration (or activity -- in a more general case) of the components Ox and Red in solution, and $\mu_{ox,red}^O$ are constants which are independent of $c_{ox,red}$ and ϕ, Eq. (2.14) can be rewritten as follows

$$E^O = E_o^O + \frac{RT}{nF} \ln \frac{c_{ox}}{c_{red}} \qquad (2.16)$$

where $E_o^O = V + \mu_{ox}^O - \mu_{red}^O$ is a standard potential which in this particular case is equal to the value of E^O when $c_{ox} = c_{red}$.

It has already been assumed that equilibrium at the semiconductor/electrolyte solution interface is achieved by the redox reaction (2.8) in which the material of the semiconductor electrode plays no part; the semiconductor behaves as an inert electrode.

Another important case is when the material of the electrode itself participates in the potential determining reaction, for example, when there is equilibrium between ions in solution and neutral particles (atoms) of the same element in the solid phase. Here the reversible electrode reaction is of the type

$$\{SC\}^{z+} + ze^- \rightleftharpoons \{SC\} \qquad (2.17)$$

where $\{SC\}$ and $\{SC\}^{z+}$ stand, respectively, for the species in the semiconductor and its ion in solution. Similar calculations to those above lead, in this case, to the following expression for the equilibrium electrode potential

$$E^O = E^O_O + \frac{RT}{zF} \ln c_{z+} \qquad (2.18)$$

where c_{z+} is the concentration (activity) of the ions $\{SC\}^{z+}$ in solution, whilst E^O_O is the standard electrode potential for the reaction (2.17). In the standard state the concentration of potential-determining ions in the solution is usually equal to 1 mol/dm^3.

The constant V in Eq. (2.14), and, consequently, the values of the standard potentials, depends on the nature of the reference electrode, that is, on its electrode reaction. On the other hand, there is no dependence on the nature of the metal, since the metals at the terminals of a properly disconnected electrochemical circuit are the same. Thus, it is necessary to introduce a relative scale of electrode potentials based on a particular standard reference electrode. For a reference electrode, whose potential is taken to be zero, one often chooses the normal hydrogen electrode (NHE) for which equilibrium at the electrode/solution interface is achieved by the reversible reaction

$$H^+ + e^- \rightleftharpoons \frac{1}{2} H_2 \qquad (2.19)$$

provided that the activity of H^+ ions in solution is 1 mol/dm^3, and the pressure exerted by gaseous hydrogen over the solution is 1 atm. It is assumed for this electrode that $E^O_{NHE} = 0$.

Many of the measured potential values cited below are related to the saturated calomel electrode (SCE);

its potential against the normal hydrogen electrode is
$E^O_{SCE} = 0.241$ V.

§ 2.3. The Electrochemical Potential
of Electrons in Solution

If there is a redox couple in the electrolyte
solution and equilibrium is established in accordance
with reaction (2.8), it is possible to introduce the
electrochemical potential of electrons in the solution,
$\tilde{\mu}_e^{el}$. From the thermodynamic viewpoint the detailed
mechanism for the establishment of equilibrium can be
ignored, and in particular, one can assume that electron
exchange between Ox and Red occurs without the participation of an electrode. The magnitude of $\tilde{\mu}_e^{el}$ is determined by a relation similar to (2.9). Hence,

$$\tilde{\mu}_e^{el} = \frac{1}{n} (\tilde{\mu}_{red} - \tilde{\mu}_{ox}) \qquad (2.20)$$

where $\tilde{\mu}_{red,ox}$ are taken at the potential in the bulk of
the solution $\phi(el)$, whereas the condition for electron
equilibrium of the redox couple in solution with an
electrode is as follows

$$\tilde{\mu}_e^{el} = \tilde{\mu}_e^{sc} \qquad (2.21)$$

From Eqs. (2.15) and (2.20) it follows that

$$\tilde{\mu}_e^{el} = (\tilde{\mu}^o)_e^{el} - \frac{RT}{n} ln \frac{c_{ox}}{c_{red}} \qquad (2.22)$$

It should be pointed out that the magnitude of $(\tilde{\mu}^o)_e^{el}$
does not depend on c_{ox} and c_{red}. In view of the fact
that the electrochemical potential $\tilde{\mu}_e^{el}$ at equilibrium
is constant (and equal to its value in the bulk of the
solution), its magnitude is clearly solely determined
by the properties of the Ox and Red species and the
solvent. In other words, the quantities $\tilde{\mu}_e^{el}$ and
$(\tilde{\mu}^o)_e^{el}$ have nothing to do with the nature of the electrode and do not depend on the structure of the inter-
face. Moreover, at thermodynamic equilibrium it is
precisely the quantity $\tilde{\mu}_e^{el}$ which determines the electrochemical potentials of electrons in the semiconductor,
$\tilde{\mu}_e^{sc}$, and metal electrodes, $\tilde{\mu}_e^{met}$, so that the values of
$\tilde{\mu}_e$ are the same for all the electrodes in contact with a
solution of a definite redox couple.

The calculation of $\tilde{\mu}_e^{el}$ requires the introduction of additional concepts. We shall consider a "redox couple" in solution which contains only one particle of the species, Red. Let us denote the most probable electron energy level of Red at thermodynamic equilibrium as E_{red}^o, and the analogous quantity for the oxidized species, Ox as E_{Ox}^o. In order to establish a connection between these two quantities and $\tilde{\mu}_e^{el}$, we shall consider the following thermodynamic cycle (see Fig. 2.6), which is similar to those analyzed in the literature (42-44); for simplicity we have taken n = 1:

a) the electron is transferred instantaneously from the particle Red in solution, which is in its normal sol-vated state, to the level E_{vac} (the potential energy level of an electron in vacuum at a point close to the electrolyte surface interface yet beyond the limits of surface forces) without changing the state of the solvent round the particle;

b) the state of the surrounding solvent near the ionized particle Red, which has now converted into an Ox particle, changes so that the thermodynamic equilib-rium state of the surrounding molecules of the solvent corresponding to the Ox particle is established;

c) an electron is transferred instantaneously from the level E_{vac} onto the thermodynamically equilibrated Ox particle in solution without introducing any changes in the structure of the solvent;

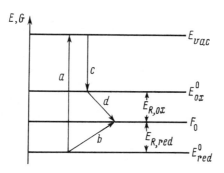

Fig. 2.6. The thermodynamic cycle used to calculate the electron energy levels for a solution containing a redox couple. E is the internal energy and G the free energy.

d) the Red particle which is formed as a result of the electron transfer triggers the reorganization of the solvent, and the system returns to the initial thermo-dynamic equilibrium state.

In this cycle the change of internal energy of the system in the step (a) characterizes the depth of the level E^O_{red} with respect to the vacuum level E_{vac}, while the change in internal energy in step (c) characterizes the depth of the level E^O_{ox} with respect to E_{vac}. (These quantities are also equal to the ionization energy of the Red component and the electron affinity of the Ox com-ponent of a redox couple in solution respectively.)

The energy changes in steps (b) and (d) are called the solvent reorganization energies, which we shall de-note as $E_{R,red}$ and $E_{R,ox}$ respectively. For the process under discussion the quantities $E_R > 0$ are the absolute values of energy (or more precisely, free energy) neces-sary to change the state of the medium (the solvent) from the initial equilibrium condition to the final equilibrium condition when the charge of the particle is changed by unity. Physically the reorganization energy is connected with the change in free energy of the en-semble of the solvent molecules surrounding the particle, as a result of changing its charge.

There is no reliable information about the values of E_R. According to (45), in aqueous solutions at room temperature, the magnitude of E_R for different ions lies approximately within the range 0.5 - 2 eV. The problems arising in model calculations of the dependence of the reorganization energy on the properties of the solvent and the characteristics of the ion has been examined (46) (see also Section 4.2).

Thus, it follows from the analysis of the thermo-dynamic cycle shown in Fig. 2.6 that

$$E^O_{red} + E_{R,red} - E^O_{ox} + E_{R,ox} = 0 \qquad (2.23)$$

The quantity F_O given from Eq. (2.23) as

$$F_o = E^O_{red} + E_{R,red} = E^O_{ox} - E_{R,ox} \qquad (2.24)$$

is obviously the free energy of a redox couple containing only one pair of Ox/Red particles. In accordance with

what has been said above, F_O is equal to the change of
free energy of such a system when its only electron is
removed (from Red to vacuum), and equilibrium is maintained. In polar liquids, water in particular, it may
be assumed that $E_{R,ox} = E_{R,red} = E_R$, since it is the
effect of the repolarization of the solvent, due to
electrostatic interactions, that makes the most significant contribution to E_R. The contribution made by
the change in short-range forces, in this case, is
small. If we assume that the reorganization energies
$E_{R,ox}$ and $E_{R,red}$ are equal to each other, we obtain from
Eqs. (2.23) and (2.24)

$$F_O = \frac{1}{2} (E^O_{ox} + E^O_{red}), \quad E^O_{ox} - E^O_{red} = 2E_R \qquad (2.25)$$

Thus, the characteristic level F_O is located centrally
between the levels E^O_{ox} and E^O_{red}, while the distance
between these levels is equal to twice the reorganization
energy (see Fig. 2.6). Note that this result can be obtained directly from an examination of the thermodynamic
equilibrium electron transfer from a Red particle to an
Ox particle within the solution.

It follows from the definition of F_O (2.24) and
Eq. (2.22) for the electrochemical potential of electrons in an ideal solution that the equality
$(\tilde{\mu}^O)_e^{el} = N_A F_O$ holds, and so finally we obtain (for n = 1)

$$\tilde{\mu}_e^{el} = N_A F_O - RT \ln \frac{c_{ox}}{c_{red}} \qquad (2.26)$$

Equation (2.26), in conjunction with Eqs. (2.24) and
(2.25), establishes the connection between $\tilde{\mu}_e^{el}$ and the
characteristics of the redox couple.

In view of the fundamental importance of Eq.
(2.26), let us obtain it by another method: using statistical mechanics and without using Eqs. (2.20) and
(2.15). With this aim in view, we shall examine a model
system. This system is a set of cells, whose average
number per unit volume is N; in the cells electrons are
placed, with no more than one in each, so that their
average number per unit volume is n $\leq$ N. The free energy
of the system provided there is one electron in a cell
is F_O. The free energy, G, of unit volume of such system
containing n electrons is given by (27)

$$G = F_o n - TS \tag{2.27}$$

where S, the configurational entropy, is given by

$$S = -kln C_N^n = -kln \frac{N!}{(N-n)!n!} \tag{2.28}$$

where C_N^n is the number of configurations: the number of ways of placing n particles in N cells. The electro-chemical (chemical) potential of electrons, related to 1 mole of substance, is equal to $\tilde{\mu}_e = N_A \partial G/\partial n$. From Eqs. (2.27), (2.28) and using the relation $dlnq!/dq = lnq$ for $q \gg 1$ we obtain

$$\tilde{\mu}_e = N_A F_o - RT ln \frac{N-n}{n} \tag{2.29}$$

Note that this model system simulates the redox couple in solution. In the redox system the electrons, whose number is equal to c_{red}, where c_{red} is the concentration of the reduced species (Red), are placed in the cells, whose number is $c_{red} + c_{ox}$ (c_{ox} is the concentration of the oxidized species (Ox)). Thus substituting $n = c_{red}$ and $N = c_{red} + c_{ox}$ in Eq. (2.29) we obtain

$$\tilde{\mu}_e = N_A F_o - RT ln \frac{c_{ox}}{c_{red}} \tag{2.30}$$

which is identical to Eq. (2.26) for $\tilde{\mu}_e^{el}$ derived above.

Thus, the electrochemical potential of electrons in solution can be introduced strictly and consistently within the framework of both a purely thermodynamic model or a statistical-mechanical approach.

The quantity $\tilde{\mu}_e^{el}$ is often called the Fermi level of electrons in solution. By analogy with the Fermi level of electrons in a semiconductor, F, let us introduce a quantity F_{redox}, which can be determined from the re-lation $F_{redox} = \tilde{\mu}_e^{el}/N_A$ and represents the electrochem-ical potential per particle of electrons in solution. The position of the level F_{redox} is determined by the presence in solution of the redox couple Ox/Red.

At the same time, it should be pointed out that the term "Fermi level of electrons in solution" (or worse, "Fermi level of solution") is inadequate and often

misleading. In this connection it must be emphasized
that the position of F_{redox} in solution is in no way
linked with the existence of "free" electrons, and is
determined by the presence of a redox couple including
only "bound" electrons. The magnitude of F_{redox} does
not characterize the energy of any actual particle.
Moreover, the solution does not contain electrons of
energy F_{redox} (or of energy F_o) alone. Hence, at the
interface of such a solution with the electrode, electrons are unable to move either from the level F_{redox}
to the electrode or from the electrode to that level:
the transitions occur to and from the levels E^O_{red} and
E^O_{ox} (see Section 4.2).

At the same time, it is precisely the position of
the level F_{redox} that determines the thermodynamic
properties of the redox couple, as well as the thermodynamic properties of the interface between a semiconductor and a solution containing such a couple. For
example, proceeding from the condition for equilibrium
(cf. Eq. (2.21))

$$F = F_{redox} \qquad\qquad (2.31)$$

the conditions for the onset of an electrochemical reaction can be written (47) as

$$F > F_{redox} \qquad \text{(cathodic reaction)} \qquad (2.32a)$$

$$F < F_{redox} \qquad \text{(anodic reaction)} \qquad (2.32b)$$

Similar expressions are used when studying the
photoelectrochemical reactions on illuminated semiconductor electrodes.

2.4. Equilibrium Electrode
   Potentials of Major
   Semiconductor Materials

To be able to conveniently utilize the notion of
the electrochemical potential, F_{redox}, of electrons in
solution, whilst characterizing electrochemical processes, it is necessary to establish a connection between
the "physical" scale of energies on which the energy of
an electron in a vacuum, E_{vac}, is taken as zero, and the
"electrochemical" scale of electrode potentials on which
the potential of some reference electrode is assumed to
be zero. This connection in its general form is

$$F_{redox} = -eE^{o} + const \qquad (2.33)$$

where E^{o} is the potential of an electrode which is in equilibrium with a solution containing a redox couple which determines the value of F_{redox}. Suppose a particular couple Ox/Red is chosen, for which the equilibrium potential, E^{o}, is conventionally regarded as zero. This solution in contact with an inert conductor is, by definition, a reference electrode. Then the constant, const, in Eq. (2.33) coincides with F_{redox} and is the free energy change in the course of the electron transfer

$$Ox + e^{-}_{vac} \longrightarrow Red \qquad (2.34)$$

Note that this reaction (2.34) is simply reaction (2.8) carried out from left to right with the consumption of an electron from a point in the vacuum near the solution surface. The change of free energy in such a process (see Section 2.3) is the sum of the energies of steps (c) and (d) in the cycle shown in Fig. 2.6. It is therefore equal to the value of F_{redox} (which coincides with F_{0} when $c_{ox} = c_{red}$). The method of determining F_{redox} is clear from Fig. 2.7 and is quite similar, in view of the comments in Section 2.1, to the determination of the work function of a solid by means of measuring the contact potential difference. Let us assume that the metal is in equilibrium with a solution containing a redox couple. Then, as is seen from the cycle in the figure

$$-F_{redox} = w_{T} + e\Delta\psi \qquad (2.35)$$

where w_{T} is the work function of a non-charged metal in a vacuum and $\Delta\psi$ is the difference in Volta potentials (the contact potential difference) in the metal/solution

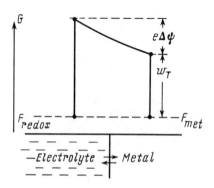

Fig. 2.7. To establish the relationship between the "physical" and the "electrochemical" energy scales.

system, emerging as a result of mutual charging of the outer (non-contiguous) surfaces of the electrode and solution when they come into contact. The Volta potential, naturally, depends (as well as F_{redox}) on the electrode potential E. The magnitude F_{redox} is determined by the particular redox couple and, therefore, changes from one reference electrode to another.

For a standard hydrogen reference electrode in aqueous solution, Eq. (2.34) takes the form

$$H^+_{aq} + e^-_{vac} \longrightarrow \tfrac{1}{2} H_2 \qquad (2.36)$$

and the free energy of this reaction is independent of the nature of the inert conductor used to make the hydrogen electrode. (Note that the Fermi levels of electrodes which have the same potential coincide irrespective of their chemical composition.) It is therefore possible to make calculations in accordance with Fig. 2.7 for, for example, an electrode made of mercury (although it is not used practically as a hydrogen electrode) as there are reliable literature values for its work function and Volta potential: $w_T = 4.48$ eV, $\Delta\psi = -0.07$ V at the normal hydrogen electrode potential (50). For the constant in Eq. (2.33) we obtain const = F (NHE) = -4.4 eV. Thus, the electrochemical potential of electrons in the metal at the potential of the normal hydrogen electrode is 4.4 eV lower than the potential energy level of an electron in vacuum at a point close to the solution surface. In other words, the potential of the hydrogen electrode "measured" relative to vacuum is -4.4V; this is sometimes called the absolute potential (see, e.g., (51)). Note that in one widely cited reference (52) an unnecessarily complicated way has been used to determine F (NHE) (using the cyclic process of evaporation of the metal electrode, ionization of the atoms, and solvation of ions thus obtained) with the result that some inaccuracy has been introduced into the calculated value.

Thus, knowing the value of E^o (for example from thermodynamic data) and using Eq. (2.33) with the calculated value for the constant, it is possible to determine F_{redox} for a given reaction.

The experimental measurement of the equilibrium potentials of semiconductor electrodes in contact with solutions of redox couples encounters certain difficulties. The stationary potential of the semiconductor electrode usually differs from the equilibrium value.

This is due to the fact that, as will be shown in Chapter 4, the attainment of equilibrium between the semiconductor and the solution is somewhat hindered. Hence, many semiconductor electrodes behave over a broad range of potentials as ideally polarizable electrodes. Remember that an electrode is termed ideally polarizable when a shift in its potential (polarization) is not accompanied by the transfer of charge across the electrode/solution interface. The stationary potential of such electrodes is determined by factors which are inadequately controlled (adsorption from solution, electrostatic charging from accidential sources, and so on).

The measurements of equilibrium potentials at semiconductor electrodes also encounters difficulties due to the fact that the semiconductor materials are often inclined to corrode, especially in aqueous solutions. The "current-less" potentials of germanium, silicon and other electrodes are therefore usually mixed, and not equilibrium, potentials.

In some cases, nevertheless, it is possible to realize the reversible potential described by Eqs. (2.16) or (2.18) at semiconductor electrodes in contact with a solution containing a redox couple. Examples of such a situation are the cadmium chalcogenide electrodes in solutions of polychalcogenides, for example, CdS in a solution containing a mixture of S^{2-} and S_2^{2-} ions. These electrodes are becoming increasingly important in connection with the conversion of solar into electrical energy (see Chapter 8).

The values of the equilibrium potentials of electrode reactions may be calculated from thermodynamic data (see, for example, (53)). Values for some reactions of practical importance are cited in Section 7.3.

To depict the thermodynamics of semiconductor electrodes in solution, it is convenient to use potential-pH diagrams (which are often called "Pourbaix diagrams" (54)); these graphically illustrate the dependence of the equilibrium potential of various reactions, in which the electrode material and solution species take part, on solution pH. Such dependences occur when, apart from Red and Ox reagents, H^+ or OH^- ions take part in potential-determining equilibria at the interface. For a number of practically important semiconductor materials, such diagrams, calculated in (55), are given in Appendix 2.

Chapter 3

THE STRUCTURE OF THE ELECTRIC DOUBLE LAYER
ON SEMICONDUCTOR ELECTRODES

In Chapter 2, based on a thermodynamic approach, the Galvani potential at the semiconductor electrode/ electrolyte solution interface was calculated, and its dependence on the characteristics of the semiconductor and the solution was examined. The very existence of a Galvani potential testifies to the fact that, as in the case of a metal/solution interface, an electric double layer emerges at the semiconductor/solution interface, which is formed by charges of opposing signs in the two contacting phases. The existence of this double layer exerts a tangible, and sometimes decisive, influence on the physical and chemical processes occurring at the interface. The structure of this electric double layer, its mathematical description, and some techniques for its experimental study will be examined in detail below.

Ξ 3.1. Model of the Electric Double Layer

The electric double layer consists of two "plates" containing charges of opposite sign, each "plate" being located in one of the contacting phases. The charge in the double layer may be either concentrated directly on the surface of the phase, that is, in a sheet whose thickness is of the order of atomic dimensions, or distributed diffusely near the surface over a distance that sometimes exceeds the atomic size by many orders of magnitude. In the semiconductor the diffuse charge near the surface is formed as a result of a redistribution of the electrons and holes, while in the electrolyte solution it is formed as a result of a redistribution of the ions which make up the so-called ionic part of the double layer. From the electroneutrality condition it follows that at equilibrium the absolute values of charge on the two "plates" of the double layer will be equal to each other.

The physical consequences of the formation of the electric double layer are of a rather general character. Firstly there is the movement of charges across the interface leading to the establishment of thermodynamic equilibrium between the two contacting phases. Secondly there are the charging processes which, generally speaking, are independent of charge transfer at the interface. Examples of such processes are the charging of surface states (for details see Section 3.5), certain kinds of adsorption and so on.

In accordance with the generally accepted model (5,56,57), three regions are distinguished within the electric double layer: the space charge region in the solution, an intermediate layer called the Helmholtz layer and the space region in the semiconductor.

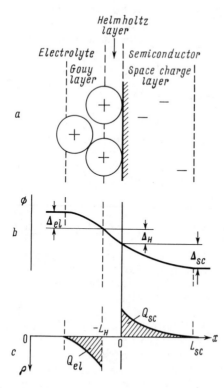

Fig. 3.1. The schematic structure of the electric double layer at the semiconductor/electrolyte interface (a), the potential distribution within this layer (b), and the charge distribution (c);
$\Delta_{el}=[\phi(-L_H)-\phi(-\infty)]=\phi_{el}$;
$\Delta_H=[\phi(0)-\phi(-L_H)]=\phi_H$;
$\Delta_{sc}=[\phi(\infty)-\phi(0)]=-\phi_{sc}$.

Fig. 3.1 shows schematically (not to scale) the
spatial distribution of charge and of the Galvani potential ϕ(el/sc) at the semiconductor/electrolyte interface.
The semiconductor occupies the area to the right of the
vertical solid line which denotes the interface (x=0).
The x axis is perpendicular to the plane of the interface
and is directed towards the semiconductor bulk. The
Helmholtz layer (58), or the dense part of the ionic double
layer, is located to the left of this line. In the
simplest model the Helmholtz layer is formed by ions
attracted to the electrode surface, and also by molecules
of the solvent. Its thickness, L_H, is of the order of
the size of an ion. Finally, the space charge layer in
the electrolyte solution is adjacent to the Helmholtz
layer starting at x = $-L_H$. This is called the diffuse
part of the ionic layer, or the Gouy layer (also called the
Gouy-Chapman layer) (59-61). A developed Gouy-Chapman
layer is formed only when the electrolyte is moderately
dilute ($\leq 10^{-1}$ mol/dm^3).

Thus, the full potential drop at the interface, the
Galvani potential, is (see Fig. 3.1)

$$\phi(el/sc)=\phi(sc)-\phi(el)=[\phi(-L_H)-\phi(-\infty)]+[\phi(0)-\phi(-L_H)]+[\phi(\infty)-\phi(0)] \quad (3.1)$$

Where the first term in square brackets is the potential
drop in the space charge layer of the electrolyte solution
(the Gouy-Chapman layer), $\phi_{el}=\phi(-L_H)-\phi(-\infty)$ and is usually
referred to as ψ', the "psi-prime potential." The second
term in square brackets in Eq. (3.1), $\phi(0)-\phi(-L_H)$, represents the potential drop in the Helmholtz layer, and
will be denoted here as ϕ_H. Finally the last term in
square brackets represents the potential drop in the
space charge layer of the semiconductor.

Here attention should be drawn to the following circumstance which is methodically important. It is an
established tradition in electrochemistry that the
positive direction of electrode potential is chosen to
correspond to increasing positive charge on the electrode.
Furthermore, the potential of the reference electrode which
is related, within the constancy of the interfacial potential drop, to the potential in the bulk of the solution,
$\phi(el)=\phi(-\infty)$, is taken as the zero of the potential scale.
On the other hand, in semiconductor surface physics potentials are read from the value in the bulk of the
semiconductor, $\phi(sc)=\phi(\infty)$. Enrichment of the space
charge layer with electrons, a negative space charge,
then corresponds to a positive potential at the surface
in accordance with the Boltzmann distribution (1.60).

Thus, the directions of the accepted potential axes of electrochemistry and semiconductor physics are directly opposed. It is for this reason that the potential of the semiconductor surface, ϕ_{sc} (read from $\phi(sc)=\phi(\infty)$) turns out to be opposite in sign to the third term in square brackets in Eq. (3.1)

$$\phi_{sc}=-[\phi(\infty)-\phi(0)]=\phi(0)-\phi(\infty)$$

Note that the positive potential direction in the semiconductor conforms with the direction chosen as positive in the "physical" scale of energies (see Section 2.3). This is reflected by the fact that in Eq. (2.33) the terms characterizing the position of the Fermi level reckoned from E_{vac} and the contribution of the electrode potential have opposite signs.

When an electrode is polarized, generally, all three components of the Galvani potential change. Taking account of what has been said,

$$\Delta\phi(el/sc) = \Delta\phi_{el} + \Delta\phi_{H} - \Delta\phi_{sc}$$

The ratio of the potential changes in the different parts of the double layer depends on the conditions at the interface.

It should be pointed out that dipoles close to the interface also contribute to the formation of the double layer. Here we mean the oriented adsorption of the solvent molecules and the dipole moment produced by the formation of polar bonds between the surface atoms of the semiconductor and adatoms. Moreover, a certain dipole moment, determined by the distribution of electron density, is sure to exist on the surface of the solid (this is particularly true of semiconductors) even when it is in contact with a vacuum. This initial "surface dipole" is deformed by the contact with the condensed phase (the electrolyte solution), and this is equivalent to the emergence of an additional (as compared with the "vacuum" value) dipole moment at the electrode surface.

The aforesaid dipoles make no contribution to the charge on the electrode but they do give rise to some potential drop, $\phi_{H,dip}$, (often included in ϕ_{H}). As experiments show (see Section 3.8) this dipole potential drop may alter to a rather small degree when the potential of the electrode changes, but it is usually highly sensitive to the type of preliminary processing of the electrode's surface and to the composition of the electrolyte.

Ξ 3.2. S p a c e   C h a r g e   i n   S e m i c o n d u c t o r : F r e e   C a r r i e r s   a n d   P o t e n t i a l D i s t r i b u t i o n

The formation of a space charge and the electric potential drop close to the semiconductor surface associated with it leads to a change in potential energy of the electrons with distance, that is to the bending of the energy bands (Fig. 3.2). Since the charge on the electron is negative, the bands are bent downwards if $\phi_{sc} > 0$, and upwards if $\phi_{sc} < 0$, with a total magnitude of bending equal to $|e\phi_{sc}|$.

In the special case when $\phi_{sc} = 0$, the bands remain unbent ("flat") right up to the surface. The potential, E, of the semiconductor electrode (measured against some reference electrode) which corresponds to this special case, $\phi_{sc} = 0$, is called the flat band potential, E_{fb}. At the flat band potential, by definition, the potential on the semiconductor surface coincides with the value of the potential in the semiconductor bulk.

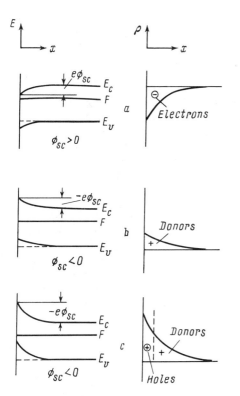

Fig. 3.2. The band bending and charge distribution near the surface of an n-type semi-conductor for various values of the potential drop, ϕ_{sc}:
a – accumulation layer;
b – depletion layer;
c – inversion layer.

The flat band potential in the electrochemistry of semiconductors is equivalent to the potential of zero charge (pzc) in the electrochemistry of metals. According to Frumkin (62), it is necessary to distinguish between the "free" (formed by electrons) and the "bound" (in the form of chemisorbed particles) charge on the electrode. Correspondingly, there exist potentials of zero free charge and of zero full charge, which is made up of free and bound charges. In conformity with this definition, the space charge of a semiconductor is free and, consequently, the flat band potential is the potential of zero free charge.

Let us emphasize that at $E = E_{fb}$ (when $\phi_{sc} = 0$) the potential drop in the Helmholtz layer, ϕ_H, may be non-zero, due, for example, to specific ionic adsorption and the contributions of surface dipoles.

When the bands are bent downwards, electrons "roll down" along the bottom of the conduction band towards the surface, and a layer enriched with electrons emerges near the surface (Fig. 3.2a). When the bands are bent upwards, electrons, on the other hand, "roll away" into the depth of the crystal (Fig. 3.2b,c), while the holes "roll down" to the surface. Considering an n-type semiconductor as a specific example, a layer enriched with majority carriers (electrons) is formed when the bands are bent downwards. When they are bent upwards, two situations are possible. If the bands are not bent upwards by very much, the number of electrons near the surface will already be small, yet the number of holes will still be insignificant. In this case the space charge near the surface consists, in the main, of immobile ionized donor atoms. This emerging space charge layer is called a Mott-Schottky or depletion layer (Fig. 3.2b). When the bands are bent upwards to an even greater extent, the charge created by minority carriers (holes) becomes dominant; this layer is then called an inversion layer (Fig. 3.2c). In a p-type semiconductor an accumulation layer is found when the bands are bent upwards ($\phi_{sc} < 0$), whilst the depletion and inversion layers occur when the bands are bent downwards ($\phi_{sc} > 0$).

Let us now move on to the quantitative description of the properties of the semiconductor space charge layer. The charge density ρ is due to both mobile carriers (electrons and holes) and immobile ones (ionized donor and/or acceptor type impurities). At equilibrium, in a non-degenerate semiconductor the local concentrations of electrons and holes are described by the Boltzmann

distribution (1.60), the potential, $\phi(sc)$, in the bulk
of semiconductor being chosen as zero. The transposition
to another origin is simply achieved by replacing
$\phi(x)$ by $\phi(x) - \phi(sc)$.

The potential distribution, $\phi(x)$ in Eq. (1.60),
can be obtained by solving Eq. (1.61). If we assume
that both donors and acceptors are fully ionized and
use Eqs. (1.60) and (1.62), this equation can be
written as follows

$$\frac{d^2\phi}{dx^2} = - \frac{e}{\varepsilon_{sc}\varepsilon_o} (p_o e^{-e\phi(x)/kT} - n_o e^{e\phi(x)/kT} + N_D - N_A) \quad (3.2)$$

The above equation is called the self-consistent Poisson-
Boltzmann equation.

From the condition of electro-neutrality it follows
that in the bulk of semiconductor $(x \to \infty)$,

$$p_o + N_D - n_o - N_A = 0 \quad (3.3)$$

Taking into account Eq. (3.3) the boundary conditions
for Eq. (3.2) are

$$\phi(0) = \phi_{sc}, \quad d\phi/dx \big|_{x\to\infty} = 0$$

In addition, as $x\to\infty$, we have $\phi\to0$ from our choice of the
zero of potential.

If the potential drop ϕ_{sc} is sufficiently small, so
that $|e\phi_{sc}/kT| \ll 1$, the exponential terms on the right
hand side of Eq. (3.2) can be expanded as a series and
then using Eq. (3.3), it is possible to transform Eq.
(3.2) to give

$$\frac{d^2\phi}{dx^2} = \frac{e^2}{\varepsilon_{sc}\varepsilon_o kT} (n_o + p_o)\phi(x) \quad (3.4)$$

The analytical solution of Eq. (3.4) is easily found,

$$\phi(x) = \phi_{sc} \, exp(-x/L_D) \quad (3.5)$$

where

$$L_D = \left[\frac{\varepsilon_{sc} \varepsilon_o kT}{e^2 (n_o + p_o)} \right]^{\frac{1}{2}}$$

In accordance with Eq. (3.5), the potential tends exponentially to zero on moving away from the interface. It is evident that the electric field in the space charge layer

$$\xi = - \frac{d\phi}{dx} = \phi(x)/L_D$$

as well as the concentrations of carriers

$$n(x) = n_o\left(1 + \frac{e\phi_{sc}}{kT} e^{-x/L_D}\right), \quad p(x) = p_o\left(1 - \frac{e\phi_{sc}}{kT} e^{-x/L_D}\right) \quad (3.6)$$

will also both change exponentially. Thus, the characteristic length L_D, which is called the Debye length, or the screening length, scales the variation of these physical parameters within the space charge layer of the semiconductor. Assuming that, for example, $n_o = 10^{15}$ cm^{-3}, $n_o \gg p_o$ and $\varepsilon_{sc} = 10$, we find that at room temperature $L_D = 10^{-5}$ cm.

Another important case for which Eq. (3.2) can be solved analytically is that of a relatively heavily doped n-type semiconductor with a negative (or for p-type a positive) surface potential so that a depletion layer is formed. As a specific example let us assume that $N_D \gg n_i$, N_A, so that $n_o \simeq N_D$ and $p_o \ll N_D$. In this case there apparently exists a range of negative potentials, $\phi_{inv} < \phi_{sc} < -kT/e$, where $\phi_{inv} < 0$ and

$$\phi_{inv} = - \frac{kT}{e} \ln \frac{N_D}{p_o} \quad (3.7)$$

when the N_D term in brackets on the right-hand side of Eq. (3.2) dominates. In this case Eq. (3.2) will be

$$\frac{d^2\phi}{dx^2} = - \frac{eN_D}{\varepsilon_{sc}\varepsilon_o} \quad (3.8)$$

Equation (3.8) holds only when there is a depletion layer. This is why the boundary conditions for Eq. (3.8), strictly speaking, should be taken not at $x \to \infty$ but at the boundary of the depletion layer at $x = L_{sc}$ (where it is necessary to match the found solution to that of Eq. (3.2) which contains additional terms on the right hand side). However, for $x > L_{sc}$ the field is small, so that it can be assumed to be negligible. Making this assumption, let us impose the following boundary conditions on Eq. (3.8)

$$\phi(L_{sc}) = 0, \quad d\phi/dx \big|_{x=L_{sc}} = 0 \tag{3.9}$$

which, together with the condition $\phi(0) = \phi_{sc}$, determine the two integration constants in Eq. (3.8) and the thickness of the space charge layer, L_{sc}. Integration then gives

$$\phi(x) = \begin{cases} -\dfrac{eN_D}{2\varepsilon_{sc}\varepsilon_o}(x - L_{sc})^2, & 0 \leqslant x \leqslant L_{sc} \\ 0, & x \geqslant L_{sc} \end{cases} \tag{3.10}$$

where $L_{sc} = (2\varepsilon_{sc}\varepsilon_o |\phi_{sc}|/eN_D)^{\frac{1}{2}}$

The approximate parabolic solution of Eq. (3.2) in Eq. (3.10) accurately describes the variation of the potential with distance, $\phi(x)$, with the exception of the usually insignificant "tail" in the depth of the space charge layer. Let us also note that inasmuch as

$$p_o + n_o \simeq N_D, \text{ then } L_{sc} = L_D\left(\frac{2e|\phi_{sc}|}{kT}\right)^{\frac{1}{2}}$$

Thus, the thickness of the depletion layer, L_{sc}, somewhat exceeds the Debye length, L_D.

In the general case Eq. (3.2) is not integrated, although its first integral can be calculated exactly, thus making it possible to obtain a number of useful relations. Multiplying the two parts of Eq. (3.2) by $d\phi/dx$ and using the relationship

$$\frac{d^2\phi}{dx^2}\frac{d\phi}{dx} = \frac{d}{dx}\left[\frac{1}{2}\left(\frac{d\phi}{dx}\right)^2\right]$$

we obtain, as a result of integrating the left-hand side of Eq. (3.2) from $x = 0$ to ∞, and the right-hand side from $\phi = \phi_{sc}$ to 0,

$$\left(\frac{dY}{dx} \right)^2 = - \frac{2e^2}{\varepsilon_{sc}\varepsilon_o kT} \left[(N_D - N_A) Y - p_o (e^{-Y}-1) - n_o (e^Y-1) \right] \quad (3.11)$$

where $Y \equiv e\phi_{sc}/kT$. In calculating the integration constant, we have used the fact that in the depth of semiconductor, when

$$x \to \infty, \quad \phi = 0 \text{ and } \frac{d\phi}{dx} = 0$$

Introducing the dimensionless parameter $\lambda = (p_o/n_o)^{\frac{1}{2}}$ and bearing in mind that, in accordance with Eqs. (1.16) and (1.17)

$$N_A - N_D = p_o - n_o = n_i (\lambda - \lambda^{-1}) \quad (3.12)$$

the expression for the electric field at the semiconductor surface

$$\xi_{sc} = - \frac{d\phi}{dx} \bigg|_{x=0}$$

in accordance with Eq. (3.11), can be written as (63)

$$\xi_{sc} = \pm \frac{kT}{eL_D^i} F(Y,\lambda) \quad (3.13)$$

where

$$L_D^i = \left(\frac{\varepsilon_{sc}\varepsilon_o kT}{2e^2 n_i} \right)^{\frac{1}{2}}$$

is the Debye length for the intrinsic semiconductor ($p_o = n_o = n_i$) and

$$F(Y,\lambda) = [\lambda (e^{-Y}-1) + \lambda^{-1}(e^Y-1) + (\lambda-\lambda^{-1})Y]^{\frac{1}{2}} \quad (3.14)$$

The function $F(Y,\lambda)$ has been 'tabulated: see, for example, (64) (and also Appendix 3); graphs of it can be found in (65-67). The sign in Eq. (3.13) should be chosen to be the same as that of Y, as can be seen from our previous solution for small ϕ_{sc}, Eq. (3.5).

Using the function $F(Y,\lambda)$ the full charge on the semiconductor per unit surface area, Q_{sc}, is readily calculated

$$Q_{sc} = \int_0^\infty \rho(x)\,dx = e\int_0^\infty [p(x)-n(x) + N_D - N_A]\,dx \quad (3.15)$$

According to Eqs. (1.61), (3.13) and (3.14), we then have

$$Q_{sc} = -\varepsilon_{sc}\varepsilon_0\xi_{sc} = \pm\frac{kT\varepsilon_{sc}\varepsilon_0}{eL_D^i}F(Y,\lambda) \quad (3.16)$$

The quantities

$$\Gamma_n = \int_0^\infty (n - n_0)\,dx \quad \text{and} \quad \Gamma_p = \int_0^\infty (p - p_0)\,dx$$

are called the surface excesses of electrons and holes respectively. It is obvious that $Q_{sc} = e(\Gamma_p - \Gamma_n)$. The surface excesses, Γ_n and Γ_p, are expressed through integrals containing $F(Y,\lambda)$. The results of numerical integration can be found in the literature as nomograms and tables (see, for example, (65)).

The calculations above were made with the assumption that the free carriers obey Boltzmann statistics throughout the space charge region. However, when an accumulation or inversion layer emerges near the interface, this assumption is not always fulfilled. For sufficiently strong fields the concentration of carriers in this layer may become so large that, to calculate it, Fermi-Dirac statistics may be required. In this case the carriers near the surface are degenerate. The corresponding condition, as for degeneracy in the bulk, is given by the inequality of Eq. (1.23), when the concentrations of carriers in the semiconductor bulk are replaced by their surface concentrations n_s and p_s. Looked at in another way, from the conditions in Eq. (1.24), surface degeneracy sets in when the band bending

is so large that near the surface one of the band edges
($E_{c,s}$ or $E_{v,s}$) comes close ($\lesssim 3kT$) to the Fermi level F.

In the simplest approach surface degeneracy is taken
into account by solving a self-consistent Poisson's
equation of the type given in Eq. (3.2), but using the
Fermi-Dirac distribution to calculate the charge density
(see (56, 68, 69)). The electric field at the surface
in this case is described by a formula of the type of
(3.13) but with a different function, also tabulated in
the works cited. This approach, when all remaining
quantum effects in the space charge layer are neglected,
has not been fully investigated. The problem is that
under the conditions discussed above the de Broglie wavelengths, λ_B, of electrons and holes (see p.15) turn out
to be comparable not only with the mean distances between carriers, but also with the thickness of the
accumulation or inversion layer. To provide a more
correct account of the quantum effects, it is necessary
to solve the Schrödinger equation for electrons in the
space charge region. In recent years such an approach
has been developed by a number of workers (70-75). Calculations are usually based on a model of free electrons
possessing an effective mass and moving in a one-dimensional electric field. Proceeding from the one-dimensional Schrödinger equation it is possible to find wave
functions for the electrons (holes) ψ_k (the index k denotes the entire set of quantum numbers describing the
given state). The charge density ρ is related to the
wave functions ψ_k by

$$\rho = - e \sum_k |\psi_k|^2 f(E_k) \qquad (3.17)$$

where $f(E_k)$ is the Fermi-Dirac distribution function.
Depending on the details of the calculations (the use of
various boundary conditions for the Schrödinger equation,
the methods used to find self-consistent potentials and
wave functions ψ_k, etc.), the results of calculations
made by different authors are somewhat different. In
the most advanced calculations of this type (75) it is
possible to achieve a good quantitative agreement with
experiments.

Let us now consider some of the model-independent
peculiarities of the near-surface region of semiconductors when there is degeneracy. We shall assume that,
as a specific example, the bands are bent heavily downwards. Then in the deep and narrow potential well for
electrons, which is formed close to the surface, discrete

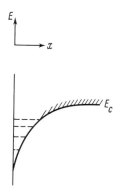

Fig. 3.3. Schematic
view of the potential
well near the surface
in which discrete
electron energy levels
occur (dotted lines).

quantum levels emerge, so that surface quantization
takes place (see Fig. 3.3). This means that the energy
of electrons, bound with their motion in the x-direction
(i.e., perpendicularly to the surface), can only assume
definite discrete values. At the same time, the spectrum of energies for motion in the y and z directions
remains continuous and so electrons in the quantum states
can move freely along the surface, remaining inside the
potential well. This is evidently true of the holes as
well. The only difference is that for them the potential
well occurs when the bands are bent upwards. Studies of
the specific behavior of quasi-two-dimensional (2D) electron
and hole gases in the narrow layers formed when the bands
are heavily bent near the surface of a semiconductor have
been conducted in recent years (76-78). The emergence of
discrete levels for carriers in the space charge layer
may also influence the kinetics of the electrochemical
and photoelectrochemical processes (see Section 6.4).

If at a given value of the surface charge, Q_{sc}, (or
of field ξ_{sc}) the carriers are degenerate at the surface,
the magnitude of the band bending, $|e\phi_{sc}|$, grows. Con-
versely, for a given band bending the corresponding
charge proves to be smaller than that calculated without
taking into account quantum effects. This also leads to
an increase (at given Q_{sc} and ξ_{sc}) of the thickness of
the space charge layer as compared to the "classical"
value (the so-called quantum expansion of the space charge
layer).

In view of the fact that degeneracy occurs under
conditions of sufficiently high concentration of carriers
near the surface, it is usually accompanied by a peculiar
"metallization" of the near surface region, which changes,
among other things, the optical and electrochemical properties of the semiconductor electrode. Thus, as a result

of degeneracy, a shift in the intrinsic absorption edge
for electromagnetic radiation takes place as is the case
in the bulk of a degenerate semiconductor (see p.26).
Finally, there is another important effect on the electrode, namely the redistribution of electric potential
in the system. In particular, in the course of electrode polarization, under the conditions of the surface
degeneracy, the additional potential difference is mainly
concentrated in the Helmholtz layer as is the case for
metallic electrodes.

Ξ 3.3. S p a c e   C h a r g e   i n   S e m i c o n d u c t o r :
        D i f f e r e n t i a l   C a p a c i t y

The differential capacity of the space charge layer
in a semiconductor, related to unit area of the interface, is given by $C_{sc} = dQ_{sc}/dE$ or, using the fact that
$dE = - d\phi_{sc}$ (see Section 3.1), by $C_{sc} = - dQ_{sc}/d\phi_{sc}$.
Using Eq. (3.16), we find that

$$C_{sc} = \frac{\varepsilon_{sc}\varepsilon_o}{2L_D^i} \frac{|\lambda^{-1}(e^Y-1)-\lambda(e^{-Y}-1)|}{F(Y,\lambda)} \qquad (3.18)$$

where, as before, $Y = e\phi_{sc}/kT$. Equation (3.18) describes
the dependence of C_{sc} on Y at an arbitrary doping level
over the entire range of the potential Y provided there
is no surface degeneracy of the carriers.

The expression for C_{sc} is substantially simplified
for an intrinsic semiconductor. Assuming that $n_o = p_o = n_i$,
or $\lambda = 1$, we obtain from Eq. (3.18)

$$C_{sc} = \frac{\varepsilon_{sc}\varepsilon_o}{L_D^i} \cosh\left(\frac{Y}{2}\right) \qquad (3.19)$$

The dependence of $lg(C_{sc})$ on Y is shown schematically in
Fig. 3.4a. The function $lg\, C_{sc} - Y$, as well as that of
$C_{sc} - Y$, is a symmetrical curve with the minimum at $Y=0$.
The position of this minimum on the electrode potential
scale coincides with the flat band potential. The value
of the differential capacity at the minimum $C_{sc}^{min}=\varepsilon_{sc}\varepsilon_o/L_D^i$
coincides with the capacity of a parallel-plate condenser
of unit area with the distance between plates being equal
to the Debye length, L_D^i, filled with a material having a
dielectric constant, ε_{sc}, equal to that of the semiconductor.

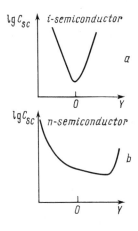

Fig. 3.4. The differential capacity of the space charge region in the semiconductor, C_{sc}, as a function of the potential drop, $\phi_{sc}=YkT/e$: a) an intrinsic semiconductor; b) an n-type semiconductor.

For extrinsic semiconductors the curve C_{sc} - Y becomes asymmetric, and the minimum shifts to the region where Y > 0 for n-type semiconductors, and to Y < 0 for p-type semiconductors (see Fig. 3.4b). For the entire function C_{sc} - Y there exist three characteristic intervals of Y. In particular, for an n-type semiconductor, in accordance with Eq. (3.18)

$$C_{sc} = \frac{\varepsilon_{sc}\varepsilon_o}{L_D\sqrt{2}} \begin{cases} e^{Y/2}, & Y \gg 1 \quad (3.20a) \\[2ex] (|Y|-1)^{-1/2}, & Y_{inv} < Y < -1 \quad (3.20b) \\[2ex] (p_o/n_o)^{1/2}e^{|Y|/2}, & |Y| \gg 1, \ Y < 0 \quad (3.20c) \end{cases}$$

Here $L_D = (\varepsilon_{sc}\varepsilon_o kT/e^2 n_o)^{1/2}$ which from the general definition, Eq. (3.5), is the Debye length for an n-type semiconductor (where $n_o \gg p_o$) and $Y_{inv} < 0$ is given by the relation $Y_{inv} = e\phi_{inv}/kT$ (see Eq. (3.7)). Equations (3.20) describe the space charge capacity of a semiconductor when an accumulation (a), depletion (b), or inversion (c) layer is formed.

Of greatest interest is the case when a depletion layer is formed; from Eq. (3.20b), we have

$$C_{sc}^{-2} = \frac{2L_D^2}{\varepsilon_{sc}^2\varepsilon_o^2} (|Y|-1) \quad (3.21)$$

and consequently, the C_{sc}-Y dependence may be linearized by plotting C_{sc}^{-2}-Y. Such a plot is called a "Mott-Schottky plot."

Let us note in this connection that in the case of wide-bandgap sufficiently doped semiconductors (so that $\lambda \gg 1$ or $\lambda^{-1} \gg 1$, see Eq. (3.12)), the interval of potential ϕ_{sc}, over which a depletion layer is formed, is rather large; for example, for an n-type semiconductor it is $\phi_{inv} < \phi_{sc} < -kT/e$. This turns out to be much wider than the potential intervals over which accumulation and inversion layers are found. This is illustrated by the energy diagram in Fig. 3.2: non-degenerate accumulation and inversion layers exist within band bending intervals of the order of E_c-F, while for the depletion layer the range is of the order of $E_g - 2(E_c-F)$, which in this case is much more than E_c-F. It is precisely for this reason that, irrespective of the conductivity type, a depletion layer is usually formed on electrodes made from wide-bandgap semiconductors. This fact largely determines the electrochemical properties of wide-bandgap semiconductor materials.

The thickness of the space charge layer, L_{sc}, can be determined in the general case from the condition $C_{sc} = \varepsilon_{sc} \varepsilon_0 / L_{sc}$. For the magnitude of L_{sc} defined by Eq. (3.20) we get

$$L_{sc} = L_D \sqrt{2} \begin{cases} e^{-Y/2}, & Y \gg 1 \quad (3.22a) \\[2ex] (|Y|-1)^{1/2}, & Y_{inv} < Y < -1 \quad (3.22b) \\[2ex] (n_o/p_o)^{1/2} e^{-|Y|/2}, & |Y| \gg 1, \ Y < 0 \quad (3.22c) \end{cases}$$

It follows from Eq. (3.22) that $L_{sc} < L_D$ for the cases in which an accumulation (Eq. (3.22a)) or inversion (Eq. (3.22c)) layer is formed whereas $L_{sc} > L_D$ when a depletion layer is formed (Eq. (3.22b)). The magnitude of L_{sc} changes over several orders of magnitude and is particularly sensitive to variations in the surface potential. However, one should bear in mind that the effects connected with degeneracy, which were mentioned above, can become very important.

In the case of the incomplete ionization of donors and acceptors the calculations become somewhat more difficult, although basically the problem is solved in a similar way. Examples of such calculations can be found in the literature (68,72).

Ξ 3.4. S p a c e   C h a r g e   i n   E l e c t r o l y t e :
        P o t e n t i a l   D i s t r i b u t i o n   i n   t h e
        D o u b l e   L a y e r

The description of the potential distribution,
$\phi(x)$, in the space charge region of the solution, the
diffuse part of the ionic layer, is particularly simple
for binary 1:1 electrolytes. Starting from the calculations above, let us assume that in Eq. (3.2) $N_A=N_D=0$
(so that, in accordance with Eq. (3.3), $n_o = p_o$) and make
the following substitutions: ε_{el} for ε_{sc}, where ε_{el} is
the static dielectric constant of the electrolyte; c_o
for n_o, where c_o is the concentration of cations (or
anions) in the bulk of the electrolyte solution ($x\to-\infty$);
ϕ for $\phi-\phi(el)$, where $\phi(el)$ is the value of the electric
potential in the bulk of the solution (see Fig. 3.1).
This gives in place of Eq. (3.2) a self-consistent
equation describing the potential distribution, $\phi(x)$,
in the electrolyte solution (79-82)

$$\frac{d^2\phi}{dx^2} = \frac{2ec_o}{\varepsilon_{el}\varepsilon_o}\, sinh\left[\frac{\phi(x)-\phi(el)}{kT}\right] \tag{3.23}$$

where $d\phi/dx = 0$ at $x\to-\infty$ and $\phi|_{x=-L_H} - \phi(el)=\phi_{el}=\psi'$ (note
that ψ' is the magnitude of the electric potential at
the outer Helmholtz plane with respect to the potential
in the bulk of the solution) are the boundary conditions
for Eq. (3.23).

Unlike Eq. (3.2), Eq. (3.23) is integrated to
the end. Its accurate solution, using the boundary conditions given, is as follows

$$\ln\, tanh\left\{\frac{e[\phi(x)-\phi(el)]}{4kT}\right\} - \ln\, tanh\frac{e\psi'}{4kT} = \frac{x+L_H}{L_G} \tag{3.24}$$

where $L_G = (\varepsilon_{el}\varepsilon_o kT/2e^2 c_o)^{\frac{1}{2}}$ is the characteristic length
determining the spatial scale of the diffuse ionic layer
in the electrolyte. The length L_G introduced by Gouy is
analogous to the Debye length, L_D, of the semiconductor.
(From the historical viewpoint, it would be more correct
to call both lengths, L_D and L_G, after Gouy, who was the
first (59,60) to examine expressions of this type.)

The electric field, $\xi_{el} = - d\phi/dx$, at the interface between the diffuse ionic layer in the electrolyte and the dense layer is calculated from Eq. (3.23) in the same way as was done previously for the calculation of the field ξ_{sc} at the semiconductor surface or directly from Eq. (3.14) using the same substitutions. As a result we obtain

$$\xi_{el} = \sqrt{\frac{8c_o kT}{\varepsilon_{el}\varepsilon_o}} \; sinh \; \frac{e\psi'}{2kT} \tag{3.25}$$

Formulae of the type given in Eq. (3.25) which establish a relationship between the field and the potential can also be obtained for the more general case of solutions of asymmetric electrolytes. Thus, for an electrolyte consisting of singly, doubly and triply charged ions ($z = - 1$, $+2$ and $+3$, respectively: for example a solution of $FeCl_2$ + $FeCl_3$), instead of Eq. (3.25) we get

$$\xi_{el} = \pm \left\{ \frac{2kT}{\varepsilon_{el}\varepsilon_o} \left[c_3^o \left(e^{-3e\psi'/kT} - 1 \right) + c_2^o \left(e^{-2e\psi'/kT} - 1 \right) + c_1^o \left(e^{e\psi'/kT} - 1 \right) \right] \right\}^{\frac{1}{2}} \tag{3.26}$$

where the c_i^o ($i=1,2,3$) are the concentrations of ions with the corresponding charge numbers in the bulk of the solution. The sign in front of the brackets should be chosen to be the same as the sign of ψ'. Equation (3.26) is the generalization of Eq. (3.25) and it turns into Eq. (3.25) if we assume that $c_3^o = 0$, $c_2^o = c_1^o$, replace 2e with e in the second term, and use the formula

$$cosh \; t = 2 \; sinh^2 \; (t/2) + 1$$

Using an expression of the type given in Eq. (3.14) for the semiconductor electrode and of the type given in Eqs. (3.25) and (3.26) for the electrolyte solution, it is possible to relate ϕ_{el}, ϕ_H, and ϕ_{sc}. Using ξ_H to denote the electric field strength in the Helmholtz layer, and ε_H for its dielectric constant, we find, after taking into account the conditions for electric induction continuity and the difference in the directions of the potential axes for ϕ_{sc} and ϕ_{el}, ϕ_H, that

$$-\varepsilon_{el} \left(\frac{d\phi}{dx} \right)_{x=-L_H} = \varepsilon_H \xi_H = \varepsilon_{sc} \left(\frac{d\phi}{dx} \right)_{x=0} \tag{3.27}$$

In particular, on the formation of a depletion layer in a semiconductor

$$\left(\phi_{sc} < 0, \ \frac{kT}{e} < |\phi_{sc}| < |\phi_{inv}|\right)$$

using Eqs. (3.10), (3.25) and (3.27) we have

$$\phi_{el} = \frac{2kT}{e} \ arc \ sinh \left(\frac{e\varepsilon_{sc} N_D |\phi_{sc}|}{4c_o \varepsilon_{el} kT} \right)^{\frac{1}{2}}$$

$$\phi_H = \frac{L_H \sqrt{2}}{\varepsilon_H} (e\varepsilon_{sc} \varepsilon_o N_D |\phi_{sc}|)^{\frac{1}{2}}$$

(3.28)

Similarly ϕ_{el} and ϕ_H can be found as functions of ϕ_{sc} at the arbitrary value of ϕ_{sc}.

In conjunction with Eq. (3.1), this makes it possible to find the dependences of ϕ_{el}, ϕ_H, and ϕ_{sc} on the Galvani potential $\phi(el/sc)$. In other words, the problem of the distribution of the full potential drop between the semiconductor and the electrolyte solution may be completely solved with sufficiently general assumptions.

The differential capacity, C, of the electric double layer is determined by the ratio $C = dQ/dE$, where E is the electrode potential, and Q is the electrode charge. For $dE/dQ = C^{-1}$, since $dE = d\phi(el/sc)$, we obtain

$$C^{-1} = C_{sc}^{-1} + C_H^{-1} + C_{el}^{-1}$$

(3.29)

where $C_H = dQ/d\phi_H$ and $C_{el} = dQ/d\phi_{el}$ are the differential capacities of the Helmholtz layer and the space charge layer in the electrolyte (or diffuse ionic layer) respectively. Thus, the resultant differential capacity of the semiconductor/electrolyte interface is made up of three capacities: C_{sc}, C_H, and C_{el} connected in series (see Fig. 3.5).

In view of the discussion above the following important point should be considered. When the calculations were made, it was assumed that the space

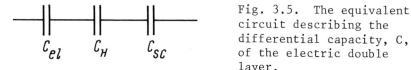

Fig. 3.5. The equivalent circuit describing the differential capacity, C, of the electric double layer.

charge region of both the semiconductor and the electrolyte were in equilibrium. When a current flows equilibrium may be perturbed and, consequently the potential
distribution can be distorted. However, the resultant
distortion in the diffuse layer of the electrolyte
solution is insignificant, so long as the current flowing
is much smaller than the limiting diffusion current in
solution (see, for example, (80-83)). Such distortions
may also be ignored in the semiconductor provided that
the current is smaller than the limiting current for
minority carriers (this is dealt with in detail in
Section 4.5).

Moreover, even the formulated conditions need not
actually be fulfilled. The experiment is usually
carried out in such a way that transport limitations in
the electrolyte are nonexistent. As far as the distortions arising from the current flow in the semiconductor
are concerned, when there is a depletion layer (which is
the particular case that is most often of practical importance) these distortions are insignificant even when
the current is large. Physically this stems from the
fact that the charge in the depletion layer of the semiconductor is mainly due to the immobile donors. That is
why the additional charge brought in (or taken out) by
the current exerts no influence on the potential distribution. Mathematically this can be seen from the
fact that Eq. (3.8) does not contain any terms derived
from the equilibrium Boltzmann distribution, and it is
precisely these terms that are modified by the flow of
current.

Similarly, when an accumulation layer is formed,
due to the excess of majority carriers, the equilibrium
potential distribution may only be upset when rather
large currents are passed with the participation of
these carriers. In this case the effect of the current
on the distribution of those carriers depleted from the
space charge region is easily taken into account if the
equilibrium potential distribution is known (see, for
example, (5), Section 29). Thus, the relations which have
already been obtained for the spatial distribution of
the field and the potential can be used in many cases
for the description of non-equilibrium processes.

Note that the problem of the potential distribution
is considerably simplified, if the concentration of the
electrolyte is high enough. It can be seen, in particular, from Eq. (3.29) that formally the value of ϕ_{el}
tends to zero as $c_o \to \infty$. This means that there are

definite conditions under which the potential drop in
the diffuse layer in the electrolyte solution can be
neglected. The necessary inequalities directly follow
from the expressions obtained before. Thus, if
$|e\phi_{sc}/kT| \lesssim 1$ then $|\xi_{sc}| = |\phi_{sc}|/L_D$, $|\xi_{el}| = |\phi_{el}|/L_G$ and the
condition of $|\phi_{el}| << |\phi_{sc}|$ is as follows

$$\varepsilon_{sc} L_G / \varepsilon_{el} L_D << 1 \qquad\qquad (3.30)$$

In a more general case, the following approximate
ratio $\varepsilon_{sc}|\phi_{sc}|/L_{sc} \approx \varepsilon_{el}|\phi_{el}|/L_{el}$ can be used as an es-
timate; this follows from (3.27) and leads to the con-
dition

$$\varepsilon_{sc} L_{el} / \varepsilon_{el} L_{sc} << 1 \qquad\qquad (3.30')$$

where L_{el} and L_{sc} are the effective dimensions of the
space charge regions in the electrolyte and the semiconductor, which depend, generally speaking, on ϕ_{el} and ϕ_{sc}.
Expressions for L_{sc} are given by Eqs. (3.22); expressions for L_{el} are obtained in a similar manner by
means of substituting $\lambda=1$ and $n_i=c_0$ in Eqs. (3.14) and
(3.18). If $|\phi_{el}|$ and $|\phi_{sc}|$ are small ($\lesssim kT/e$), then
$L_{sc}=L_D$, $L_{el}=L_G$, and Eq. (3.30') coincides with Eq.
(3.30).

Similarly the potential drop in the Helmholtz layer
is small when compared with $|\phi_{sc}|$ if the following in-
equality is observed

$$\varepsilon_{sc} L_H / \varepsilon_H L_{sc} << 1 \qquad\qquad (3.31)$$

Note that the conditions expressed by the inequalities
given in (3.30) and (3.31) are actually equivalent to

$$C_{sc}/C_{el} << 1 \qquad C_{sc}/C_H << 1 \qquad\qquad (3.32)$$

Estimates show that with reasonable values of the
system parameters, the conditions expressed in the inequalities (3.30) and (3.31) or (3.32) are observed. At
the same time, it should be borne in mind that even when
(3.31) holds, a relatively small potential drop in the
Helmholtz layer (or its change when E changes) may play
a rather significant role since it is precisely this
quantity that often exerts a decisive influence over the
rates of electrochemical reactions at the interface.

Thus, the following general conclusion can be drawn: in view of the fact that out of the three characteristic lengths - L_{sc}, L_H and L_{el} - in the majority of cases the first one is much bigger than the two others, the absolute value of the potential drop $|\phi_{sc}|$ proves to be much bigger than the absolute values of $|\phi_H|$ and $|\phi_{el}|$, and, consequently, comprises the main part of the interfacial potential drop ϕ(el/sc). In a similar way, as is seen from Eq. (3.32), the basic contribution to the resultant capacity, C, of the electric double layer is made by the capacity C_{sc} of the space charge layer of the semiconductor.

This, however, is not observed in the following cases:

(i) for strongly doped semiconductors in which the Fermi level in the bulk of the material is close to the edge of the majority carrier band or is even located within that band (bulk degeneracy);

(ii) under very heavy charging of the electrode, when the Fermi level on the surface comes close to the edge of the conduction or valence band (surface degeneracy);

(iii) when the concentration of surface states is very high: the change in their degree of filling leads to a redistribution of the charge and the potential. In particular, their charging may substantially increase the relative contribution of the ϕ_H component to the total potential drop, ϕ(el/sc).

Finally, it should be pointed out that the contribution from the capacity of the space charge layer in the electrolyte, C_{el}, and also the potential drop, ϕ_{el}, can be neglected when the solution concentration is sufficiently high; this can be achieved by introducing an indifferent (background) electrolyte into the solution.

All of this can equally be applied to the potential distribution at the interface when the electrode is polarized by an externally applied voltage. The change in the electrode potential, E, is accompanied by a change in the various components of the Galvani potential, ϕ(el/sc). If we assume that $\Delta E = \Delta\phi$(el/sc) then (see p. 66)

$$\Delta E = \Delta\phi_{el} + \Delta\phi_H - \Delta\phi_{sc} \qquad (3.33)$$

Let us examine two important extreme cases.

(i) We will assume that when the potential of the electrode is changed, $|\Delta\phi_{sc}| >> |\Delta\phi_H|$, $|\Delta\phi_{el}|$. That is to say, the potential drops in the Helmholtz and the diffuse ionic layers remain practically unchanged when the electrode is polarized and thus $|\Delta E| = |\Delta\phi_{sc}|$. This is why the positions of all the energy levels at the surface and, in particular, of the band edges, $E_{c,s}$ and $E_{v,s}$, remain constant with respect to the positions of the energy levels in the electrolyte solution and to the reference electrode (Fig. 3.6b). In this case the energy band edges are said to be "pinned" at the surface.

(ii) At the other extreme, we will assume that, due to some of the effects mentioned above, when the electrode is polarized, $|\Delta\phi_H| >> |\Delta\phi_{sc}|$, $|\Delta\phi_{el}|$. That is to say, it is the potential drop in the Helmholtz layer that changes with changing polarization. Under these conditions (See Fig. 3.6c), the energy levels of the surface are shifted with respect to the energy levels in solution by the value of $\Delta E = \Delta\phi_H$. With respect to the Fermi level in the semiconductor the band edges, $E_{c,s}$ and $E_{v,s}$, at the surface maintain the same relative position as the electrode potential changes since ϕ_{sc} remains constant. In order to emphasize the difference between this situation and the previous one, it is said that the bands are

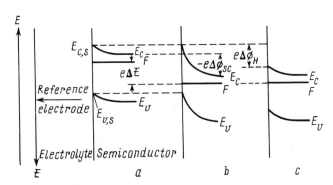

Fig. 3.6. The energy scheme for a semiconductor/electrolyte junction to which an external voltage is applied. It illustrates the pinning of the band edges (transition from a to b) and Fermi level pinning (transition from a to c) at the surface of a semiconductor electrode.

"unpinned" (the same phenomenon is, not very aptly, called Fermi level pinning).

For real systems an intermediate case is often observed in which, when the electrode is polarized, the two potential drops (ϕ_{SC} and ϕ_H) change simultaneously. Hence, neither the band edges nor the Fermi level are pinned at the surface.

Ξ 3.5. <u>S u r f a c e   S t a t e s</u>

It has already been mentioned in the preceding sections that the surface states may play an important role in the processes at the interface[*]. The surface states influence many of the electrical, optical and electrochemical properties of semiconductors and these can thus change substantially when the surface is processed (on metal electrodes, due to the very large number of "free" electrons in the metal, the surface states do not usually play an important role).

We shall now deal in greater depth with the nature and properties of surface states. The general reason for the emergence of electron interface (surface) energy levels is that in a bounded crystal in addition to the electronic states corresponding to electrons moving in the bulk of the crystal (the delocalized states), other states also necessarily emerge. In these states electrons are localized at the very surface. Thus, in addition to the bulk energy levels forming the energy bands, there are localized energy levels at the surface proper.

The existence of localized surface energy levels has two very important effects. Firstly, electrons and holes may be trapped at the surface, producing a surface electric charge and thus inducing a charge of opposite sign in the bulk. The influence of the surface on the equilibrium properties of the semiconductor is intimately linked with this phenomenon. Secondly, the surface energy levels are capable of radically changing the kinetics of all the processes involving electrons and holes: on the one hand, they provide additional centers

---

[*]When a solid/liquid interface is examined, it would be more precise to talk not of surface but of interface electron states. When using the generally accepted term "surface states" below we shall take it to mean "interface states."

for the recombination and generation of current carriers, while, on the other, they may mediate charge transfer across the interface. Thus non-equilibrium electron and hole processes, and in particular photoprocesses, are affected by interface states.

Tamm (84) was the first to show that one of the causes of surface states is the discontinuity in the periodic potential at the crystal surface. It is for precisely this reason that, for a bounded crystal, additional solutions of Schrödinger's equation for the electron arise, which sharply attenuate on moving away from the surface in either direction (this problem is dealt with in greater detail in (85)). The existence of surface states can be explained by an analysis of the chemical bonds in the crystal; this was first done by Shockley (86) for the simplest one-dimensional model. When described in these terms, the surface electron levels arise as a result of dangling electron bonds (due to the break in the lattice) at the surface.

The interface electron states present on the pure surfaces of a crystal are usually called intrinsic states. In recent years, considerable progress has been made in both theoretical and experimental methods of studying intrinsic surface states (see, for example, (87-89)). For a number of semiconductors, it was possible to obtain the surface electron states density distribution, $N_{ss}(E)$, for the different single crystal faces. In some cases it is possible to establish, by interpretation of the experimental data, that the distribution $N_{ss}(E)$ has one or more sharp and narrow maxima. In such cases the surface levels are discrete (or quasi-discrete) and, to a certain extent, are similar to the levels in one-dimensional crystals. The maximum integral density of the surface states does not exceed $10^{15} cm^{-2}$, in other words, it approximately coincides with the number of lattice points per unit surface area of the solid. Additional intrinsic surface states may emerge under the influence of an electric field (90). Furthermore specific surface localization of electrons occurs when a magnetic field is imposed parallel to the surface (91-92).

Above it was assumed that the surface of the crystal was atomically pure (the ideal case). Under normal conditions, that is to say, on real surfaces, and in particular, on those in contact with electrolyte solutions, the situation is complicated by the presence of adsorbed atoms (adatoms) and even layers on the

surface, as well as by structural defects. These can
exchange electrons with the bulk of the semiconductor.
As a result, surface electron energy levels of a different nature and with different properties from the
intrinsic levels mentioned above are formed. These
"extrinsic" levels play an important part in the processes of adsorption and catalysis (93,94).

Thus, the real surface of a semiconductor possesses
various types of electronic states and their corresponding electron energy levels. These levels are
characterized by a complex energy spectrum; they may
be both donors and acceptors, and their concentration depends on the surface treatment (in particular, on
chemical etching) and may be of the order of
10^{14}-10^{15}cm^{-2}.

Non-stationary experimental methods make it possible
to assess the characteristic relaxation time of the
surface states. The processes determining the relaxation of charge in the surface states are, in general,
diverse (interaction with phonons and electrons of the
solid, adsorption-desorption, diffusion along the
surface and through oxide and other films covering the
semiconductor, and so on). Moreover, in most cases there
are sets of levels with different characteristic relaxation times on the surface.

Let us use a simple monoenergetic model to evaluate
the influence of the surface electron levels on the potential and charge distributions in the electric double
layer. Following the description of the bulk levels in
Section 1.4, let us assume that all the electron surface
levels are characterized by the same energy E_{ss} and can
lose or capture only one electron (see Fig. 3.7). As a
specific case, we shall consider donor levels which are
positively charged when empty and neutral when filled

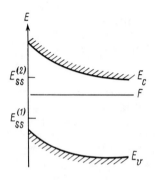

Fig. 3.7. Monoenergetic
electron levels at a
semiconductor surface:
the level with energy
$E_{ss}^{(1)}$ is filled; the
level with energy $E_{ss}^{(2)}$
is empty.

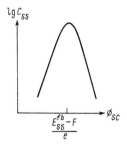

Fig. 3.8. Differential
surface states capacity,
C_{ss}, as a function of
potential drop, ϕ_{sc}.

(occupied by an electron). Using the Fermi distribution
(1.2), for the density of charge, Q_{ss}, on the surface
levels we obtain the following expression

$$Q_{ss} = eN_{ss} \left[1 + exp \left(\frac{F - E_{ss}^{fb} + e\phi_{sc}}{kT} \right) \right]^{-1} \qquad (3.34)$$

where N_{ss} is the full number of donor levels per unit of
interfacial area, and E_{ss}^{fb} is the energy of the surface
level at the flat band potential, so that $E_{ss}=E_{ss}^{fb}-e\phi_{sc}$.

The differential capacity connected with the surface level, C_{ss}, is calculated in a similar way to the
capacity, C_{ss}, of the space charge layer: $C_{ss}=dQ_{ss}/dE$.

Using Eq. (3.34), we obtain

$$C_{ss} = \frac{e^2N_{ss}}{kT} \frac{exp(Y_{ss} + Y)}{[1 + exp(Y_{ss} + Y)]^2} \qquad (3.35)$$

where $Y_{ss}=(F-E_{ss}^{fb})/kT$ and, as before, $Y=e\phi_{sc}/kT$. According to Eq. (3.35), the dependence of C_{ss} on Y, as well
as of $lg\ C_{ss}$ on Y, is described by a curve with a maximum (see Fig. 3.8) at $\phi_{sc}=(E_{ss}^{fb}-F)/e$, where
$C_{ss}^{max}=e^2N_{ss}/4kT$. On both sides of this maximum C_{ss}
tends exponentially to zero, so that

$$C_{ss} \simeq \frac{e^2N_{ss}}{kT} e^{-|Y_{ss} + Y|} \qquad \text{when} \quad |Y| \rightarrow \infty \qquad (3.36)$$

If the number of surface levels is sufficiently large
($N_{ss} \gtrsim 10^{14} cm^2$), then in the vicinity of the maximum

(within several kT/e) the capacity of the surface levels becomes comparable to the capacity of the Helmholtz layer $C_H = \varepsilon_H \varepsilon_0 / L_H$.

Note that since the full charge on the semiconductor "plate" of the double layer, Q, is equal to $Q = Q_{sc} + Q_{ss}$, the differential capacity of that "plate" is equal to the sum of the capacities, $C_{sc} + C_{ss}$. The resultant capacity of the interface, C, can be represented by the circuit shown in Fig. 3.9. It is clear from the estimates above that, under certain conditions, the magnitude of C_{ss} may prove to be larger than C_{sc}.

It has already been pointed out that the presence of a large number of charged surface states may lead to a redistribution of the potential drop between the semiconductor and the solution. Indeed, taking into account the charge in the surface states, the relation between the electric field in the semiconductor and in the Helmholtz layer, ξ_H, is given by the relation (cf. Eq. (3.27)):

$$\varepsilon_H \varepsilon_0 \xi_H = - \varepsilon_{sc} \varepsilon_0 \xi_{sc} + Q_{ss} \tag{3.37}$$

Of most interest here is the particular case when the first term on the right-hand side of Eq. (3.37), connected with the contribution from the space charge in the semiconductor (see Eq. (3.16)), can be neglected when compared to the second term determined by the charge in the surface states. Then the change in the potential drop in the Helmholtz layer, $d\phi_H = L_H d\xi_H$, due to a change in the potential drop in the semiconductor, $d\phi_{sc}$, is described by the ratio

$$- \frac{d\phi_H}{d\phi_{sc}} = - \frac{L_H}{\varepsilon_H \varepsilon_0} \frac{dQ_{ss}}{d\phi_{sc}} = \frac{C_{ss}}{C_H} \tag{3.38}$$

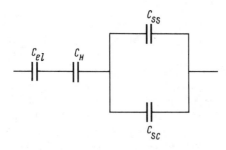

Fig. 3.9. The equivalent circuit describing the capacity, C, of the semiconductor electrode taking into account the surface states capacity.

From Eq. (3.35) it follows that the right-hand side of
Eq. (3.38) may exceed unity provided that

$$N_{ss} > \frac{4\varepsilon_H\varepsilon_o kT}{e^2 L_H^2}$$ (3.39)

Under these circumstances, when the electrode is polarized,
in the vicinity of the potential $\phi_{sc} \simeq (E_{ss}^{fb}-F)/e$ we have
$|d\phi_H| > |d\phi_{sc}|$, and thus the change in the potential drop
in the Helmholtz layer turns out to be bigger than the
change in the potential drop in the semiconductor. If a
set of electronic levels with different values of E_{ss}^{fb}
exists at the interface, as opposed to the single energy
case, then the effective "metallization" of the semiconductor electrode may occur over a rather broad potential
range. When the opposite inequality to (3.39) applies,
the potential distribution at the interface is described
by the ratios discussed in Section 3.4.

Ξ 3.6. S u r f a c e   R e c o m b i n a t i o n

Surface levels, like local impurity levels in the
bulk, may take part in the recombination and thermal
generation of electrons and holes and, consequently,
exert a strong influence on their lifetime. The kinetics
of charge carrier recombination and generation on the
surface are quantitatively characterized by the surface
recombination velocity.

This quantity can be introduced in the following
way. Let $R_{p,s}$ be the resultant rate of recombination of
holes on the surface, that is, the difference between the
number of acts of their capture and of their thermal
ejection by the surface levels per unit surface area per
unit time. At equilibrium, when $p_s = p_s^o$, we have
$R_{p,s} = 0$. Under non-equilibrium conditions, as was the
case for bulk recombination (see Eq. (1.42)), it is formally possible to write, for the holes

$$R_{p,s} = s_p'(p_s - p_s^o)$$ (3.40)

and, by analogy, for the electrons

$$R_{n,s} = s_n'(n_s - n_s^o)$$ (3.41)

The coefficients s_p' and s_n' have the dimensions of velocity, and Eqs. (3.40) and (3.41) may serve as
formal definitions of the surface recombination velocity.
However, the expressions for $R_{p,s}$ and $R_{n,s}$ are usually
written down in a somewhat different way. Let us consider a plane at the boundary of the space charge and
quasi-neutral bulk regions of the semiconductor where
the bands are no longer bent. We will denote the concentrations of holes and electrons at this plane as
$p(L_{sc})$ and $n(L_{sc})$, respectively. We then assume that:

(i) the bulk recombination in the space charge
region can be neglected (the observance of the inequalities $L_{sc} \ll L_p$, L_n is *sine qua non*);

(ii) the rate of surface recombination determining
the fluxes of carriers to the surface is not too large.

In this case, between p_s and $p(L_{sc})$ and also between n_s
and $n(L_{sc})$ the same unambiguous relations exist as at
equilibrium. In particular, for a non-degenerate semiconductor

$$p_s = p(L_{sc})e^{-Y}, \quad n_s = n(L_{sc})e^{Y} \qquad (3.42)$$

Then, instead of Eqs. (3.40) and (3.41), we can write
as follows

$$R_{p,s} = s_p[p(L_{sc}) - p_o], \quad R_{n,s} = s_n[n(L_{sc}) - n_o] \quad (3.43)$$

where p_o and n_o are the thermodynamic equilibrium concentrations of carriers in the semiconductor bulk. It is
precisely the coefficients s_p and s_n in Eqs. (3.42) and
(3.43) that are called, respectively, the surface recombination velocities for holes and electrons. In the general case the magnitudes of s_p and s_n (like s_p' and s_n') are
not equal and may depend on δp and δn. Experiments have
shown that the numerical values of s_p and s_n, depending
on the conditions of the surface, may vary over a rather
broad range, from 1 to 10^5 cm/s, at room temperature.

Let us examine in detail the process of recombination
through monoenergetic surface levels. As in the case of
bulk recombination (see Section 1.4) it is necessary to
take into consideration the four types of elementary acts
shown schematically in Fig. 3.10. Working from the same
principles, we obtain

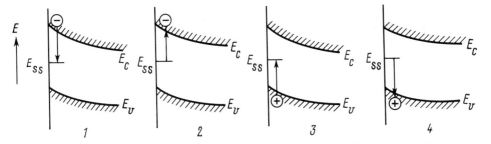

Fig. 3.10. The elementary stages in recombination via surface levels: 1) capture of an electron; 2) ejection of an electron to the conduction band; 3) capture of a hole; 4) ejection of a hole to the valence band.

$$R_{n,s} = a_{n,s} N_{ss} [n_s (1-f_{ss}) - n_1 f_{ss}] \qquad (3.44)$$

$$R_{p,s} = a_{p,s} N_{ss} [p_s f_{ss} - p_1 (1 - f_{ss})] \qquad (3.45)$$

The significance of the symbols is clear from a comparison with Eqs. (1.48) and (1.49). The only difference is that now N_{ss} is the surface (rather than bulk) density of recombination centers; the quantities n_1 and p_1 are given by the same relations as before, with the value of E_t replaced by E_{ss} (the energy of the surface level). The parameters $a_{n,s}$ and $a_{p,s}$ in Eqs. (3.44) and (3.45) have precisely the same meaning as in Eqs. (1.48) and (1.49) and can vary within the same limits.

In the steady state $R_{n,s} = R_{p,s} = R_s$ and so f_{ss}, the non-equilibrium occupancy of the surface levels, is given from Eqs. (3.44) and (3.45) by

$$f_{ss} = \frac{a_{n,s} n_s + a_{p,s} p_1}{a_{n,s} (n_s + n_1) + a_{p,s} (p_s + p_1)} \qquad (3.46)$$

Substituting this expression in Eqs. (3.44) or (3.45), we obtain an expression similar to Eq. (1.53):

$$R_s = N_{ss} a_{n,s} a_{p,s} \frac{p_s n_s - p_1 n_1}{a_{n,s} (n_s + n_1) + a_{p,s} (p_s + p_1)} \qquad (3.47)$$

where $p_1 n_1 = n_o p_o = n_i^2$. Assuming that

(i)   the semiconductor is non-degenerate;

(ii)   the quasi-neutrality condition is ful-filled, $\delta n(L_{sc}) = \delta p(L_{sc})$;

(iii) the perturbation from equilibrium is in-significant, $\delta n \ll p_o$, n_o,

then from Eq. (3.47) for the surface recombination velocity, $s_p = s_n = s = R_s/\delta n$, we get (cf. the expression for τ_t^{-1}, ensuing from Eq. (1.54))

$$s = N_{ss} a_{n,s} a_{p,s} \frac{n_o + p_o}{a_{n,s}(n_o e^Y + n_1) + a_{p,s}(p_o e^{-Y} + p_1)} \tag{3.48}$$

Equation (3.48), in accordance with experiment, predicts that s passes through a maximum at a particular value of $Y = Y^{max}$. Differentiating Eq. (3.48) with respect to Y and putting it equal to zero we find, from the denominator, that

$$Y^{max} = \ln\left(\frac{p_o}{n_o} \frac{a_{p,s}}{a_{n,s}}\right)^{1/2} \tag{3.49}$$

Knowing the value of Y^{max} from experiment and knowing the ratio $p_o/n_o = n_i^2/n_o^2 = p_o^2/n_i^2$, it is possible to find the ratio $a_{p,s}/a_{n,s}$.

An analysis of the experimental data shows (93,94) that of the whole set of surface levels only a part of them participates in the processes of recombination. The other levels can effectively exchange charge carriers with only one of the bands and are therefore traps for electrons or holes.

In conclusion, let us note that the process of surface recombination of electron-hole pairs occurring via the surface states is basically similar at both the semiconductor/electrolyte and the semiconductor/vacuum (gas) boundaries. At the same time, at the semiconductor/electrolyte interface there is an additional specific recombination mechanism; this will be examined in Section 7.2.

Ξ 3.7. <u>M e t h o d s f o r S t u d y i n g t h e E l e c t r i c</u> <u>D o u b l e   L a y e r   :   M e a s u r e m e n t   o f</u> <u>F l a t   B a n d   P o t e n t i a l</u>

The flat band potential (like the potential of zero charge in the electrochemistry of metals) is an important characteristic of the semiconductor/electrolyte system. Its magnitude is used in the quantitative description of the double layer structure and of the kinetics of electrochemical reactions on semiconductor electrodes. A number of experimental methods to determine the flat band potential and study the potential distribution at the interface have been elaborated. Some of the essential details of the experimental study of the surface properties of semiconductor electrodes in the absence of any net current across the interface, that is, under conditions of ideal polarizability, are briefly examined below.

Historically the first method of studying the distribution of the potential drop at the semiconductor/solution junction and, in particular, for determining the flat band potential was the surface conductivity method (see, for example, (95-97)). This technique is based on the fact that the local electronic conductivity in the space charge region differs from that of the noncharged semiconductor bulk and depends (63) on the potential drop, ϕ_{sc}, in the semiconductor. In recent years, however, the surface conductivity method has declined in importance, due to a number of practically unavoidable experimental complications and uncertainties.

Today, the most common technique is the measurement of the differential capacity of the electric double layer. By comparison of the theoretically calculated curve for the differential capacity, C_{sc}, originating from the existence of a space charge in the semiconductor (see Section 3.3), with that measured experimentally, it is possible to determine the potential drop in the semiconductor, ϕ_{sc}.

We shall restrict ourselves to the most important case of electrodes made from semiconductors with sufficiently wide bandgaps. On such an electrode a depletion layer is formed over a wide range of potentials. The relationship between the differential capacity, C_{sc}, and the potential drop in the semiconductor, ϕ_{sc}, is given by Eq. (3.21); this can also be written in the following form:

$$C_{sc}^{-2} = \frac{2}{\varepsilon_{sc}\varepsilon_o eN_D} (|\phi_{sc}| - kT/e) \qquad (3.50)$$

Let us assume that the electrolyte concentration is sufficiently high so that the contribution made by the diffuse part of the ionic double layer to the resultant capacity C can be neglected. In this case if the experimentally determined dependence of the capacity, C, on the electrode potential, E, yields a linear plot for C^{-2} against E, in accordance with Eq. (3.50), it is often taken as proof that the two conditions

$$C_{sc} \ll C_H, \quad |\Delta\phi_{sc}| \gg |\Delta\phi_H| \qquad (3.51)$$

have been observed. In other words, it is concluded that:

(i) the measured capacity of the electrode, C, is fully determined by the capacity of the space charge layer in the semiconductor,

(ii) when the potential of the electrode is changed, it is only the potential drop in the semiconductor that changes, the potential drop in the Helmholtz layer remaining constant; in particular, $\phi_{sc} = -(E - E_{fb})$ where E_{fb} is the flat band potential.

If the conditions in (3.51) are satisfied, then using Eq. (3.50) it is possible from the intercept of the Mott-Schottky plot with the potential axis to determine the flat band potential E_{fb}, and from its slope the donor concentration N_D.

The differential capacity method is widely applied in electrochemical studies of semiconductor electrodes. At the same time, it should be stressed that the direct use of Eq. (3.50) to experimentally determine both E_{fb} and N_D is based on a number of suppositions (which are often tacitly accepted) about the properties of the semiconductor/electrolyte interface. Let us enumerate the principal suppositions.

(i) It is maintained that the measured capacity C is not distorted by any leakage current across the interface, by the ohmic resistance of the electrode and

electrolyte, etc. A proper account of these effects is
a separate problem and is usually taken within the framework of the assumptions about the electrode's equivalent
circuit, in which the unknown quantity C_{sc} is singled
out.

(ii) It is supposed that donors (acceptors) in the
semiconductor are, firstly, fully ionized at the temperature of the measurements and, secondly, that they are
homogeneously distributed in the sample, at least
within the limits of the space charge region. (Inhomogeneity, on a scale which is large when compared with
the thickness of the space region, can be determined by
the capacity method in combination with layer-by-layer
etching of the semiconductor material.) If the concentrations N_D, N_A depend on the x-coordinate, then, instead
of Eq. (3.50), for the dependence of C_{sc} on ϕ_{sc} more
complicated relations are obtained. Deviations from Eq.
(3.50) are also observed in the presence of deep donors
(acceptors) which are not ionized in the semiconductor
bulk at the measurement temperature but become ionized
in the electric field of the space charge layer, thereby
contributing to the capacity (99).

(iii) It is held that the measured capacity represents precisely the capacity of the space charge
region and does not include, for example, the capacity
of fast surface states, adsorption capacity, and so on.
In some cases this prerequisite is naturally observed,
for example, on zinc oxide electrodes (see below), but
more often than not the contribution from surface state
capacity turns out to be considerable.

As to the effect of surface levels on the measured
capacity, it often proves possible to diminish this contribution by increasing the frequency of the alternating
current when measuring the capacity. In this case the
levels with a comparatively long relaxation time, whose
concentration is usually relatively large, become gradually less important until at sufficiently high frequencies the measured capacity practically represents the
space charge capacity.

(iv) It is supposed that the Helmholtz layer
capacity C_H exerts no influence on the measured capacity
C. A more detailed analysis will show, however, that
such a conclusion, in particular when dealing with rather
heavily doped semiconductors, may prove to be incorrect
(98).

Indeed, when we include, for the time being, the influence of the dense part of the double layer, we have

$$E - E_{fb} = - \phi_{sc} + \Delta\phi_H, \quad C^{-1} = C_{sc}^{-1} + C_H^{-1} \quad (3.52)$$

where $\Delta\phi_H$ is the change in the potential drop in the Helmholtz layer when the potential of the electrode changes from E_{fb} to E [*]. Assuming that the charge on the surface levels can be neglected, so that $Q_{sc}=-Q_H$, and that the capacity, C_H, of the Helmholtz layer does not depend on the potential, we get

$$\Delta\phi_H = \frac{Q_{sc}}{C_H} \quad (3.53)$$

From Eq. (3.16) it follows that Q_{sc} for the depletion layer is given by the expression

$$Q_{sc} = (2e\varepsilon_{sc}\varepsilon_o N_D)^{\frac{1}{2}} \ (|\phi_{sc}| - kT/e)^{\frac{1}{2}} \quad (3.54)$$

Substituting Eq. (3.54) in Eq. (3.53) and Eq. (3.53) in the first equation in (3.52), we obtain a quadratic expression in the unknown $(|\phi_{sc}|-kT/e)^{\frac{1}{2}}$, whose solution is

$$(|\phi_{sc}|-kT/e)^{\frac{1}{2}}=-b+[b^2+(E-E_{fb})-kT/e]^{\frac{1}{2}} \quad (3.55)$$

where $b=(2e\varepsilon_{sc}\varepsilon_o N_D)^{\frac{1}{2}}/C_H$. Finally, expressing $|\phi_{sc}|-kT/e$ in terms of C_{sc} using Eq. (3.50) and C_{sc} in terms of C and C_H using the second equation in (3.52), we obtain from Eq. (3.55) the relationships between C, C_H and the electrode potential

$$C^{-2} = C_H^{-2} + \frac{2}{e\varepsilon_{sc}\varepsilon_o N_D} (E-E_{fb}-kT/e) \quad (3.56)$$

From Eq. (3.56) it can be seen that C^{-2} is linearly dependent on E just as it was when we assumed that

---

[*] Remember that the potential drop in the Helmholtz layer, ϕ_H, at flat band potential in the general case is non-zero unlike ϕ_{sc} (see Section 3.2).

ϕ_{sc}=-$(E-E_{fb})$ in Eq. (3.50); moreover, the slope of the corresponding straight line is the same as it would have been if there were no influence from the Helmholtz capacitance. Now, however, the potential

$$E' = E_{fb} + kT/e - e\varepsilon_{sc}\varepsilon_o N_D/(2C_H^2) \qquad (3.57)$$

describes the intercept of the plot on the electrode potential axis in accordance with Eq. (3.56). With respect to the potential determined from Eq. (3.50), where it was assumed that inequalities in (3.51) were valid, the intercept is shifted by an amount

$$-e\varepsilon_{sc}\varepsilon_o N_D/(2C_H^2)$$

This shift, especially in the case of rather heavily doped semiconductors, is far from negligible[*].

Thus, an experimentally observed linear dependence of C^{-2} on E does not necessarily mean that the inequalities in (3.51) hold. Looked at in another way, deviations from linearity do not by themselves invalidate the inequalities in (3.51). Finally, it should be pointed out that in many cases uncertainties are introduced by the frequency dependence of the Mott-Schottky plots. (An increase in the measuring a.c. frequency is often used, as was shown above, to decrease the contribution of the surface states to the measured capacity.) When the frequency is changed, these plots are deformed in either of two ways:

(i) More often than not, only the slope changes whilst the intercept on the potential axis remains constant.

(ii) Sometimes, along with the change in slope, there is a shift of the entire line with respect to the potential axis, and the point of intersection on that axis is a function of frequency.

No comprehensive explanation of these trends has as yet been offered. Despite this fact, plots of type (i) are used in evaluating flat band potentials by extrapolation to the intersection on the potential axis; however, the value of $N_D(N_A)$ cannot apparently be found

---

[*]For example, for TiO$_2$ (ε_{sc} = 173) at N_D = 10^{19} cm^{-3} and C_H = 10^{-6} F/cm^2 the shift amounts to approximately -0.12 V.

from their slope. Type (ii) plots are considered to be unfit even for determining the magnitude of E_{fb}.

It follows from above that in order to use the linear dependence of C^{-2} on E obtained experimentally as proof of the validity of the inequalities in (3.51), it is necessary to use some additional criterion, for example the coincidence of the donor concentration, N_D, calculated from Eq. (3.50) with the concentration measured by an independent method (for example, making use of the Hall effect (15)).

Other methods of determining the flat band potential are based on the fact that during the generation of electron-hole pairs in the semiconductor, for example, under illumination, the bands unbend (for details see Section 6.1). In this case the potential of the electrode tends toward the flat band potential and, if the illumination is strong enough, attains it (100). The same thing occurs (101) when the carriers are injected through a p-n junction located on the back side of a thin plate electrode so that it is close to the electrode/solution interface. The type of electrode used in this case is described in Section 4.6.

Of interest among the newer methods is that based on the measurement of the surface stress in a solid when its potential is modulated (102). This method may be efficiently applied to semiconductor electrodes (103).

In conclusion, let us briefly deal with some of the basic details of the experimental techniques used to study the electrochemistry of semiconductor materials. As has already been mentioned, both the bulk and surface properties of semiconductors are highly sensitive to imperfections in the crystal structure and the presence of impurities. For this reason single-crystal semiconductor electrodes with their surfaces correctly oriented relative to the main crystallographic axis are preferred for experimental studies. The relatively high resistance of semiconductors gives rise to tangible Ohmic potential drops. This makes the choice of an optimal geometry for the electrodes important.

One should bear in mind that for semiconductors with bandgaps broader than ca. 1 eV the intrinsic conductivity is negligibly small. In fact they are insulators and therefore are practically unsuitable for electrochemical measurements. As electrode materials they are used only when doped with donor or acceptor type admixtures.

Stringent demands are made on the preparation of the electrode surface before measurements. After machining (cutting, polishing) the surface region of the crystal contains many structural defects (the so-called damaged layer). The sample is subjected to chemical or electrochemical etching in order to remove this layer and expose the undamaged crystal structure beneath it. Special attention is given to the purification of reagents. Higher demands are also made on the quality of the electrical contacts; they should be both Ohmic and of low resistance. For details of the techniques used in the experimental studies of semiconductor electrodes see (5,104).

3.8. Selected Experimental Results

3.8.1. Potential Distribution, Semiconductor Space Charge and Flat Band Potential.

Up until the present time most of the information about the structure of the double layer on semiconductor electrodes has been obtained by the differential capacity method. To use capacity measurements for the determination of ϕ_{sc}, it is necessary, as has already been mentioned, to get rid of the influence of fast surface states, and take into account the capacity of the Helmholtz layer, and thus isolate the capacity of the space charge layer, C_{sc} (cf. Figs. 3.5 and 3.9).

The experimentally measured (105) dependence of the capacity of a germanium electrode on its potential, E, is shown in Fig. 3.11 (dots). The full line represents the variation of C_{sc} with ϕ_{sc} calculated using Eq. (3.18). The E and ϕ_{sc} scales are identical. The good agreement between the experimental and theoretical curves shows that under these experimental conditions (105) $|\Delta E| = |\Delta \phi_{sc}|$ and, consequently, the potential drop in the Helmholtz layer does not change in the course of electrode polarization.

There is a more frequent case in which the experimental "capacity vs. electrode potential" curve is broader than that calculated from Eq. (3.18), the "capacity vs. potential drop in the space charge layer" curve. This means that as the potential of the electrode, E, changes, the potential drop in the Helmholtz layer, ϕ_H, also changes, $\Delta \phi_{sc}$ and $\Delta \phi_H$ being comparable in magnitude. In accordance with the comments in Section 3.4, a change in the Helmholtz potential drop frequently testifies to a high density of surface states, whose charge

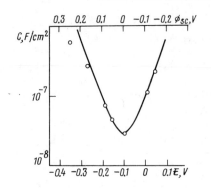

Fig. 3.11. Plot of the electrode capacity as a function of potential for an intrinsic germanium electrode in 0.1 N H_2SO_4: the points are from experiment; the curve is calculated in accordance with Eq. (3.18) (105).

gradually changes over the entire range of potentials, i.e., these levels are far from being mono-energetic. However, while exerting an influence on the potential distribution at the interface, these states make no contribution to the differential capacity measured at a sufficiently high frequency. Their relaxation time is 1-10 s. When the potential of the electrode changes at a faster rate, the charge on these states has insufficient time to relax, whilst when the potential changes slowly, the potential drop in both the space charge and Helmholtz layers changes simultaneously so that it is possible to estimate the value of $\Delta\phi_{sc}/\Delta\phi_H$. Usually when the potential of a germanium electrode in aqueous solution, and also in some non-aqueous solutions, is slowly changed, the change in the Helmholtz potential drop amounts to between 30 and 50 per cent of the change in the electrode potential (96,106). This is illustrated by the ϕ_{sc} vs. E curve for a germanium electrode (see Fig. 3.12).

Quite different behavior, as compared with germanium, is displayed by zinc oxide electrodes (107). As discussed above, wide-gap semiconductors like zinc oxide

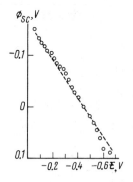

Fig. 3.12. The dependence of the potential drop in the space charge region on the potential for an n-type germanium electrode, 30 ohm· cm (the electrolyte was KBr in N-methylformamide) (96).

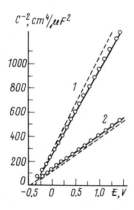

Fig. 3.13. The dependence of the square of the inverse capacity on the potential for ZnO electrodes in 1 N KCl at pH 8.5. Conductivity: 1 - 0.59 $ohm^{-1} \cdot cm^{-1}$; 2 - 1.79 $ohm^{-1} \cdot cm^{-1}$. The dotted lines are calculated in accordance with Eq. (3.50) (107).

(E_g = 3.2eV) are characterized by the formation of a depletion layer when put in contact with an electrolyte solution. On this type of semiconductor it has proved scarcely possible to observe an inversion layer because the transition from depletion into inversion is beyond the range of the ideal polarizability of the electrode.

The above effect of Helmholtz capacity, C_H, on the resultant capacity, C, of not too highly doped zinc oxide electrodes can be neglected and the capacity vs. potential curves are described by Eq. (3.50) (see Fig. 3.13). In practice the capacity, C, does not depend on the measuring a.c. frequency over a broad range of frequencies (from 50 Hz to 100 kHz). This is a rather rare case, inasmuch as the capacity of semiconductor electrodes usually changes, to a greater or lesser extent, with frequency. When calculating "theoretical" C_{sc}^{-2} vs. ϕ_{sc} curves (Fig. 3.13), the values of the donor concentration N_D found from Hall effect measurements were used. The difference between the calculated and measured values for the capacity does not exceed 2 per cent. Hence, two conclusions (107) can be drawn:

(i) On the zinc oxide electrode the density of fast surface states which would contribute to the measured capacity is very low (not more than 10^9 cm^{-2}).

(ii) Even with a slowly sweeping electrode potential (this is precisely how the curves in Fig. 3.13 were taken), it is only the potential drop in the space charge layer that changes, whereas the Helmholtz potential drop, although it is far from being small (see below), remains constant.

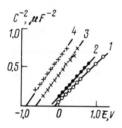

Fig. 3.14. The dependence of the square of the inverse capacity on the potential for a TiO$_2$ electrode. A.c. frequency: 62 kHz; solution pH: 1 - 0.7; 2 - 3.0; 3 - 9.1; 4 - 12.9. The potentials are cited with respect to the Saturated Calomel Electrode (113).

Capacitive characteristics qualitatively similar to those for zinc oxide are found, with small variations, for many other wide bandgap semiconductors: cadmium sulfide (108), potassium tantalate (109), silicon carbide (110), tin oxide (111), titanium dioxide (112, 113) (Fig. 3.14), gallium phosphide (114) (Fig. 3.15), etc.

As a rule, the measured capacity of silicon electrodes is much higher than that of the space charge layer, C_{sc}. Apparently, the native oxide formed on silicon in contact with aqueous solution is a powerful source of fast surface states. Only in rare cases, for example, in concentrated hydrofluoric acid solutions, is the capacity of the electrode close to C_{sc}, and the change in potential during polarization completely located within the space charge layer of the semiconductor (115). Flat band potentials have been determined for a number of semiconductors using the methods mentioned in Section 3.7. Appendix 4 contains some of the most reliable values.

Bearing in mind the concepts discussed in Section 2.2, let us examine the dependence of the flat band potential on the concentration of free carriers in the semiconductor. For this purpose we shall examine the potential distribution in a cell with a semiconductor electrode (Fig. 3.16, cf. Fig. 2.5) at flat band potential. Since the concentration of the doping impurity

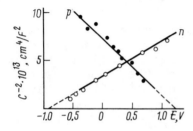

Fig. 3.15. The dependence of the square of the inverse capacity on potential for n- and p-type GaP electrodes in 1 N H$_2$SO$_4$ (114). (Reprinted by permission of the Publisher, The Electrochemical Society, Inc.)

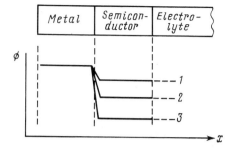

Fig. 3.16. The potential distribution in a cell with a
semiconductor electrode at
flat band potential: 1,2,3 -
samples with different bulk
concentrations of conduction
band electrons.

determining the concentration of majority carriers is
negligibly small, its presence does not alter the chemical properties of the semiconductor. Hence, it can be
assumed that impurities exert no influence on the Helmholtz potential drop at the interface, inasmuch as it
is controlled by the chemical interaction of the electrode material with the electrolyte solution. At the
flat band potential $\phi_{SC} = 0$. Thus the measured potential difference in the circuit, that is, the flat band
potential, only depends on the concentration of electrons in the sample through the potential drop at the
contact between the semiconductor and the metal contact
lead. This drop (as well as the work function of the
semiconductor) is determined by the chemical potential
of electrons in the semiconductor. In an extrinsic
semiconductor, according to Eq. (1.22a), it changes by
2.3 kT/e when the concentration of electrons changes by
a factor of 10. Indeed, as is seen from Fig. 3.17, the
experimentally measured flat band potential shifts by
precisely this value ($\approx$60 mV at room temperature) when
the concentration of electrons in the sample increases
by an order of magnitude (107). This is a manifestation
of the well-known proposition (116) that the potentials
of zero charge in a series of electrodes differ by
approximately the same amount as the work functions. For
this reason for a particular semiconductor material the
flat band potential of a p-type sample is more positive
than that of an n-type sample (cf. Fig. 3.15).

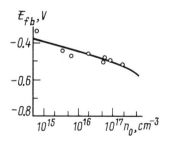

Fig. 3.17. The dependence
of the flat band potential
of zinc oxide on the bulk
concentration of conduction
band electrons in the sample.
1 N KCl, pH 8.5 (107).

It should be stressed that in the case of metal electrodes the Helmholtz potential drop at the potential of the zero charge differs for different metals due to differences in their interactions with the solution, for example, the degree of solvent adsorption. This introduces uncertainties when an attempt is made to establish a quantitative correlation between the work function and zero charge potential. It is precisely semiconductor rather than metal electrodes which provide an opportunity to investigate such a correlation without the complications introduced by differences in the chemical nature of the electrodes being compared.

3.8.2. T h e   H e l m h o l t z   L a y e r. The flat band potential of a semiconductor electrode, E_{fb}, corresponding to zero space charge in the semiconductor is not a characteristic constant of the semiconductor material alone. The value of the potential, $E=E_{fb}$, measured against a certain reference electrode depends, in particular, on the composition of the solution, the state of the semiconductor surface, its crystallographic orientation, adsorption, and so on. Changes in E_{fb} reflect changes, depending on the factors enumerated, in the potential drop in the Helmholtz layer. Hence, the measurement of the flat band potential has become a major tool for the study of the Helmholtz layer on semiconductor electrodes, whereas many traditional methods developed for the electrochemistry of metals find only limited application due to interference from the space charge in the semiconductor.

On electrodes which are subject to oxidation in solution or readily adsorb oxygen, a strong influence is exerted on the Helmholtz potential drop and, hence, on the flat band potential by the degree of surface oxidation. For example, on a germanium electrode at the stationary potential in a solution of an electrolyte in which the germanium oxides are easily dissolved (e.g., hydrofluoric acid), there are no oxide phases on the surface but chemisorbed oxygen can be found. The coverage of the surface by oxygen, or hydroxyl groups, which is the same thing, is close to unity. Under strong cathodic polarization, the adsorbed oxygen is reduced and the surface is covered by adsorbed hydrogen (or by germanium hydride). The Ge-O bond is polar, so that a certain dipole potential drop $\phi_{H,dip}$ is associated with chemisorbed oxygen. Therefore the transition from a "hydroxide" to a "hydride" type surface is accompanied by a negative shift in the flat band potential of approximately 0.6 V (117). A similar shift is also

observed in the course of surface reduction-oxidation processes at electrodes made from some other semiconductors, although the detailed mechanism of the surface
reactions is not fully clear (118).

On oxide semiconductors (TiO_2, ZnO and others),
and also on semiconductors whose surface is usually
oxidized (for example, germanium), the concentrations
of OH^- and H^+ ions chemisorbed from aqueous solutions
are not equal to each other and this fact is necessarily
observed as a contribution to ϕ_H and, hence, to the magnitude of E_{fb}. In this case the flat band potential is,
as a rule, a linear function of pH (see Fig. 3.18).
This may stem from the influence exerted by the pH on
the equilibrium between the chemisorbed ions on the
electrode surface and those in the solution as well as
on the dissociation of surface compounds that, in their
turn, alter the value of ϕ_H (119). For example, hydroxyl groups on the oxidized germanium surface dissociate in alkaline solutions in accordance with the
equation

$$\text{--Ge--OH} \rightleftarrows \text{--Ge--O}^- + H^+ \tag{3.58}$$

This leads to the emergence of an additional potential
drop in the Helmholtz layer. When the solution pH is
changed, the dissociation equilibrium shifts. That is
why the concentration of the charged surface groups and
the associated potential drop changes. In the simplest
case

$$\Delta\phi_H = -\frac{2.3kT}{e}(\text{pH}) \tag{3.59}$$

This potential drop is practically independent of the
potential at the semiconductor surface and is, therefore,

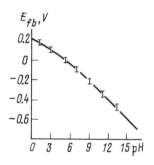

E_{fb}, V

Fig. 3.18. The influence
of solution pH on the
flat band potential of
germanium (with intrinsic
conductivity) (119). The
potentials are cited with
respect to the Normal
Hydrogen Electrode.

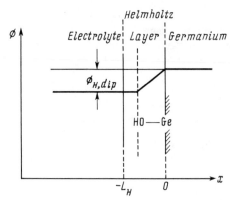

Fig. 3.19. The potential distribution at the germanium/electrolyte interface at the flat band potential and at the isoelectric point of the surface oxide.

also present at the flat band potential. Here the charge due to the surface anionic GeO⁻ groups is entirely neutralized by the charge on the electrostatically adsorbed cations in the ionic layer; the charge on the surface states can often be neglected.

The surface oxides on germanium (and many other semiconductors) are amphoteric, so that they dissociate in alkaline solutions like acids (see Eq. (3.58)), and in acidic solutions like bases. At a certain pH value the dissociation of the surface oxide groups is practically suppressed and hence the charge caused by it in the ionic layer is absent. Similarly, in the case of adsorbed OH⁻ and H⁺ ions, at a certain pH value, depending on the nature of the electrode, the overall charge of adsorbed ions will be zero. This pH value is called the isoelectric point, or, which is nearly the same thing, the point of zero zeta potential (pzzp). By definition (62), this is the potential of zero full charge corresponding to the isoelectric point when $E = E_{fb}$. At the same time, even under these conditions, the value of $\Delta\phi_H$ may be non-zero due to the adsorption at the interface of dipole molecules of the solvent, and also (see above) to the polar character of the chemical bond between germanium and oxygen (see Fig. 3.19).

The isoelectric point can be determined by titrating a suspension of the semiconductor material (120) or by applying other methods. For example, the isoelectric point of the germanium surface oxide is 2.5 (121), a value which is very close to that for the crystalline material GeO_2.

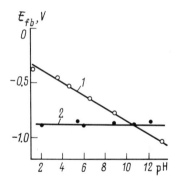

Fig. 3.20. The influence
of the solution pH on the
flat band potentials of
TiO_2 (I) and CdS (2) electrodes (112,186).

The concepts applied above to the structure of the
Helmholtz layer on the germanium electrode may be
applied to many other semiconductors. In particular, a
similar dissociation of the surface oxygen-containing
groups is observed on oxide semiconductors, for example,
TiO_2, WO_3, etc., and this is manifested in the dependence of the flat band potential on pH (Fig. 3.20, curve
1).

Thus, for many semiconductor electrodes the flat
band potential depends upon the extent of surface oxidation, and it is possible to distinguish two more or
less definite states of the surface: "oxidized" and
"reduced" (cf. the "hydroxide" and "hydride" surfaces
of the germanium electrode). Coming back to the table
of flat band potentials (Appendix 4), it should be
pointed out that, although the state of the surface was
not accurately defined in all cases, taking into account
the dependence of the flat band potential on the semiconductor's conductivity type (cf. Fig. 3.15), it can
be assumed that in the case of p-type specimens we are
actually dealing with the "oxidized" surface, while in
the case of n-type specimens, with the "reduced" surface.

For some semiconductor materials, for example, CdS,
the flat band potential does not depend on pH (Fig. 3.20,
curve 2). However, it may change when substances which
are specifically adsorbed on the semiconductor are
added to the solution, for example sulfides in the
case of CdS (123,124) (see Fig. 3.21). These shifts are
due to polar adsorption bonds, as well as the ionic
charge associated with the adsorbed species.

Alongside the experimental approach, in (122) a
method of evaluating the magnitude of the flat band potential, E_{fb}, on the basis of the semiconductor and solution
properties was proposed. The essence of the method is

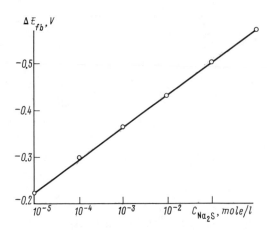

Fig. 3.21. The shift in the flat band potential of a CdS electrode as a function of the Na_2S concentration (with respect to the Na_2S free solution) (124). (Reprinted by permission of the Publisher, The Electrochemical Society, Inc.)

as follows: in accordance with Eq. (3.1), with a sufficiently high electrolyte concentration at thermodynamic equilibrium we have

$$e(\phi_H^O - \phi_{sc}^O) = F - F_{redox(ref)} = -\chi + F - E_c - F_{redox(ref)} \qquad (3.60)$$

Here $F_{redox(ref)}$ relates to the reference electrode and χ is the electron affinity of the semiconductor; see Section 2.1. The value of $(F - E_c) < 0$ can be found from Eqs. (1.20) and (1.22), and χ with the help of the semi-empirical formulae linking χ with the first ionization potential I_1, the electron affinity A_f of the elements forming the semiconductor compound, and its bandgap E_g; for example, in the simplest case of a binary semiconductor MX we have $\chi = (A_f + I_1 + E_g)/2$. The value of E_g, if it is not known directly from experiments can, in its turn, be estimated on the basis of the properties of the elements making up the semiconductor compound (see Section 1.1). If we assume approximately that at the isoelectric point (pzzp) the value of ϕ_H^O is equal to zero, then from Eq. (3.60) proceeding from the known value of $F_{redox(ref)}$ we can calculate ϕ_{sc}^O and hence the flat band potential.

The value of the pH corresponding to pzzp can be calculated if we know the standard electrochemical potential of the reactions in which adsorbing ions take part or, as has already been mentioned, it may be determined experimentally. This makes it possible to verify Eq. (3.60) experimentally. It turns out, in particular, that for a great number of oxide semiconductors, as might be expected, there exists a linear relationship between ϕ_{sc}^O and the pH. A linear relationship

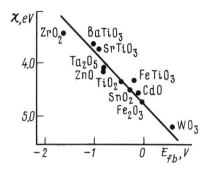

Fig. 3.22. The calculated electron affinity as a function of the flat band potential (measured against the Saturated Calomel Electrode) for a series of semiconductors at their isoelectric points (122).

has also been found empirically between the calculated values for χ and the measured values of E_{fb} (see Fig. 3.22). Using this plot, it is possible to approximately find the flat band potential by calculating the value of χ for the given semiconductor.

In conclusion, note that the existence of a linear relationship between χ and E_{fb} is quite similar to the well known relationship between the work function and the potential of zero charge in the electrochemistry of metals (62). This analogy becomes particularly clear when we consider that for wide gap doped semiconductors (Fig. 3.22) the value of E_C-F is usually very small ($\lesssim 0.1$ eV), hence $\chi \approx w_T$. In the preceding section we pointed out that in the case of metals some of the experimentally observed deviations from the linear dependence mentioned above stem from the differences in dipole potential drops produced by solvent adsorption on the different metals (62). A similar effect (although apparently less significant for a homologous series of semiconductor electrodes) may also occur in the electrochemistry of semiconductors.

3.8.3. E n e r g y   D i a g r a m   o f   t h e   S e m i c o n d u c t o r / E l e c t r o l y t e   C o n t a c t. It is convenient to discuss the details of the thermodynamic and kinetic behavior of semiconductor electrodes with the aid of an energy diagram for the interface on which electrochemical potentials, along with the energies of the band edges of the semiconductor, E_C and E_V, are displayed. In the semiconductor this is the Fermi level, F; in the solution it is the electrochemical potential, F_{redox}, of the redox couple which exchanges electrons with the semiconductor.

Let us now examine the method of construction of such an energy level diagram on the basis of experimental

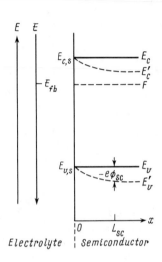

Fig. 3.23. The energy level
diagram for a semiconductor/
electrolyte junction at the
flat band potential (full
lines) and at a more positive
potential (dotted lines).

data. The determination of the flat band potential, E_{fb},
is the point of departure. If the flat band potential is
known, then we also know the position of the electrochemical
potential(Fermi level) of the semiconductor, F, on the
electrode potential scale, that is, against the chosen
reference electrode. Again note that the Fermi level of
the semiconductor electrode coincides with the Fermi
level of the reference electrode when the electrodes are
at the same potential. With the help of Eq. (2.33) it
is possible to determine the position of the Fermi level,
F, on the "physical" scale of energies in which the energy
of the electron in vacuum at a point close to the solution surface is taken as the reference point (Fig. 3.23).
Then, using the known value of the majority carrier concentration and the relations in (1.22) describing the
position of the Fermi level with respect to the majority
carrier band edge (the conduction band for an n-type
material, or the valence band for a p-type material)
it is possible to find the majority carrier band
edge in the bulk of the semiconductor and, consequently, also at the surface when $E=E_{fb}$. Since we know
the bandgap, $E_g=E_c-E_v$, it is also easy to find the
position of the minority carrier band edge. Finally, proceeding from the equilibrium potential value for the
redox couple and again using (2.33), one can find the
level of F_{redox} in solution. Energy level diagrams of
semiconductor/aqueous solution junctions for a number
of the most widely used materials are shown in Fig.
3.24 (125).

For any electrode potential differing from E_{fb}, the
band edges in the semiconductor bulk (in Fig. 3.22 these
are E_c' and E_v' respectively) shift by the value of ϕ_{sc}

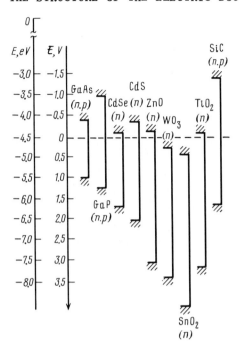

Fig. 3.24. The position of semiconductor band edges with respect to the E_{vac} level in aqueous solutions of surface-inactive electrolytes (pH 1). Potentials are cited with respect to the Normal Hydrogen Electrode (125). Note: in keeping with (125) we have used a value of −4.5 eV for F_{NHE} (the more accurate value is −4.4 eV; see Section 2.4).

with respect to their position at the surface. The position of all the levels in the semiconductor with respect to the solution levels depends on whether the potential drop in the Helmholtz layer, ϕ_H, changes with changing electrode potential (this drop is not explicitly marked in the drawings but it is automatically taken into consideration when determining E_{fb} and thus in the positioning of F on the scale of electrode potentials).

At this point it is necessary to distinguish between the two extreme situations mentioned in Section 3.4, band edge pinning and Fermi level pinning at the surface of the semiconductor electrode (cf. Fig. 3.23 and 3.6). Let us emphasize that pinning of the bands can be observed not only when polarizing the electrode with an external voltage source, but also in some cases upon illumination of the electrode if photoadsorption and photodesorption processes do not occur*, so that the structure of the

---

*The phenomenon of photoadsorption and photodesorption, that is, the change in the surface coverage of adsorbate under illumination of the adsorbent, has been observed on the "dry" surface of some semiconductor materials (see, for example, (94,126,127)). It can be assumed that such processes also take place at the semiconductor/ electrolyte interface but this problem has not yet been studied.

Helmholtz layer and the potential drop, ϕ_H, within it remain unchanged. Such a situation is observed, for example, on the TiO_2 electrode (128,129) and also probably on some other oxide electrodes.

An intermediate case often occurs, where $\Delta\phi_{SC}$ and $\Delta\phi_H$ are of comparable size. For example, on electrodes made of Ge (96), GaP (130) and $MoSe_2$ (131) the values of both ϕ_H and ϕ_{SC} change considerably with both polarization and illumination of the electrode. Sometimes over a certain potential range the band edges of the semiconductor are pinned at the surface ($\phi_H \approx$ const), while outside this range the value of ϕ_H begins to change significantly and the bands are unpinned. Such details of the behavior of semiconductor electrodes have a large effect, particularly on the performance of photoelectrochemical energy convertors.

Chapter 4

THE KINETICS OF ELECTROCHEMICAL REACTIONS ON SEMICONDUCTOR ELECTRODES

Electrochemical reactions are heterogeneous chemical reactions which occur at the electrode/electrolyte interface and are accompanied by the transfer of electrical charge across that interface. It is only natural that electrode reactions proceed in accordance with the fundamental laws of chemical kinetics, although they have their own specific features which are determined by the peculiarities of the transfer of charge across the interphase boundary. The energetics and, consequently, the rate of charge transfer from one phase to the other depend on the electric potential difference between the phases. As a result, in addition to the "usual" variables of chemical kinetics - concentration, temperature, and so on - electrochemical kinetics is characterized by an additional independent variable, the electrode potential. It is significant that this quantity, as well as the rate of the electrode reaction, may vary considerably as a result of electrochemical polarization.

The features specific to electrode processes on semiconductor electrodes as compared with those on metallic electrodes are connected, first, with their electronic structure, in particular with the bandgap and the two types of charge carriers and, second, with the essential influence of the applied electric field on the concentration of these carriers near the surface. It is precisely on these points that we shall concentrate our attention.

It should be noted that prior to the 1920's the concept that the charge transfer across the interface was always a fast process prevailed, the measured rate of the electrode process being determined, on the whole, by other processes, for example, the transport of reagents to the electrode surface or subsequent chemical reactions. More recently, however, the concept of the "slowness" of the electrochemical reaction proper (80-82) has been firmly established in electrochemical kinetics. This concept is applicable to a diverse range of electrode reactions, including those occurring on semiconductor electrodes.

4.1. Phenomenological Description of Electrode Reactions

Let us examine a semiconductor electrode on which a redox reaction of the type in Eq. (2.8) occurs. Electrons from both the conduction and valence bands can, in general, take part in the electrode process. Thus the four different electron transfers shown in Fig. 4.1. correspond to the reversible reaction under discussion. The transfers consisting of the removal of electrons from the semiconductor or the injection of holes contribute to the cathodic current, and those consisting of the injection of electrons and removal of holes contribute to the anodic current. Thus, the total current is made up of four currents, i_n^c, i_p^c, i_n^a, i_p^a, and the reaction in question is the sum of two partial reactions. Assuming that in Eq. (2.8) n=1, these reactions can be written as follows

$$Ox + e^- \rightleftarrows Red \qquad\qquad (4.1a)$$

$$Ox \rightleftarrows Red + h^+ \qquad\qquad (4.1b)$$

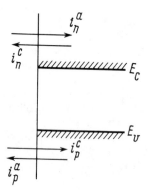

Fig. 4.1. The electron and hole currents at a semiconductor electrode in a solution containing a redox couple.

representing reaction through the conduction band or valence band respectively (here the symbol h^+ signifies a positively charged hole in the valence band). On a metallic electrode only the first of these two reactions can take place.

Let us examine the expressions for the currents corresponding to the two reactions (4.1). (It should be borne in mind that the term current here really refers to the current density.) At thermodynamic equilibrium the current crossing the interface due to the "hole" reaction (4.1b), corresponding to the transfer of electrons from the solution species to the electrode, may be written as follows

$$(i_p^a)^o = e \left\{ k_p^a \; e^{-E_p^a/kT} \right\} c_{red,s}^o \, p_s^o \qquad (4.2)$$

where $c_{red,s}^o$ is the equilibrium concentration of the reduced species, Red, in solution near the electrode surface and p_s^o is the equilibrium concentration of holes in the valence band near the surface of the semiconductor (equivalent to the number of vacant places for electrons arriving from the solution species). The concentration p_s^o, in the absence of degeneracy, can be written as $p_s^o = p_o \; exp(-e\phi_{sc}^o/kT)$ where ϕ_{sc}^o is the equilibrium potential drop in the semiconductor. The product of the concentration, p_s^o and $c_{red,s}^o$, is the sort of factor which is common in chemical kinetics. The quantity in brackets is, in contrast, a factor peculiar to electrochemical systems, which characterizes the rate of the process. Here E_p^a is the activation energy of the process (strictly, one should speak of the free energy of activation) and k_p^a is the pre-exponential factor.

Similarly, the equilibrium current for the same reaction (4.1b) but in the opposite direction, resulting from the transfer of electrons from the valence band to the solution (or of holes to the valence band) is given by the equation

$$(i_p^c)^o = e \left\{ k_p^c \; e^{-E_p^c/kT} \right\} c_{ox,s}^o \, n_v \qquad (4.3)$$

where $c_{ox,s}^o$ is the equilibrium concentration of the oxidized species, Ox, close to the surface and $n_v \approx N_v$ is the concentration of electrons in the valence band near the surface; the latter quantity, unlike p_s^o, can be regarded as constant and independent of potential. At equilibrium, in accordance with the principle of detailed balance (see, for example, (132)), the currents in Eq. (4.2) and Eq. (4.3) are of equal absolute magnitude, and represent, by definition, the exchange current i_p^o for the electrode reaction occurring through the valence band

$$i_p^o = (i_p^a)^o = (i_p^c)^o \qquad (4.4)$$

Let us now assume that the equilibrium conditions at the interface are upset, for example by switching on a constant voltage source. As a result of the polar-ization, the electrode potential adopts a certain value E which differs from the equilibrium value E^o. The quantity

$$\eta = E-E^o \qquad (4.5)$$

is called the overvoltage. In the general case, the change in the Galvani potential, which is equal to η, is distributed between the semiconductor, the Helmholtz layer and the electrolyte solution. Below, however, we shall assume that the electrolyte concentration is suf-ficiently high so that the potential drop in the elec-trolyte solution can be neglected. This is usually the case for most experiments:

$$\eta = \eta_H + \eta_{sc} \qquad (4.6)$$

where $\eta_H = \phi_H - \phi_H^o$ and $\eta_{sc} = \phi_{sc}^o - \phi_{sc}$ are the components of the overvoltage associated with the Helmholtz layer and the semiconductor respectively. These quantities exert rather substantial influences on the processes of electron transfer. Note that, as a result of the dif-ference in the direction of the potential axes (Section 3.1) adopted in electrochemistry and in semiconductor surface physics, $\eta_H = \Delta\phi_H$ but $\eta_{sc} = - \Delta\phi_{sc}$.

The activation energy is a function of the dropping potential in the region where the discharge occurs, that is, in the Helmholtz layer. As is the case with metallic electrodes (80,82), it can be assumed that the dependence of the activation energies, $E_p^{a,c}$, on the overvoltage, η_H, is expressed by the equations

$$E_p^a(\eta_H) = E_p^a - e\alpha_p\eta_H \quad ; \quad E_p^c(\eta_H) = E_p^c + e(1-\alpha_p)\eta_H \qquad (4.7)$$

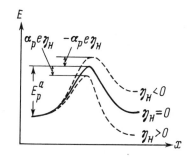

Fig. 4.2. The dependence of the activation energy for electrochemical reaction (hole current) on the overvoltage in the Helmholtz layer η_H.

The quantity $0 \leqslant \alpha_p \leqslant 1$ is called the transfer coefficient and characterizes the extent of the potential's influence on the activation energy of the electrode process. According to Eq. (4.7) the height of the corresponding energy barrier changes linearly with the energy difference between the initial and final states. Fig. 4.2 shows how this happens. The magnitude of the transfer coefficient, and also the limits of applicability of this notion depend on the microscopic characteristics of the electrode reaction, and are examined below.

Taking into account Eqs. (4.4) and (4.5) and bearing in mind that the value of k_p^a does not depend on η, the expression for the current i_p^a can be written as follows*

$$i_p^a = i_p^0 \left\{ \frac{p_s}{p_s^o} \times \frac{c_{red,s}}{c_{red,s}^o} \right\} e^{e\alpha_p \eta_H / kT} \qquad (4.8)$$

where $c_{red,s}$ and p_s are the concentrations (generally speaking, non-equilibrium concentrations) of the reagents at the interface. Similarly, taking into consideration the fact that the value of n_v is constant, we get for the current i_p^c

$$i_p^c = i_p^0 \frac{c_{ox,s}}{c_{ox,s}^o} e^{-e(1-\alpha_p)\eta_H / kT} \qquad (4.9)$$

---

*Taking into account that the discrete nature of the charge in the electric double layer may formally lead to the dependence of k_p^a on η_H. Tunnelling of electrons through the narrow barrier of the space charge layer, which may become essential for sufficiently strong fields (see Section 4.4), results in an effective dependence of k_p^a on η_{sc}.

The resultant current for the electrochemical reaction occurring through the valence band is equal to $i_p = i_p^a - i_p^c$. According to Eqs. (4.8) and (4.9), we obtain

$$i_p = i_p^o \left\{ \frac{p_s}{p_s^o} \times \frac{c_{red,s}}{c_{red,s}^o} e^{e\alpha_p \eta_H /kT} - \frac{c_{ox,s}}{c_{ox,s}^o} e^{-e(1-\alpha_p)\eta_H/kT} \right\} \quad (4.10)$$

Using a similar argument it is possible to show that the resultant current, i_n, associated with the electrochemical reaction occurring through the conduction band (reaction (4.1a)), is given by

$$i_n = i_n^o \left\{ \frac{c_{red,s}}{c_{red,s}^o} e^{e\alpha_n \eta_H /kT} - \frac{n_s}{n_s^o} \times \frac{c_{ox,s}}{c_{ox,s}^o} e^{-e(1-\alpha_n)\eta_H/kT} \right\} \quad (4.11)$$

while the exchange current, i_n^o, for this process is

$$i_n^o = e \left\{ k_n^a e^{-E_n^o/kT} \right\} c_{red,s}^o \quad p_c = e \left\{ k_n^c e^{-E_n^C/kT} \right\} c_{ox,s}^o n_s^o$$

The meaning of the notation used here is obvious from a comparison with the notation in Eqs. (4.2) and (4.3): in particular, $p_c \approx N_c$. Here, however, it is p_c - the concentration near the surface of unoccupied electronic levels in the conduction band - that should be regarded as constant. The value of n_s^o, in contrast, depends on the equilibrium value of the potential drop, ϕ_{sc}^o: in the absence of degeneracy $n_s^o = n_o exp(e\phi_{sc}^o/kT)$.

The exchange currents i_p^o and i_n^o and the transfer coefficients, which are included in the expressions obtained, are phenomenological parameters of the theory, and can be found directly from the experiment. The bigger the exchange current, the "faster" the electrode reaction. According to Eqs. (4.10) and (4.11), the latter means that the faster the reaction, the lower the overvoltage required to achieve any given current density.

Equations (4.10) and (4.11) describe under sufficiently general assumptions the electron currents through the valence and conduction bands of a semiconductor when an electrochemical reaction occurs at the interface. The full electric current i (the resultant charge flux) per unit interface area is equal to

$$i = i_p + i_n \qquad (4.12)$$

It often happens that exchange currents via the two bands, i_p^o and i_n^o, differ considerably so that in Eq. (4.12) one of the terms plays a decisive role. It should also be noted that for small overvoltages, according to Eqs. (4.10)-(4.12), there is a linear relationship between the magnitude of i and η. In this case the equivalent "resistance" depends on the values of the exchange currents, and by measuring this resistance it is possible to determine these currents.

For certain important cases the relations mentioned above may be considerably simplified. If the electrode reaction currents (Eqs. (4.10) and (4.11)) are sufficiently small, so that the magnitudes of i_p and i_n are less than the transport limiting currents for the reagents Red and Ox, the effects of concentration polarization in the solution may be neglected (83). In other words, the current flow does not perturb the concentration distributions c_{red} and c_{ox}, hence,

$$c_{red,s} = c_{red,s}^o = c_{red} \; ; \; c_{ox,s} = c_{ox,s}^o = c_{ox} \qquad (4.13)$$

where c_{red} and c_{ox} are the concentrations of Red and Ox in the bulk of the solution.

When the relations in (4.13) hold, Eqs. (4.10) and (4.11) assume the following form

$$i_p = i_p^o \left(\frac{p_s}{p_s^o} e^{\alpha_p \eta_H/kT} - e^{-e(1-\alpha_p)\eta_H/kT} \right) \qquad (4.14)$$

$$i_n = i_n^o \left(e^{\alpha_n \eta_H/kT} - \frac{n_s}{n_s^o} e^{-e(1-\alpha_n)\eta_H/kT} \right) \qquad (4.15)$$

It should be emphasized at this point that the currents i_p and i_n are not solely determined by the change in the Helmholtz potential drop η_H, as would be the case for an electrochemical reaction at a metallic electrode. For semiconductor electrodes the currents are linked with the potential drop in the space charge layer of the semiconductor, ϕ_{sc}, through the factors p_s/p_s^o and n_s/n_s^o in Eqs. (4.14) and (4.15). If the whole of the

applied potential drop occurs within the Helmholtz layer
(for example, due to the presence of a high density of
charged surface states), then $\eta = \eta_H$ and hence $p_s/p_s^o = 1$,
$n_s/n_s^o = 1$. For the current i_n through the conduction
band in such a case, we get the following expression

$$i_n = i_n^o \left(e^{e\alpha_n\eta/kT} - e^{-e(1-\alpha_n)\eta/kT} \right) \qquad (4.16)$$

which coincides with the corresponding expression in the
electrochemistry of metals (80-82).

In the opposite extreme case the entire potential
drop occurs within the semiconductor, and $\eta = \eta_{sc}$.
Using the Boltzmann distribution to find n_s and assuming
that $\eta_H = 0$, we obtain in this case from Eq. (4.15) (in
particular instead of Eq. (4.16))

$$i = i_n^o \left[1 - exp(-e\eta/kT) \right] \qquad (4.17)$$

At sufficiently large deviations from equilibrium when
$e|\eta|/kT \gg 1$, differentiation of Eq. (4.17) with respect
to the electrode potential, E, and taking account of
the fact that $dE = d\eta_{sc} = d\eta$ gives

$$dln|i_n|/dE = -e/kT \qquad (4.18)$$

The formally similar relation for the case of large
overvoltages ($e|\eta|/kT > 1$) is also found in the electro-
chemistry of metals (Tafel law). Equation (4.18) is
entirely analogous with the Tafel law (and also with the
corresponding relationship from Eq. (4.15), if we
assume that in Eq. (4.15) the transfer coefficient, α_n,
is equal to zero and suppose that $dE = d\eta_H = d\eta$). However, despite this formal coincidence, the physical
sense of Eq. (4.18) in these two cases (that is, when
$\eta = \eta_{sc}$ and when $\eta = \eta_H$ respectively) is totally different, as can be seen from the discussion above.

If we assume that the exchange currents are rather
large ($i_n^o \gg i_n$ and $i_p^o \gg i_p$) the expressions in parentheses
in Eqs. (4.14) and (4.15) are equal to zero. When
$\eta = \eta_{sc}$, so that $\eta_H = 0$, we obtain

$$n_s = n_s^o \quad ; \quad p_s = p_s^o \qquad (4.19)$$

The physical meaning of these equations is clear enough: if the electrode reaction is sufficiently fast (corresponding to a large exchange current) then the surface concentrations of the reagents remain equal to their equilibrium values, even when there is a current flowing. Equations (4.14)-(4.17) and (4.19) may be used as boundary conditions in the calculation of the mobile carrier distribution inside the semiconductor.

In conclusion, let us note that if the concentrations of the reagents Red and Ox in solution are relatively small, then effects connected with the transport of reagents to the electrode may become important. As a result of this the values of $c_{red,s}$ and $c_{ox,s}$ will no longer be equal to c_{red} and c_{ox}. Modification of Eqs. (4.10) and (4.11), to take into account the effects of transport in the assumption that $\eta_H = 0$, has been carried out in the literature (133). These effects, however, are insignificant if the corresponding currents remain less than the limiting currents for the transport of Red and Ox in solution. In order to make an assessment of the effect one may assume that

$$i_{red,ox}^{lim} \simeq e z_{red,ox} \cdot D_{red,ox} \cdot c_{red,ox} / \delta_D$$

where $D_{red,ox}$ is the diffusion coefficient, $z_{red,ox}$ is the charge number of the Red or Ox species, and δ_D is the thickness of the diffusion layer in solution.

Ξ 4.2.   F u n d a m e n t a l s   o f   t h e   T h e o r y   o f   t h e   E l e m e n t a r y   A c t

The previous section contained a phenomenological approach to the description of electrochemical reactions on semiconductor electrodes. This approach is based on fundamental kinetic principles such as the law of mass action and the principle of detailed balance. In some cases, however, the use of these general relations alone proves to be insufficient.

The clearest understanding of the physical nature of the charge transfer process is attained when it is described on a microscopic level. Such an approach makes it possible, to a certain extent, to reveal the meaning of the parameters discussed above, to establish the limits of applicability of the phenomenological approach, and also to chart the ways towards further generalization. In particular this latter aspect is especially important due to the need for the development of a theory for the photoelectrochemical reactions on semiconductor electrodes. In this section we examine in brief the fundamentals of the microscopic theory of electron transfer (the theory of the elementary act). A detailed and all-round account, along with the various applications and history of the problem, is given in (46, 134-138).

It should be pointed out that the approach charac- terized by the above mentioned works is not the only possible one. It is based on a number of assumptions some of which may be invalid. Alternative approaches, however, are as yet insufficiently developed, although in recent years there has been certain progress (see, for example, (139)).

As an explicit example let us first examine the flux of electrons from a semiconductor into solution, that is, the cathodic current i^C which apparently consists of two components i_n^C and i_p^C (see Fig. 4.1); the anodic current may be analyzed in a similar way. Assuming that the particles (ions or molecules) from the solution enter into the electrode reaction independently, we get for the magnitude of i^C

$$i^C = e \int_0^\infty \int\int_{-\infty}^\infty c_{ox}(x,E')k(E,E',x)\rho(E)f(E)dEdE'dx \quad (4.20)$$

where $\rho(E)$ is the density of electron states in the semiconductor and $f(E)$ is the function describing the energy distribution of the electrons (see Eq. (1.2)). Hence $\rho(E)f(E)dE$ is the number of electrons with an energy in the interval from E to E + dE.

The function $c_{ox}(x,E')$ is the partial concentration of ions or molecules to be reduced characterized by an energy E' and being at a distance x from the electrode. We can therefore write

$$\int_{-\infty}^\infty c_{ox}(x,E')dE' = c_{ox}(x) \quad (4.21)$$

where $c_{ox}(x)$ is an "ordinary" bulk concentration for the reacting particles. Finally, the function $k(E,E',x)$ describes the rate of transition of an electron with an energy E from the semiconductor onto a particle in

solution which is characterized by the energy level E'
and is at a distance x from the electrode. Integrating
with respect to all possible distances x (that is, summing
over all particles) and over the energies of the reacting
particles and the electrons, both in the valence band and
in the conduction band, will obviously produce the current i^C. Note that since the integration with respect
to E and E' in Eq. (4.20) is carried out from $-\infty$ to $+\infty$
the choice of the energy zero may be ignored for the
time being.

In order to calculate the current from Eq. (4.20)
additional simplifying assumptions are necessary. In
particular, it is only natural to assume that the probability of electron transition sharply decreases as
the distance between the reacting particle and the electrode surface is increased. Hence, it is possible to
introduce the concept of a reaction layer within which
the act of discharge takes place. The reacting particles
are at a distance δ from the electrode, where the value
of δ is the most probable distance at which electron
transfer between the electrode and the reagent occurs.
Taking into consideration what we know of the interface
structure, we may conclude that the value of δ is close
to the thickness of the Helmholtz layer (if not coincident with it). In other words, $c_{ox}(-\delta)$ is the concentration of reacting (reducible) particles near the
surface which was previously denoted as $c_{ox,s}$.

The expression for the partial concentration
$c_{ox}(-\delta,E')$ can obviously be written as

$$c_{ox}(-\delta,E') = c_{ox,s} W(E')$$

where $W(E')$ is a function characterizing the distribution
of the energy levels, E', of the particles in the plane
of discharge (at a distance $-\delta$ from the electrode surface).

Taking this into account, we can write the general
expression (4.20) in the following simplified form

$$i^C = ec_{ox,s} \int\int\limits_{-\infty}^{\infty} W(E')k(E',E)\rho(E)f(E)\,dEdE' \qquad (4.22)$$

where $k(E,E')$ is a function describing the rate of electron transfer from the electrode onto the particles in

the plane of discharge, and depends on the magnitude of δ.

Note that we can obtain relations corresponding to the phenomenological expressions for electron and hole currents from Eq. (4.22). Indeed, in order to calculate, for example, the value of i_p^C (see Eq. (4.3)), the integration with respect to E in Eq. (4.22) should be carried out over the semi-infinite interval $(-\infty; E_{v,s})$. Using the mean value theorem and keeping in mind the fact that

$$\int_{-\infty}^{E_{v,s}} \rho(E) f(E) dE = n_v \qquad (4.23)$$

the expression resulting from Eq. (4.22) can be written as follows

$$i_p^C = ek_p c_{ox,s} n_v \qquad (4.24)$$

Here

$$k_p = \int_{-\infty}^{\infty} W(E') k(\bar{E}, E') dE' \qquad (4.25)$$

where $k(\bar{E}, E')$ is the value of $k(E, E')$ at a certain mean point $\bar{E}$ on the interval $(-\infty; E_{v,s})$. Equation (4.24) formally coincides with Eq. (4.3), and k_p from Eq. (4.25) may be identified with $k_p^C \exp(-E_p^C/kT)$.

Any further advance in the calculations demands additional ideas and concepts qualitatively different from those given above. The most important idea of this kind, which was formulated by Gurney (40), is the utilization of the Franck-Condon principle (79) in the description of electron transfer between the electrode and the reagent in solution. According to this principle, the electron transfer is most probable when the energy levels of the initial and the final states of the system coincide. Let us emphasize that at equilibrium it is the free energies of the initial and the final states of the system that are equal to each other, and not the internal energies. However, for each elementary act of electron transfer to proceed in conformity with the Franck-Condon principle it is necessary that the internal energy levels be equal. This may be achieved by

fluctuations in the internal energy. The key mechanism
in the electrochemical systems, which ensures this
"levelling" of the energies, is based on the fluctuations
of the polar solvent interacting with the charged reagent.

Physically, the Franck-Condon principle ensues from
the fact that the electron transfer occurs rather quickly:
characteristic times for electron motion are $\approx 10^{-15}$ s.
During the transition time the system, consisting of an
ion together with its surrounding solvent molecules,
turns out to be "frozen"; the magnitude of all the internal coordinates are fixed in so far as the movements
of these particles are much slower ($\approx 10^{-13}$ s). For
this reason energy exchange between the system and the
electron during the short time span of the electron
transfer is practically impossible. Consequently, electrons are transferred when the configuration of the "slow"
(heavy) particles is such that the transfer is not accompanied by any change in the energy of the system. Under
these conditions the law of conservation of energy, which
should be observed in the course of the transfer, takes
the form E = E'; thus it follows that the electronic
levels of the initial and the final states should be
equal.

It should be stressed that the condition of rapid
electron transfer is important. In the case of a slow
transfer, part of the electron subsystem energy may, in
the course of the transfer, be transmitted, for example,
to the internal degrees of freedom of the ion-solvent
system. In such a case the simple form of the law of
conservation of energy (E = E') will not apply. In the
final analysis, the small mass of the electron when compared with the masses of the reacting ion and the solvent
molecules provides the physical basis for the rapidity
of transfer.

It is also assumed that the act of transfer is not
accompanied by the absorption or radiation of a light
quanta (radiationless transition), or by any other interactions which tangibly upset the energy balance in
the electron-reagent system in solution (for example,
interactions with surface oscillating modes and plasmons,
foreign adsorbed particles, and so on).

The condition E = E' derived from the Franck-Condon
principle is expressed mathematically by the fact that
the function k(E',E) has an extremely sharp maximum when
its arguments are equal and can therefore be approximately written as follows

$$k(E', E) = k(E)\delta(E - E') \qquad (4.26)$$

where $\delta(u)$ is the Dirac delta function determined by the relations (see, for example, (132), Section 5):

$$\delta(u) = \begin{cases} 0, \text{ if } u \neq 0 \\ \\ \infty, \text{ if } u = 0 \end{cases} \qquad \int_{-\infty}^{\infty} \delta(u)\,du = 1 \qquad (4.27)$$

In accordance with Eq. (3.32), the integrals containing delta functions are calculated on the basis of the following formula

$$\int_{-\infty}^{\infty} f(u')\delta(u - u')\,du' = f(u) \qquad (4.28)$$

Using Eq. (4.26), and also Eqs. (4.27) and (4.28), the expression for the cathodic current i^c can be written as follows

$$i^c = ec_{ox,s} \int_{-\infty}^{\infty} W(E)k(E)\rho(E)f(E)\,dE \qquad (4.29)$$

Another important problem is the calculation of the probability $W(E)$ in Eq. (4.29). For this it is necessary to consider the energy characteristics of charged particles in liquid media and in particular ions in electrolyte solutions in polar solvents. In crystalline solids the density of states and the characteristic energy levels (for example, the position of the bottom of the conduction band in a semiconductor) can be defined unambiguously. When we consider liquid media, we only have the mean values of the characteristic energies (for example, for an ion or electron in a liquid). It is precisely these mean values which are used in the many thermodynamic relations and which are determined on the basis of the analysis of thermodynamic cycles of the type examined in Sections 2.3 and 2.4.

At the same time, the energy characteristics of even identical particles cannot be taken as equal due to the fluctuating interaction of the particles with their surroundings. This applies to the energy levels of particles in solution (ions, atoms, molecules) acting

as reagents in electrochemical reactions. These particles interact with dielectric polarization of the bulk
of the polar liquid which surrounds them, and also
directly with their close surroundings due to the existence of short-range forces. Fluctuations in these interactions result in shifts in the electronic energy
levels of the particles in both directions with respect
to their relatively most probable (mean) energy level
which we shall denote as E^O.

The deviation of any quantity from a most probable
value under the influence of accidental external impacts
is described by the Gauss distribution given certain
general assumptions (see, for example, (26,27)). Hence,
the probability, $W(E)$, of deviation of E from E^O can be
written as follows

$$W(E) = \frac{1}{\sqrt{2\pi}\sigma} e^{-(E-E^O)^2/2\sigma^2} \qquad (4.30)$$

The parameter σ, included in the Gauss distribution
(Eq. (4.30)) is called the dispersion and characterizes
the mean square deviation of the value of E from the
most probable value (see Fig. 4.3). With the help of
Eq. (4.30) it becomes evident that

$$\sigma^2 = \int_{-\infty}^{\infty} (E - E^O)^2 W(E)\, dE \qquad (4.31)$$

The distribution function (4.30) satisfied the normal-
ization conditions for all E^O and σ:

$$\int_{-\infty}^{\infty} W(E)\, dE = \frac{1}{\sqrt{2\pi}\sigma} \int_{-\infty}^{\infty} e^{-(E-E^O)^2/2\sigma^2}\, dE = 1 \qquad (4.32)$$

As the dispersion, σ, tends to zero, the distribution
(4.30) turns into a delta function

$$\lim_{\sigma \to 0} W(E,\sigma) = \delta(E - E') \qquad (4.33)$$

Equation (4.30) together with Eq. (4.29) (and the
corresponding expression to Eq. (4.29) for the current
i^a) makes it possible to obtain a number of important
results without further consideration of the parameter σ.

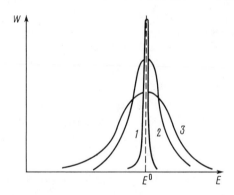

Fig. 4.3. The W(E) dependence at different values of dispersion σ; $\sigma_1 < \sigma_2 < \sigma_3$ (the indices correspond to the numbers on the curves).

For our present purposes, however, there is no need to maintain such a degree of generality. Based on model calculations (135-138), the dispersion σ in our particular physical situation turns out to be

$$\sigma = \sqrt{2kTE_R} \qquad (4.34)$$

The parameter E_R included in Eq. (4.34) is the reorganization energy which has already been discussed in Section 2.3 for a somewhat different type of process in connection with our analysis of the concept of the electrochemical potential of the electron in solution. Here the value of E_R characterizes the free energy necessary to change the state of the medium surrounding the reagent from its initial equilibrium position (corresponding to the Ox particle near the electrode) to its final equilibrium position (Red particle near the electrode). Note that the values of E_R near the electrode surface may be different from the values in the solution bulk. This is mainly because in interfacial charge transfer processes the rearrangement only takes place in the semi-space occupied by the solvent, whereas in the case of the homogeneous charge transfer process it involves the entire space. Thus, the magnitude of E_R for homogeneous and heterogeneous electron transfer reactions is generally not the same. In aqueous solutions (45) the reorganization energy is approximately $E_R \simeq 0.5$-2 eV. Hence, we obtain for the dispersion, σ, at room temperature from Eq. (4.34), $\sigma \simeq 0.1$-0.2 eV. Thus, on the one hand, $\sigma \gg kT$ and, on the other, $\sigma \ll E_g$.

Note that Eq. (4.33) has a clear physical meaning when Eq. (4.34) is taken into account: in the absence of fluctuations of the medium ($T \to 0$) or in the absence of any influence by the medium on the energy levels of the particles ($E_R \to 0$) these levels become equal to E^O for all identical particles.

Substituting Eqs. (4.30) and (4.34) in Eq. (4.29), we obtain the following expression for the electron transfer current

$$i^C = \frac{ec_{ox,s}}{\sqrt{4\pi kTE_{R,ox}}} \int_{-\infty}^{\infty} k(E)\rho(E)f(E)\exp\left\{-\frac{(E-E_{ox}^O)^2}{4kTE_{R,ox}}\right\}dE \quad (4.35)$$

where E_{ox}^O is the most probable (thermodynamic) value of the electron energy level for Ox-particles in solution (this has already been discussed in Section 2.3) and $E_{R,ox}$ is the reorganization energy for the Ox-particle.

Equation (4.35) is rather important. For this reason we shall show how it is possible to obtain an expression similar to Eq. (4.35) by means of the model calculations mentioned above before going on to analyze the equation in detail. Furthermore, the models used are necessary for the calculation of the pre-exponential factor $k(E)$, and also in the analysis of some of the details of electron transfer. In this context let us examine one such approach to the model calculation of the electron current.

Fig. 4.4 shows a simple (unidimensional) description of the potential energy, U, of a system consisting of a reacting particle and an electrode in its initial (i) and final (f) states as a function of the solvent dimensionless coordinate, q (the coordinates of "slow" particles). At the solvent coordinate q_i^o, which corresponds to the initial equilibrium state of the system, the energies (terms) U_i and U_f differ considerably from

Fig. 4.4. Potential curves U_i and U_f as functions of the solvent coordinate.

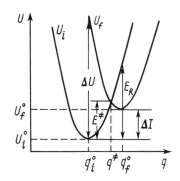

each other: $U_f(q_i^0) - U_i(q_i^0) = \Delta U$. Hence, the electron transition requires the consumption of energy ΔU. Since there is no source of energy, a transition in this state of the system cannot occur. Electron transition, in accordance with the Franck-Condon principle, is possible at the coordinate $q^{\neq}$. This corresponds to the inter-section point of the two curves for the initial and final states, when the energies U_i and U_f are equal to each other. The transition state for the system corres-ponds to the point $q^{\neq}$. This is achieved as a result of thermal fluctuations of the solvent. The attainment of the configuration on $q^{\neq}$ requires an activation energy $E^{\neq}$ (for the transitions examined in Section 4.1, the activation energies $E_p^{a,c}$ and $E_n^{a,c}$ stood for the value of $E^{\neq}$). This activation barrier is sometimes called the Franck-Condon barrier.

Taking the discussion above into account, the elec-tron current i^c can be written as follows (134)

$$i_c = ec_{ox,s} \int_{-\infty}^{\infty} T(E)\rho(E)exp(-E^{\neq}/kT)dE \qquad (4.36)$$

where $T(E)$ is the so-called transmission coefficient which is presumably insignificantly dependent on the energy E. To find the dependence of $E^{\neq}$ on E let us return to the potential energy curves. We shall assume that the fluctuations are wholly determined by the os-cillations of the solvent dipoles, which occur at a single fixed frequency ω_1 (equal, for example, to the frequency of the librational oscillations of the sol-vent molecules). In this case the curves U_i and U_f are identical parabolas differing only in the coor-dinates of their minima. The coordinate of the transition state, $q^{\neq}$, can be found from the condition

$$U_i(q^{\neq}) = U_f(q^{\neq}) \qquad (4.37)$$

where

$$U = U_{i,f}^0 + \tfrac{1}{2}\hbar\omega_1(q - q_{i,f}^0)^2 \qquad (4.38)$$

and we find that, from Eq. (4.37), this coordinate is given by

$$q^{\neq} = \tfrac{1}{2}(q_i^0 + q_f^0) + \frac{U_f^0 - U_i^0}{\hbar\omega_1(q_f^0 - q_i^0)} \qquad (4.39)$$

The energy at this point is found by substitution from
Eq. (4.39) in Eq. (4.37), to give

$$U(q^{\neq}) = U_i^o + \frac{[\hbar\omega_1(q_f^o - q_i^o)^2/2 + U_f^o - U_i^o]^2}{2\hbar\omega_1(q_f^o - q_i^o)^2}$$

(4.40)

The reorganization energy E_R, as is seen from
Fig. 4.4, is then

$$E_R = \tfrac{1}{2}\hbar\omega_1(q_f^o - q_i^o)^2$$

(4.41)

Since the potential energy of a harmonic oscillator is
$\omega_1 q^2/2$, the value of E_R described by Eq. (4.41) precisely corresponds, within the framework of the model,
to the general definition of E_R given above.

Introducing the quantity $\Delta I = U_f^o - U_i^o$, the heat of
reaction, we obtain the final result that the activation
energy $E^{\neq} = U(q^{\neq}) - U_i^o$ is, from Eqs. (4.40) and (4.41)

$$E^{\neq} = \frac{(E_R + \Delta I)^2}{4E_R}$$

(4.42)

A similar result is also found for the more general case
when the existence of many ($m \geqslant 1$) coordinates characterizing the state of the solvent (reaction coordinates)
is considered. The reorganization energy can be determined on the basis of an obvious generalization of
Eq. (4.41)

$$E_R = \tfrac{1}{2} \sum_{j=1}^{m} \hbar\omega_j(q_f^o - q_i^o)_j^2$$

(4.43)

where the index ($1 \leqslant j \leqslant m$) refers to the oscillator
corresponding to the j-th coordinate.

The heat of reaction when an electron with an
energy E passes from the electrode to the solution is

$$\Delta I = F_o - E$$

(4.44)

where F_o is the thermodynamic equilibrium energy of the
electron in a solution containing the redox system (see
Section 2.3).

Combining Eqs. (4.44), (4.42) and (4.36) and using the fact that in this particular case $E_R = E_{R,ox}$, we obtain the final expression for the cathodic current

$$i^c = ec_{ox,s} \int_{-\infty}^{\infty} T(E) \rho(E) f(E) exp \left\{ - \frac{(F_o - E + E_{R,ox})^2}{4kTE_{R,ox}} \right\} dE \quad (4.45)$$

This expression coincides fully with Eq. (4.35), if we take into account the fact that, according to Eq. (2.24), $E_{ox}^O = F_o + E_{R,ox}$, and replace $T(E)$ by $k(E)/\sqrt{4\pi kTE_{R,ox}}$.

Thus, the model based on the concept of parabolic terms, $U_{i,f}$, produces the same results as the statistical model based on the Gauss distribution (Eq. (4.30)) in which a dispersion σ is specifically chosen. One can believe that expressions of the type given in Eq. (4.45) are actually more general than the assumptions inherent in the models from which the results are derived.

Similarly, for the anodic current associated with the transition of electrons from the reduced reagent (the Red particles) to the electrode we obtain the following expression

$$i^a = \frac{ec_{red,s}}{\sqrt{4\pi kTE_{R,red}}} \int_{-\infty}^{\infty} k(E) \rho(E) [1-f(E)] exp \left\{ - \frac{(E-E_{red}^O)^2}{4kTE_{R,red}} \right\} dE \quad (4.46)$$

where $E_{red}^O = F_o - E_{R,red}$. Equation (4.46) differs from Eq. (4.35), among other things, in the replacement of $c_{ox,s}$ by $c_{red,s}$ and $f(E)$ by $1-f(E)$; this latter term is the probability that the state with energy E in the semiconductor is vacant.

Finally the currents i^c and i^a for electron transfer may be written in the following symmetrical and compact form (140):

$$i^c = e \int_{-\infty}^{\infty} k(E) \mathcal{D}_{ox}(E) \mathcal{D}_{sc}^{occup}(E) dE \quad (4.47)$$

$$i^a = e \int_{-\infty}^{\infty} k(E) \mathcal{D}_{red}(E) \mathcal{D}_{sc}^{vacant}(E) dE \quad (4.48)$$

where $\mathcal{D}_{ox}$ and $\mathcal{D}_{red}$ are functions of the density of
vacant and occupied electron states in solution, while
$\mathcal{D}_{sc}^{vacant}$ and $\mathcal{D}_{sc}^{occup}$ are the corresponding terms for the
semiconductor electrode (or more generally an arbitrary
electronic conductor). At the same time

$$\mathcal{D}_{ox} = \frac{c_{ox,s}}{\sqrt{4\pi kTE_{R,ox}}} \; exp \left\{ - \frac{(E - E_{ox}^{o})^2}{4kTE_{R,ox}} \right\}$$

$$\mathcal{D}_{red} = \frac{c_{red,s}}{\sqrt{4\pi kTE_{R,red}}} \; exp \left\{ - \frac{(E - E_{red}^{o})^2}{4kTE_{R,red}} \right\} \tag{4.49}$$

where

$$E_{ox}^{o} = F_{o} + E_{R,ox} , \quad E_{red}^{o} = F_{o} - E_{R,red} \tag{4.50}$$

and

$$\mathcal{D}_{sc}^{occup} = \rho(E)f(E), \quad \mathcal{D}_{sc}^{vacant} = \rho(E)[1 - f(E)] \tag{4.51}$$

The coefficient $k(E)$ in Eqs. (4.35), (4.46) and also in
(4.47) and (4.48) is proportional to the probability of
electron transfer between the states of equal energy
in the solid and in the solution. (This probability
is

$$\frac{2\pi}{\hbar} |\psi_{1}(E) H_{if} \psi_{f}(E)|^2$$

where the $\psi_{i,f}$ are the wave functions for the electron
in the initial and the final states and H_{if} is the
Hamiltonian describing the interactions leading to transition between these states.) The calculation of the
coefficient $k(E)$ requires a highly detailed model for
the process under discussion. However, such a detailed
treatment is not necessary here, since a number of important results can be obtained directly from the
general relation given above.

§ 4.3. The Calculation of Charge
         Transfer Currents

Let us examine the mutual disposition of the characteristic energy levels for the solution and the electrode. This disposition exerts a decisive influence
over the magnitudes of the electron transfer currents.

Figure 4.5 shows schematically the distribution of
electron energy states at the interface of a solution

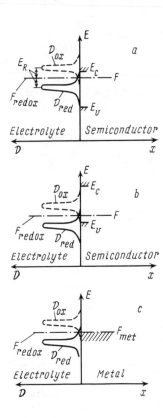

Fig. 4.5. Distribution
of energy states at the
junction between (a)
an n-type semiconductor,
(b) a p-type semiconduc-
tor, or (c) a metal and
an electrolyte which
contains a redox couple.

containing a redox couple in equilibrium with n- and
p-type semiconductors and a metal. At equilibrium the
electrochemical potentials of the solid, F, and the
solution, F_{redox}, coincide (see Section 2.3). The distributions of the vacant and occupied electron states
in the solution, $\mathcal{D}_{ox}$ and $\mathcal{D}_{red}$, are shifted with respect
to the position of F_{redox} by $E_{R,ox}$ and $E_{R,red}$. (Strictly speaking, the values of $\mathcal{D}_{ox}$ and $\mathcal{D}_{red}$ are shifted by
$E_{R,ox}$ and $E_{R,red}$ with respect to the level F_0, which
differs from F_{redox} by $kT \ln(c_{ox}/c_{red})$.) They may either
overlap or not overlap with the distributions of elec-
tron states, $\mathcal{D}_{sc}^{vacant}$ and $\mathcal{D}_{sc}^{occup}$, in the solid. According
to the Franck-Condon principle, a contribution to the
electron current is only made by those energy intervals
for which the densities of states overlap; this ensures
the possibility of electron transitions at constant
energy. These intervals correspond to the overlap (see
Eqs. (4.47) and (4.48)) of $\mathcal{D}_{ox}$ with $\mathcal{D}_{sc}^{occup}$ and of $\mathcal{D}_{red}$
with $\mathcal{D}_{sc}^{vacant}$ respectively. In the bandgap $\mathcal{D}_{sc}^{occup} =$
$\mathcal{D}_{sc}^{vacant} = 0$ (for metallic electrodes at ordinary tem-
peratures $\mathcal{D}_{met}^{vacant} = 0$ when $E < F_{met}$ and $\mathcal{D}_{met}^{occup} = 0$ when
$E > F_{met}$).

The change in electrode potential leads, firstly, to a change in the absolute magnitudes of the densities $\mathcal{D}_{sc}$ and $\mathcal{D}_{ox,red}$ and, secondly, to a change of their relative positions. Consequently, the energy intervals for their overlap also change. Depending on the particular conditions, either of these two effects may dominate.

Taking these points into account, we obtain, from Eqs. (4.46) and (4.48), the following expressions for the partial currents i_n^a and i_p^a (passing through the conduction and valence bands respectively), which make up the anodic current i^a:

$$i_n^a = \frac{ec_{red,s}}{\sqrt{4\pi kTE_R}} \int_{E_{c,s}}^{\infty} k(E)\rho(E)\,exp\left\{-\frac{(E - F_o + E_R)^2}{4kTE_R}\right\} dE \quad (4.52)$$

$$i_p^a = \frac{ec_{red,s}}{\sqrt{4\pi kTE_R}} \int_{-\infty}^{E_{v,s}} k(E)\rho(E)e^{(E-F)/kT}\,exp\left\{-\frac{(E - F_o + E_R)^2}{4kTE_R}\right\} dE \quad (4.53)$$

For simplicity we assume that $E_{R,ox} = E_{R,red} = E_R$. One should bear in mind that, in the absence of carrier degeneracy, for the conduction band $1 - f(E) \approx 1$, and for the valence band $1 - f(E) \approx exp\{(E - F)/kT\}$.

Note that the upper and lower limits of integration in Eqs. (4.52) and (4.53) are determined by the positions of the energy band edges at the surface of the semiconductor ($E_{c,s}$ and $E_{v,s}$). In accordance with Section 3.2,

$$E_{c,s} = E_c - e\phi_{sc}, \quad E_{v,s} = E_v - e\phi_{sc} \quad (4.54)$$

The integrals may be calculated analytically with the help of approximations derived from the physics of the process. The function $exp\{-(E - F_o + E_R)^2/4kTE_R\}$ decays sharply on both sides of its maximum (at $E = F_o - E_R$); the characteristic energy interval of this decay is $\sigma = \sqrt{2kTE_R}$, so that $\sigma \gg kT$ but, at the same time, $\sigma \ll E_g$. The distribution of energy levels for the case in question is shown in Fig. 4.5. Here, for simplicity, it is assumed that $c_{ox} = c_{red}$ (and therefore $F_O = F_{redox}$). The widths of the distributions of $\mathcal{D}_{ox}$ and $\mathcal{D}_{red}$ are determined by the value of the dispersion σ (see Eq. (4.34) and Fig. 4.3).

The other functions within the integrals in Eqs. (4.52) and (4.53) either change much more slowly (for example, $k(E)$ and $\rho(E)$) or decay sharply when moving away from the band edge (for example $exp\{ - (F - E)/kT\}$). Thus the energy interval of the overlap contributing to the current is in many cases quite narrow, and it can be assumed that charge transfer occurs only from energy regions close to the band edges. In view of this we can substitute $E = E_{c,s}$ and $E = E_{v,s}$ in the functions which vary slowly with energy in Eqs. (4.52) and (4.53) and take them out of the integral sign to obtain the following final expressions for the currents i_n^a and i_p^a:

$$i_n^a = ec_{red,s} \left(\frac{kT}{4\pi E_R} \right)^{\frac{1}{2}} k(E_{c,s}) N_c \, exp \left\{ - \frac{(E_{c,s} - F_O + E_R)^2}{4kTE_R} \right\} \quad (4.55)$$

$$i_p^a = ec_{red,s} \left(\frac{kT}{4\pi E_R} \right)^{\frac{1}{2}} k(E_{v,s}) p_s \, exp \left\{ - \frac{(E_{v,s} - F_O + E_R)^2}{4kTE_R} \right\} \quad (4.56)$$

Thus, the anodic current through the conduction band, i_n^a, is proportional to the density of states N_c, which does not depend on the surface potential, while the anodic current through the valence band, i_p^a, is proportional to the surface concentration of holes which depends on that potential. Similarly, for the components of the cathodic current we obtain from Eqs. (4.35) or (4.47)

$$i_n^c = ec_{ox,s} \left(\frac{kT}{4\pi E_R} \right)^{\frac{1}{2}} k(E_{c,s}) n_s \, exp \left\{ - \frac{(E_{c,s} - F_O - E_R)^2}{4kTE_R} \right\} \quad (4.57)$$

$$i_p^c = ec_{ox,s} \left(\frac{kT}{4\pi E_R} \right)^{\frac{1}{2}} k(E_{v,s}) N_v \, exp \left\{ - \frac{(E_{v,s} - F_O - E_R)^2}{4kTE_R} \right\} \quad (4.58)$$

With an eye to subsequent comparison, let us also find the expressions for electron transfer currents on metal electrodes. Assuming that charge transfer occurs within a narrow energy interval in the vicinity of the

Fermi level in the metal, F_{met}, and using similar approximations, we shall get for the anodic current

$$i_{met}^a = ec_{red,s}\left(\frac{kT}{4\pi E_R}\right)^{\frac{1}{2}} k(F_{met})\rho(F_{met})\,exp\left\{-\frac{(F_{met}-F_o+E_R)^2}{4kTE_R}\right\} \quad (4.59)$$

and for the cathodic current

$$i_{met}^c = ec_{ox,s}\left(\frac{kT}{4\pi E_R}\right)^{\frac{1}{2}} k(F_{met})\rho(F_{met})\,exp\left\{-\frac{(F_{met}-F_o-E_R)^2}{4kTE_R}\right\} \quad (4.60)$$

Equations (4.55) to (4.60) are the solution to the problem of calculating the electron transfer currents on semiconductor and metal electrodes.

Ξ 4.4 A n a l y s i s   o f   t h e   E x p r e s s i o n s   f o r   t h e   C h a r g e — T r a n s f e r   C u r r e n t s

Exchange currents are easily found from the equations in Section 4.3. The values of i_n^o and i_p^o are directly given by Eqs. (4.55) and (4.48) on substituting the value of $E_{c,s}$ and $E_{v,s}$ at the equilibrium surface potential, that is, $E_{c,s}^o = E_{c,s}$ and $E_{v,s}^o = E_{v,s}$ when $\phi_{sc} = \phi_{sc}^o$ (see Eq. (4.54)). Attention should be paid to the fact that when $\phi_{sc} = \phi_{sc}^o$, the conditions $i_n^c = i_n^a$ and $i_p^c = i_p^c$ following from the principle of detailed balance are automatically satisfied. Indeed, equating, for example, Eqs. (4.55) and (4.57) and transforming exponents we get

$$c_{red,s} N_c\,exp\left\{-\frac{(E_{c,s}^o - F_o)}{2kT}\right\} = c_{ox,s} n_s^o\,exp\left\{\frac{(E_{c,s}^o - F_o)}{2kT}\right\} \quad (4.61)$$

At thermodynamic equilibrium $F_{redox} = F$ and therefore, in accordance with Eq. (2.26), $F_o = F + kT\,ln(c_{ox}/c_{red})$ and since F_{redox} is constant throughout the entire bulk of the solution, $c_{ox}/c_{red} = c_{ox,s}/c_{red,s}$. Finally, if we substitute for F_o in Eq. (4.61) and take into consideration the first of the relations in Eq. (4.54) when $\phi_{sc} = \phi_{sc}^o$, and also the relations

$$n_o = N_c\,exp\left(\frac{F - E_c}{kT}\right) \quad \text{and} \quad n_s^o = n_o\,exp(e\phi_{sc}/kT)$$

it is evident that the equality (4.61) satisfies an identity.

For the ratio of electron and hole exchange currents, using Eqs. (4.55) and (4.58), we find

$$\frac{i_n^O}{i_p^O} = \frac{c_{red,s}}{c_{ox,s}} \frac{N_c}{N_v} \frac{k(E_{c,s})}{k(E_{v,s})} \, exp \left\{ -\left(1 + \frac{E_g}{2E_R}\right) \frac{(E_{v,s} + E_{c,s} - 2F_O)}{2kT} \right\} \quad (4.62)$$

Taking Eq. (4.54) into account, this can be rewritten to an accuracy of the pre-exponential factor in a simpler approximate form, as

$$\frac{i_n^O}{i_p^O} \simeq exp \left\{ -\left(1 + \frac{E_g}{2E_R}\right) \left(\frac{F_i - F - \phi_{sc}^O}{kT}\right) \right\} \quad (4.63)$$

where F_i is the Fermi level for the intrinsic semiconductor. In particular, from Eq. (4.63) the influence exerted by the doping of the material, and also (through ϕ_{sc}^O) by the position of F_{redox} in solution on the ratio of the exchange currents can be seen. Thus, the more positive the value of E_{redox} (that is, the lower F_{redox} and consequently, the more probable the overlap of the distribution function of the occupied levels Red in solution with the valence band of the semiconductor) the more dominant the hole exchange current becomes ($i_p^O > i_n^O$). On the other hand, redox couples with sufficiently negative values of E_{redox} exchange charges mainly with the conduction band ($i_n^O > i_p^O$) (see Fig. 4.5a and b). In accordance with Eq. (4.63), one should expect particularly large differences between i_n^O and i_p^O in the case of wide bandgap semiconductors for which the value of $|F_i - F|$ is usually large. Here practically all electron exchange with the solution occurs through only one of the two energy bands.

All that has been stated above is obviously equally valid when the potential deviates from its equilibrium value.

The exchange current on a metal electrode i_{met}^O (see Fig. 4.5c) is obtained from Eqs. (4.59) or (4.60). In view of the condition $F_{met} = F_{redox}$, from which it follows that at equilibrium $F_O = F_{met} + kT\ln(c_{ox}/c_{red})$, and also bearing in mind the inequality $kT/E_R \ll 1$, we find

$$i^O_{met} = e(c_{ox,s} c_{red,s})^{\frac{1}{2}} k(F_{met}) \rho(F_{met}) exp(-E_R/4kT) \quad (4.64)$$

From a comparison of Eq. (4.64) with (4.55) and
(4.58) it follows that the electron and hole exchange
currents on semiconductor electrodes are many orders of
magnitude smaller than on metal electrodes. Indeed,
from Eqs. (4.55) and (4.64) we find

$$\frac{i^O_n}{i^O_{met}} \simeq \frac{N_c}{\rho(F_{met})} e^{-(E_c - F)/kT} e^{-(E_c - F)^2/4kTE_R} \quad (4.65)$$

All three cofactors in Eq. (4.65) are smaller than unity,
and the first one alone amounts to $10^{-2} - 10^{-3}$. The
ratio i^O_p/i^O_{met} is evaluated in a similar way.

The above relation is supposed to manifest itself
distinctly, for example, in electrode processes on metal
electrodes which are accompanied by the formation of a
new phase on the electrode surface. In particular, we
can believe that the decrease in the degree of reversibility of electrode reactions at metal electrodes
(which is a consequence of the decrease in exchange
current) when their surface is oxidized is due to the
fact that the emerging oxide layer is a semiconductor.

Now let us examine some specific features of non-
equilibrium processes. The polarization of the electrode leads

   (i)   to a change in the concentrations of carriers
n_s and p_s near the surface due to the deviation of ϕ_{sc}
from the equilibrium value, inasmuch as $n_{sc} \neq 0$;

   (ii)   to a vertical shift of the energy distribution
$\mathcal{D}_{ox}$ and $\mathcal{D}_{red}$ in solution with respect to the distribution
$\mathcal{D}^{occup}_{sc}$ and $\mathcal{D}^{vacant}_{sc}$ in the electrode at the expense of an
additional potential drop in the Helmholtz layer because
$n_H \neq 0$;

   (iii) to a change in the concentrations of reagents
$c_{ox,s}$ and $c_{red,s}$ near the electrode as a result of the
redistribution of the ionic species in solution (the
ψ'-effect) such that $n_{el} \neq 0$.

The latter phenomenon is insignificant if the concentration of electrolyte solution is sufficiently high
(see Section 3.4) and, in any case, it is not specific
to semiconductor electrodes. If necessary, it can be
taken into account in precisely the same way as it is
dealt with in the description of the kinetics of electrochemical reaction on metals. Therefore it will not
be examined in what follows.

When the electrode is polarized, the level F_{redox}
is shifted with respect to F by an amount $e\eta$ so that
$F_{redox} - F = e\eta$ (see Fig. 4.6). The shift in F_{redox}
(and consequently F_O and $E_{Ox,red}^O$) with respect to the
semiconductor band edges at the interface is $e\eta_H$. In
general $\eta_H \neq \eta$; the difference $\eta_{sc} = \eta - \eta_H$ is reflected
in an additional band bending (again it should be pointed
out that $\eta_{sc} = \phi_{sc}^O - \phi_{sc} = -\Delta\phi_{sc}$). Correspondingly,
the dependence of the partial currents $i_{n,p}^{a,c}$ on η_H can
be found by replacing the values of the differences
$E_{c,s} - F_O$ and $E_{v,s} - F_O$ with $E_{c,s}^O - F_O - e\eta_n$ and
$E_{v,s}^O - F_O - e\eta_H$. The quantities in the exponents in
Eqs. (4.55)-(4.58) in no way depend on the value of η_{sc}.
In view of this, we obtain for the current i_p^c, in particular, from Eq. (4.58) the following expression

$$i_p^c = e c_{Ox,s} \left(\frac{kT}{4\pi E_R} \right)^{\frac{1}{2}} k(E_{v,s}) N_v \, exp \left\{ - \frac{(E_{v,s} - F_O - e\eta_H - E_R)^2}{4kTE_R} \right\} \qquad (4.66)$$

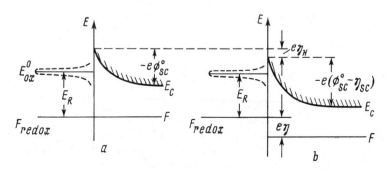

Fig. 4.6. Mutual disposition of the energy
levels in a semiconductor and electrolyte
containing a redox couple (for simplicity,
the figure shows the edge of the conduction
band alone and the level of only Ox reagent):
a) in equilibrium, b) when the electrode is
polarized.

It follows from Eq. (4.66) that in the general case there is no exponential relationship between the current and the overvoltage of the type postulated in the phenomenological theory (Eqs. (4.3) and (4.7)). At the same time, it can be assumed that in many cases the following conditions are fulfilled

$$F - E_v + e\phi_{sc} \ll E_{R'}, \quad e\eta_H \ll E_R \qquad (4.67)$$

Transforming the expression in the exponent in Eq. (4.66) using the fact that $F_0 \simeq F$, in conjunction with the inequalities in Eq. (4.67), and neglecting the squared term in $e\eta_H$, we obtain the following approximation

$$i_p^c = i_p^o \, e^{-e\eta_H/2kT} \qquad (4.68a)$$

Similarly, using the inequalities of the type in Eq. (4.67), we find from Eq. (4.56) that the current i_p^a is proportional to $exp(e\eta_H/2kT)$.

The change in the potential drop in the semiconductor by η_{sc} leads to the deviation in the concentration p_s, which is included in Eq. (4.46), from its equilibrium value. As a result, we obtain for i_p^a

$$i_p^a = i_p^o \, \frac{p_s}{p_o} \, e^{e\eta_H/2kT} \qquad (6.68b)$$

When we compare Eqs. (4.68a) and (4.68b) with the corresponding terms in the expression for the current (4.14), it is evident that they coincide if the transfer coefficient α_p in Eq. (4.14) is assumed to be equal to 1/2. Similarly, from Eqs. (4.55) and (4.57) we obtain the terms included in Eq. (4.15) provided that $\alpha_n = 1/2$.

Thus, the phenomenological expressions obtained previously under certain conditions ensue from a microscopic examination. At the same time, it should be stressed that when the inequalities of the type in (4.67) are upset, within the framework of the phenomenological relations of the type in (4.7) the apparent transfer coefficient may differ from 1/2.

If $\eta_H=0$ and $\eta = \eta_{sc}$, that is, when the entire additional potential drop is concentrated in the layer of the semiconductor near the surface, Eqs. (4.55)-(4.58)

become especially simple. In this case the magnitudes
of $F_O - E_{c,s}$ and $F_O - E_{v,s}$ are independent of the electrode potential (see Fig. 4.6), and we get the following
relations for the partial currents directly

$$i_n^a = i_n^O \quad , \quad i_n^c = i_n^O \, e^{-e\eta/kT} \tag{4.69}$$

$$i_p^c = i_p^O \quad , \quad i_p^a = i_p^O \, e^{e\eta/kT} \tag{4.70}$$

The dependence of the currents i_n and i_p on $\eta = \eta_{sc}$,
formed from the partial currents (Eqs. (4.69) and (4.70))
is shown in Fig. 4.7. As is seen from Fig. 4.7, the
current-voltage characteristics are markedly asymmetric.
The resultant electron current i_n increases exponentially under cathodic polarization whilst remaining
limited by the value of i_n^O under anodic polarization.
In contrast, the current i_p increases under anodic polarization and is limited by the value of i_p^O under cathodic
polarization. (The exponential increase in the current
i_p is actually restricted in n-type semiconductors, and
the current i_n is similarly restricted in p-type semiconductors, by the magnitude of the corresponding
limiting current of minority carriers i_p^{lim} or i_n^{lim}; see
Section 4.5.)

When $\eta = 0$, we simultaneously have $i_p^a = i_p^c$ and
$i_n^a = i_n^c$. If electrons and holes were to act as independent (or quasi-independent) reagents, then the two
equalities would correspond to different equilibrium
potentials. This is precisely the situation that can
occur under illumination of the semiconductor electrode
(see Section 6.3). From Eq. (4.69) the expression (4.17) for
the electron current follows, while from Eq. (4.70) a similar

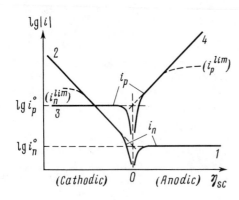

Fig. 4.7. The dependence of
$lg|i|$ on η_{sc} for electron
and hole currents: 1) i_n^a;
2) i_n^c; 3) i_p^c; 4) i_p^a (the
limiting currents for minor-
ity carriers are denoted by
dotted lines).

expression for the hole current follows. The fulfill-
ment of inequalities of the type in (4.67) is no longer
necessary, as was the case before, for these expressions
to be valid.

In conclusion, note that the equality $\eta = \eta_H$ always
applies on metal electrodes and this, in accordance with
Eqs. (4.59) and (4.60), leads to the following relations

$$i_{met}^c = i_{met}^o \, exp\left(\frac{E_R}{4kT} \right) exp\left\{ -\frac{(e\eta + E_R)^2}{4kTE_R} \right\}$$

$$(4.71)$$

$$i_{met}^c = i_{met}^o \, exp\left(\frac{E_R}{4kT} \right) exp\left\{ -\frac{(e\eta - E_R)^2}{4kTE_R} \right\}$$

If the inequality $e\eta \ll E_R$ is observed, we can ignore
terms in $(e\eta)^2$ in the exponents of Eq. (4.71) and obtain

$$i_{met}^c = i_{met}^o \, exp\left(-e\eta/2kT\right), \quad i_{met}^a = i_{met}^o \, exp\left(e\eta/2kT\right) \quad (4.72)$$

These relations (4.72) are found to hold with a sufficient accuracy within a rather broad range of η.
This may serve as a qualitative confirmation of the
accuracy of the concepts underlying the theory.

In the calculations in the preceding sections,
apart from the use of the Franck-Condon principle and
the notion of the fluctuations of electron levels of
the reagents in solution, it was also assumed that

(i)   electron transitions occurred directly between
the levels in the conduction band and/or the valence band
of the semiconductor electrode and the levels of the reagents in solution;

(ii)   energies of electrons and holes in the band
taking part in the electrode reactions were confined to
a narrow interval (of the order of several kT) in the
vicinity of the energy of the band edges at the surface,
$E_{c,s}$ and $E_{v,s}$.

Now let us briefly examine the cases when electrode
processes occur with the violation of these two conditions. The latter condition may be upset in the case
of relatively strong band bending near the surface.

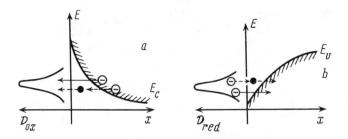

Fig. 4.8. Tunnelling of electrons through
a potential barrier formed by the space
charge region : a) tunnelling from the
conduction band; b) tunnelling to the valence band. Resonant tunnelling via localized states is denoted by the dotted line.

Under these circumstances the thickness of the space
charge layer L_{sc} becomes very small (Eq. (3.22)), and
electrons tunnelling through the space charge region
(Fig. 4.8) can make a substantial contribution to the
current of the electrode reaction. Taking the effects
of tunnelling into account may become especially important if the energy levels E_{ox}^O and E_{red}^O of the redox
couple in solution are far from the band edges $E_{c,s}$ and
$E_{v,s}$ at the surface, so that transitions from the levels
$E_{c,s}$ and $E_{v,s}$ become less probable. Quantum-mechanical
calculations of this effect with the use of a unidimensional model potential can be found in the literature
(45, 141-145). In particular, in (145) resonance tunnelling with the participation of localized states
inside the potential barrier has been examined (see
Fig. 4.8). The electron levels of impurity centers in
the space charge region can be modelled by these states.
In these cases, with certain additional suppositions,
we find, as we found previously, a linear relationship
between the logarithm of the current and the overvoltage
(Tafel law).

The first of the formulated conditions is violated
when surface states play a part in electron transfer
(146-150). The model for electron transfer with the
participation of recombination type surface levels is
represented schematically in Fig. 4.9. Under steady-
state conditions, in the absence of any electrode reaction, the degree of filling of this level is determined
by the condition of the equality of the electron and hole
fluxes, $R_{n,s}$ and $R_{p,s}$ (see Eqs. (3.44) and (3.45)). When
the electrode reaction proceeds, instead of the equality

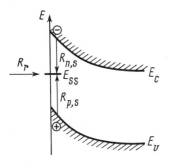

Fig. 4.9. The scheme for an electrode re-action occurring via a surface electron level of the recom-bination type.

$R_{n,s} = R_{p,s}$ we get $R_{p,s} = R_{n,s} + R_r$. Here R_r is the flux of electrons due to the electrode reaction flowing from the reagent in solution onto the surface level. The phenomenological calculation of such a process for the simplest model has been published (150), and the magnitude of the flux R_r found to depend on the relation-ship between the rate of electron transfer across the interface and the rate of surface recombination.

Electron transfer with the participation of the surface states may be important, as in the case of tunnelling effects, when the levels E_{Ox}^O and E_{red}^O are located in the bandgap; in other words, they are most significant in the case of wide-bandgap semiconductors.

Finally, a mechanism which is a kind of combination of the two preceding ones is possible, namely, the tun-nelling of electrons between energy levels in allowed bands in the bulk and the energy levels on the surface with subsequent transition to the solution (electrode reaction). In this case, the surface states play the role of a sort of a bridge for the electron transfer and can strongly increase the rate of the electrode pro-cess. Such a mechanism has been observed (111,151) on so-called "optically transparent electrodes" (films of heavily doped tin dioxide, SnO_2).

Approaches to the description of electron transfer currents using the methods of irreversible thermody-namics have been published (152, 153).

Ξ 4.5. Limiting Currents of Minority Carriers

The examination above is based on the assumption that the flux of current through the interface does not violate the equilibrium distribution of electron and hole concentrations in the semiconductor. However, if

one takes into consideration the fact that the concentration of current carriers in some semiconductor
materials is relatively small (this concerns primarily
minority carriers), the assertion that the validity of
this assumption is restricted to rather small currents
will not be unexpected. In practice the development of
an electrode reaction may substantially disturb the distribution of electrons and holes in the semiconductor
electrode. In particular, if minority carriers are involved in the electrode reaction, then its rate may be
limited by transport of these carriers from the semiconductor bulk to the surface (see Fig. 4.7). This
effect is similar to the limitation of the rate of an
electrode process by mass transport of reagents Ox and
Red in solution (see p.123). Let us now calculate the
rate of transport of minority carriers and, as a specific example, of the holes in an n-type semiconductor.

Under steady-state conditions, in the absence of
additional generation in the bulk, it is possible to
write a differential equation for the hole concentration,
in accordance with Eq. (1.68)

$$D_p \frac{d^2 p}{dx^2} + \frac{p - p_o}{\tau_p} = 0 \qquad (4.73)$$

The following expression serves as the general solution
of Eq. (4.73)

$$\delta p = A e^{-x/L_p} + B e^{x/L_p} \qquad (4.74)$$

where $\delta p = p - p_o$, $L_p = \sqrt{D_p \tau_p}$ is the diffusion length,
and A and B are integration constants. Bearing in mind
that in the bulk of the semiconductor (at $x \to \infty$) $\delta p = 0$,
we find that, from Eq. (4.74), $B = 0$, and if $L_{sc} \ll L_p$,
then

$$\delta p(x) = \delta p(L_{sc}) e^{-x/L_p} \qquad (4.75)$$

where $\delta p(L_{sc})$ is the value of δp at the boundary of the
quasi-neutral bulk with the space charge region. While
for the diffusion current

$$i_p = - e D_p \frac{dp}{dx}$$

we find from Eq. (4.75) $i_p = e D_p \delta p(L_{sc})/L_p$ at $x = L_{sc}$.
From this it follows that when the concentration of

holes at the space charge layer/quasi-neutral bulk interface tends towards zero, $\delta p = -p_o$, and the absolute magnitude of the hole current, that is, the limiting (saturation) current for the holes, is equal to

$$i_p^{lim} = eD_p \frac{p_o}{L} \qquad (4.76)$$

Physically the limiting current is the maximum value
of the hole current i_p which, under steady-state conditions, can flow from the semiconductor quasi-neutral
bulk to the space charge region at the expense of the
bulk thermal generation of electron-hole pairs (it is
assumed that the conditions $p_o \ll n_o$ and $L_{SC} \ll L_p$ are
fulfilled). Comparing Eq. (4.76) with similar expressions for the limiting diffusion currents of reagents
in solution, it is obvious that the diffusion length L_p
is an analogue of the effective thickness of the diffusion layer, δ_D. Note that the hole current in the reverse
direction, from the space charge region to quasi-neutral
bulk, is not limited in its absolute value by the magnitude of i_p^{lim}. This is evident from Eq. (4.75) in
which the positive value of δp can exceed the equilibrium
concentration p_o. The limiting current of minority carriers in a p-type semiconductor (electrons) is described
by Eq. (4.76) after the substitution of D_p by D_n, L_p by
L_n and p_o by n_o.

When comparing the limiting currents for minority
carriers, calculated by the method above, with those
measured experimentally the following two considerations
should be borne in mind. Firstly, Eq. (4.76) has been
obtained on the assumption that bulk thermal generation
is the sole source of the minority carriers when their
concentration decreases (when compared with the equilibrium value) as a result of removing them at the interface
in the course of the electrode reaction. Basically, this
is not the only possible source; generation through
levels in the space charge region (154) and generation
through surface levels may also contribute. Note that
generation through surface levels is an opposite process
to that examined at the end of Section 4.4, due to which
surface states influence the rates of electrode processes.
All of this makes an additional contribution to the hole-
limiting current that is actually measured.

Secondly, in some cases the measured limiting current proves to be greater than the "true" limiting
current for the minority carriers. This effect, which

is called "current multiplication", is usually (see, for
example, (155)) explained by the fact that the electrode
reaction includes several subsequent stages, with the
minority carriers participating in only some of them.
For instance, for the anodic dissolution of germanium it
is possible (155) to write the following scheme of reactions

$$Ge + 2h^+ \rightarrow Ge(II) \qquad\qquad (4.77a)$$

$$Ge(II) \rightarrow Ge(IV) + 2e^- \qquad\qquad (4.77b)$$

where Ge(II) and Ge(IV) denote compounds of two- and
four-valent germanium respectively. The formation of
the intermediate product, the two-valent germanium,
occurs with the consumption of holes from the valence
band. Its subsequent oxidation is accompanied by the
transition of electrons to the conduction band. According to this scheme, the ratio of the measured current
to the limiting current for the minority carriers, the
so-called current multiplication factor M, is exactly
equal to 2. Usually, however, for this and other
similar reactions $1 < M < 2$. This apparently results
from the fact that the intermediate product (for example,
Ge(II)) also enters into other reactions (disproportionation, recombination and so on) in addition to reaction
(4.77b). Taking current multiplication into account,
the value of the limiting current is determined by the
following relation

$$i_p^{lim} = eMD_p \; \frac{p_o}{L_p} \qquad\qquad (4.78)$$

It should be noted that Eq. (4.76) for the limiting
reaction current, which is determined by the supply of
minority carriers, coincides with the well known relationship for the saturation current of a p-n junction
(see, for example (156)). This is only natural, since
the physical picture of the two phenomena is similar insofar as the carrier transport in the semiconductor is
concerned. The difference in the boundary conditions
leads to the phenomenological factor M in Eq. (4.78).

Ξ 4.6. S e l e c t e d   E x p e r i m e n t a l   R e s u l t s

The major features particular to the electrochemical kinetics of semiconductor electrodes, when compared
with metal electrodes, which follow from the preceding
sections can be briefly formulated as follows:

(i)   The rate of reactions which entail the com-
sumption of minority carriers is limited by their
transport to the electrode surface.

(ii)   Basically, free carriers from both the conduction and the valence bands can take part in electrode
reactions; the predominant participation of this or that
band is determined by the equilibrium potential of the
redox reaction (see Fig. 4.5).

(iii) Due to the prevalent potential drop in the
space charge region of the semiconductor, the "effective
transfer coefficient" is close to unity, that is, the
slope of the linear $E - lg i$ plot for one-electron reactions is 2.3 kT/e = 60 mV per decade. (On metals the
slope of the linear $E - lg i$ plot at α = 0.5 is $\simeq$ 120 mV.)

Let us now examine the specific effects of the
particular processes mentioned above on the electrochemical behavior of some semiconductor materials. It is
assumed that the experimentally determined current-poten-
tial curves are not distorted by the Ohmic voltage drop
in the cell (both in the solution and in the semiconductor).

4.6.1. K i n e t i c s   w i t h   S l o w
M i n o r i t y   C a r r i e r   S u p p l y. The anodic dissolution of n-type germanium serves as a classic example
of an electrode reaction whose rate is limited by the
bulk generation of minority carriers and their transport
to the electrode surface. The study of this reaction by
Brattain and Garrett (3, 157) in 1954-1955 marked the in-
auguration of the electrochemistry of semiconductors as
an independent field of science. This reaction (see Eq.
(4.77)) consists of at least two stages. The first
occurs with the participation of holes; these are necessary for the anodic dissolution (and also for the corrosion) of the majority of semiconductor materials.
Their localization on atoms weakens the interatomic
bonds of the crystal lattice, thereby facilitating its
destruction (for details see Section 7.4).

The contribution from hole generation in the space
charge region can be neglected in samples with sufficiently large diffusion lengths for holes, L_p, such that
the condition $L_p \gg L_{sc}$ is fulfilled. Surface generation of holes is also insignificant. This is because
during the course of the anodic dissolution of germanium a kind of self-purification of the surface occurs
removing the impurities which act as generation/recom-
bination centers.

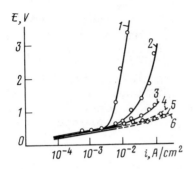

Fig. 4.10. Polarization curves of germanium anodic dissolution in 0.1 N HCl (158).
1) n-type, 1 ohm·cm:
2) n-type, 6 ohm·cm:
3) n-type, 25 ohm·cm;
4) p-type, 30 ohm·cm;
5) p-type, 6 ohm·cm;
6) p-type, 1 ohm·cm.

The shape of the anodic polarization curves is shown in Fig. 4.10 (158). It can be seen from Fig. 4.10 that while on p-type electrodes, where the majority carriers take part in the reaction, the relationship between the current and the voltage is close to an exponential one, that is, it is described by the Tafel law, on n-type electrodes there is a pronounced saturation current of minority carriers whose magnitude is determined by the electrophysical characteristics of the samples.

The quantitative test of Eq. (4.76) was carried out (159) using a series of n-type germanium samples. The value of the diffusion length of the holes, L_p, was varied in addition to their bulk concentration, p_0. As is seen from Fig. 4.11 a linear relationship really does exist between the saturation current and the p_0/L_p ratio. The anodic dissolution of other n-type semiconductor materials, for example, silicon, gallium arsenide and cadmium sulfide, which also consume holes, may be limited (see, for example, (159, 160)) by the generation of holes in the space charge region.

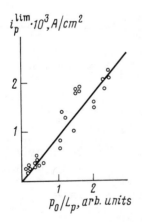

Fig. 4.11. The dependence of the limiting current density for n-type germanium anodic dissolution on the ratio p_0/L_p (159). (Reprinted by permission of the Publisher, The Electrochemical Society, Inc.)

The quantitative study of the anodic dissolution
reaction shows that the current multiplication factor
for the germanium electrode is somewhat lower than 2
(161). During the anodic dissolution of gallium arsenide the phenomenon of current multiplication is
practically nonexistant (M ≃ 1) (162).

4.6.2. P a r t i c i p a t i o n   o f   T w o
E n e r g y   B a n d s   i n   E l e c t r o c h e m i c a l
R e a c t i o n s. Let us first examine the principles of
the special experimental techniques used to determine
the contributions from the hole and the electron currents to the total current at semiconductor electrodes.
All of these methods, in one way or another, make use
of the specific feature of the electrochemical reactions
on semiconductors examined in the previous chapter; they
are based on the fact that there exists a saturation
current which is limited by the bulk generation of minority carriers.

An electrode device of the "transistor" type (13)
shown in Fig. 4.12b is widely used experimentally in
semiconductor electrochemistry. A thin plate of the
semiconductor material serves as the electrode proper.
Its thickness is such that it does not exceed the diffusion length of the minority carriers. One side of
this plate contacts with the electrolyte solution, while
the other side has a p-n junction on it. The ohmic contact to the plate is the common point coupling two
independent electrical circuits. One of these is used
to polarize the electrode with the help of an auxiliary
electrode placed in the electrolyte solution, while the
other is used to polarize (or bias) the p-n junction.
Under reverse bias, the p-n junction serves as an

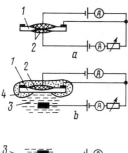
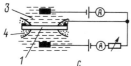

Fig. 4.12. Sectional
view of (a) an alloy
p-n-p transistor, (b)
an electrode with a
p-n junction on the
rear side and (c) a
thin bilateral elec-
trode. 1) n-Type
germanium, 2) p-type
germanium, 3) electro-
lyte, 4) insulation.

indicator for changes in minority carrier concentration
in the electrode bulk during the course of the electrochemical reaction at the electrode/electrolyte interface.
Since the thickness of the electrode is sufficiently
small, the recombination (or generation) of minority
carriers can be neglected (or, at least, it can be quantitatively taken into account). The similarity in the
behavior of such an electrode device and a planar transistor is self-evident (Fig. 4.12a). The plate with the
ohmic contact plays the role of the base, while the p-n
junction serves as a collector for the minority carriers.
For example, in the case of an n-type electrode the injection of holes into the electrode (or their extraction
from the electrode) as a result of electrochemical reaction at the electrode surface is recorded on the basis
of the rise or decay of the saturation current of the
p-n junction.

The bilateral electrode shown in Fig. 4.12c operates in a similar way. Here the p-n junction is replaced
by a second electrolytic contact at which a certain electrode reaction occurs under the regime of minority
carrier current saturation. This contact plays the role
of the collector. In all other respects the mode of
operation of the device is the same as the previous case.
The bilateral electrode technique is convenient because
it does not require the creation of a p-n junction in
the sample under study (163).

Another, methodically simpler, method (164) for de-
tecting the injection of holes into the electrode during
a cathodic reaction consists of the measurement of the
limiting current for dissolution (of the n-type semiconductor electrode) which occurs simultaneously with that
reaction. Assume that two reactions, Eqs. (4.1a) and
(4.1b), occur simultaneously at the electrode but with
quantitatively different probabilities. Then it is
possible formally to write the expression for the overall cathodic process as

$$Ox + \mu e^{-} \longrightarrow Red + \nu h^{+} \tag{4.79}$$

where μ and ν are unknown phenomenological coefficients,
the ratio μ/ν characterizing the probability of reaction
(4.1a) with respect to reaction (4.1b).

As soon as the two simultaneously occurring reactions (the anodic dissolution of the electrode material and the cathodic reduction of some reagent from

the solution) can be regarded as independent, the role of the cathodic reaction boils down to the provision of holes for the anodic reaction. Consequently, the increase in the limiting anodic current, when the reagent Ox is introduced into solution, is a measure of the "hole" component of the total current for the reaction (4.89). Subtracting this quantity from the full current, it is possible to find the electron component and, thus, to determine the coefficients μ and ν.

It should be stressed that one may expect the simultaneous participation of both the valence and conduction bands in the electrode reaction if (see Fig. 4.5) the distribution functions of the occupied $\mathcal{D}_{red}$ and non-occupied $\mathcal{D}_{ox}$ levels in the solution overlap sufficiently with the two semiconductor bands. The latter condition is satisfied by a certain relationship between the width of the bandgap of the semiconductor, the position of the electrochemical potential of the solution redox couple, and the solvent reorganization energy. In practice this condition can be expected to be more likely to occur in the case of relatively narrow bandgap semiconductors (for example, germanium and silicon).

The participation of the valence and the conduction bands in the reduction reactions of a number of oxidants at the germanium electrode has been studied (163, 164). The qualitative results are shown in Table 1. The participation of one or another band is determined, as can be seen from Table 1, by the value of the equilibrium potential of the redox couple. The more positive the potential, the greater the probability of valence band participation in the reaction, that is, the larger the contribution of the hole current to the total current (for example in the case of MnO_4^- and Ce^{4+}). On the other hand, oxidants with less positive equilibrium potentials (for example, CO_2, V^{3+}) are reduced, in the main, with the participation of conduction band electrons. (The reduction of some oxidants, for example I_3^- and H_2O_2, on germanium does not follow this rule. This is apparently due to the more complex nature of the process; see (164).)

Unlike the rather narrow bandgap materials mentioned above, for the wide bandgap semiconductors the participation of only one band in the reaction seems more probable.

Note that the phenomenon of the simplest one-electron reactions, of the type in Eq. (4.1), proceeding

TABLE   1.   The Participation of the Conduction Band and the
Valence Band in Redox Reactions at Germanium Electrodes (164)

Redox couple	Standard potential, V (vs. Normal Hydrogen Electrode)	Predominant band for electron transfer
MnO_4^-/MnO_4^{2-}	1.62	Valence
Ce^{4+}/Ce^{3+}	1.39	Valence
Quinone-hydroquinone	0.9	Valence
Fe^{3+}/Fe^{2+}	0.64	Valence
$Fe(CN)_6^{3-}/Fe(CN)_6^{4-}$	0.52	Valence
H^+/H_2	0	Conduction
V^{3+}/V^{2+}	-0.35	Conduction
$CO_2/C_2O_4^{2-}$	-0.50	Conduction

simultaneously through the conduction and valence bands
should not be confused with the phenomenon of current
multiplication mentioned above. The latter phenomenon
may be formally explained by a different mechanism with
separate stages in a complex (multistep) process.
Either electrons or holes take part in each of these
stages, while in the measured current the contributions
of all stages are summed together.

   4.6.3.   S l o w   E l e c t r o c h e m i c a l
K i n e t i c s. The kinetics of the overall process is
determined by the electrochemical step itself (the interfacial electron transfer) if all the other stages
are fast. In this case, the concentrations of reagents
(Ox or Red) in the solution and of free carriers (electrons and holes) in the semiconductor differ insignificantly from their equilibrium values. For the current
carriers in the semiconductor this is true either when
the majority carriers are consumed in the reaction or
when the current through the interface remains much
smaller than the saturation current for the minority
carriers $i_{p,n}^{lim}$.

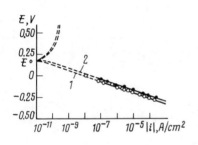

Fig. 4.13. The dependence of
$lg|i|$ on E for reaction in the
system $Fe(CN)_6^{3-}/Fe(CN)_6^{4-}$ on
ZnO electrodes with conductiv-
ities : 1) 1.7 $ohm^{-1}\cdot cm^{-1}$;
2) 0.59 $ohm^{-1}\cdot cm^{-1}$ (4). (Reprinted by permission of the
Publisher, The Electrochemical
Society, Inc.)

When these conditions are fulfilled and the entire
potential drop is concentrated in the space charge
region, we obtain experimentally linear plots of $E - lg\,i$
with slopes which are close to 60 mV/decade. An example
of such a reaction is the reduction of $Fe(CN)_6^{3-}$ of ZnO or
CdS (both n-type semiconductors). This occurs with the
participation of conduction band electrons (4, 165) (see
Fig. 4.13). It should be pointed out, however, that the
state of the electrode surface strongly influences the
slope of the $E - lg\,i$ plot. For the same electrode, but
with different preliminary treatment, it is possible to
obtain substantially different values. In all likeli-
hood, the reason for this is the redistribution of the
full overvoltage between the space charge region and the
Helmholtz layer.

The anodic dissolution of some p-type seimconductors,
which occurs with the participation of majority carriers
(holes), usually gives plots with slopes which are a bit
larger than 60 mV/decade. For example, on p-type germanium the slope has a value which is intermediate
between the characteristic values for one-electron reactions at semiconductor electrodes (60 mV) and at metal
electrodes (120 mV, see above) (166). This is in accordance with our discussion in Section 3.8; on the germanium
electrode a considerable part of the applied voltage is
usually dropped in the Helmholtz layer.

Chapter 5

ELECTROCHEMICAL PROCESSES BASED ON PHOTOEXCITATION OF REAGENTS IN SOLUTION

If photosensitive atoms or molecules are present in the solution which is in contact with the electrode, this may, as a result of irradiation, initiate electrochemical reactions with the participation of the photoexcited particles. Quanta of light then play the role of an additional reagent of sorts, which accelerates the electrode processes. The study of electrochemical reactions in which photoexcited reagents take part is of importance both in connection with applied research which is at present on the upsurge and with the development of a more profound understanding of the structure and properties of photosensitive atoms and molecules. (The notion of the "photoexcitation of a reagent in solution" is an alternative to the notion of the "photoexcitation of the electrode" which is examined in the next chapter.)

In electrode reactions which occur with the participation of photoexcited reagents, as in ordinary electrode reactions, the electron passes either from the excited species to the electrode (an anodic process) or from the electrode on to the excited species (a cathodic process). In some cases, the elementary act of charge transfer with the participation of excited reagents has much in common with ordinary (dark) electrochemical reactions (44, 167-169). This makes it possible to interpret the photoprocesses using the theoretical concepts developed earlier for the description of redox electrode reactions (see Chapter 4).

It is relatively difficult to observe electron transitions with the participation of photoexcited species
on metal electrodes. The main reason for this is that,
due to the very large density of electrons in the metal,
an electron transition in one direction is immediately
compensated for by a transition in the opposite direction. Taking this into consideration, the semiconductor/
electrolyte interface offers, in a sense, a unique
opportunity for the study of electrochemical reactions
with the participation of photoexcited reagents.

Ξ 5.1. F u n d a m e n t a l s   o f   t h e   T h e o r y
        o f   E l e c t r o d e   R e a c t i o n s   o f
        P h o t o e x c i t e d   R e a g e n t s

Let us examine the semiconductor/electrolyte interface under illumination. The quantum yield, along with
the photocurrent, may serve as the qualitative characteristic of the process. The quantum yield (quantum
efficiency) Y is defined as the ratio of the number of
electrons, which have passed through the interface
under consideration, to the number of quanta of the
exciting light striking that interface (or the number
of quanta absorbed by it).

Let the energy of the quanta be less than the
threshold of intrinsic light absorption for the semiconductor but sufficient to excite the photosensitive
ions or molecules in the solution. Thus, the electron
system of the solid may be regarded as practically uneffected, and the photoelectrochemical processes, whose
behavior now fully depends on the photoexcitation of
the reagents in solution, can be studied. We will denote
the photocurrent, determined by the occurrence of processes of this type, as i^*.

Let us start with the case where the electron passes
into the semiconductor from the photoexcited particle
which thus plays the role of a donor (D). We shall consider the situation where the energy difference between
the mean energy level of the donor particle in solution,
E_D^0, and the bottom of the conduction band in the semiconductor is large. The density distribution of the
occupied electron states in the solution, which is the
product of the donor concentration, c_D, and the distribution function for these states W_D, overlaps poorly
with the distribution of non-filled electron states in
the conduction band (see Fig. 5.1). In this case the
transfer of an electron, in accordance with Section 4.3,
is barely possible. However, illumination causes the

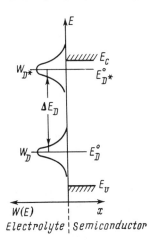

Fig. 5.1. The mutual disposition of energy levels for photoexcitation of donors in solution (photoreaction proceeds via the conduction band).

photoexcitation of the donor and can sharply increase the rate of the electrode reaction. This results from the fact that the photoexcited electron in a donor-type atom or molecule possesses an energy which is higher than E_D^O by, approximately, the value of the excitation energy, $\hbar\omega$, where ω is the frequency of the exciting radiation.

Let us assume that the lifetime of the photoexcited state is much longer than the time required to establish thermodynamic equilibrium of the photoexcited particle with its surroundings. Hence, the lifetime also turns out to be much longer than the period of the electron transition from the reagent in solution on to the electrode (see Section 4.2). Under such conditions we may introduce the concept of a mean energy level, $E_{D^*}^O$, which is the analogue of the level E_D^O and characterizes the ensemble of photoexcited donors in the medium. Moreover, it is possible, as before, to describe the probability distribution of the electron levels around the mean level, $E_{D^*}^O$, by the Gauss distribution function whose general form is given in Eq. (4.30).

If we make use of the notion of the thermodynamic ensemble of quasi-equilibrium photoexcited particles in solution with a mean electron energy level, $E_{D^*}^O = E_D^O + \Delta E_D$, then the calculation of the electron transfer rate is similar to the calculations given in Section 4.3. The current $(i_n^a)^*$ for the electron transfer from the photoexcited reagent in solution to the semiconductor conduction band, associated with the reaction

$$D^* \to D^+ + e^- \qquad (5.1)$$

is found from Eq. (4.55) and turns out to be given by
the expression

$$(i_n^a)^* = ec_{D*}T_{D*} \, exp \left\{ -\frac{(E_{c,s} - E_{D*}^O)^2}{4kTE_{R,D*}} \right\} \qquad (5.2)$$

where c_{D*} is the concentration of photoexcited donors in
the plane of discharge, E_{D*}^O is the mean electron energy
level corresponding to the excited donor, and $E_{R,D*}$ is
the reorganization energy for the photoexcited state,
which can differ from that for the ground state; finally,
the coefficient T_{D*}, by definition, is equal to

$$T_{D*} = (kT/4\pi E_{R,D*})^{\frac{1}{2}} N_c k(E_{D*}^O)$$

where the value of $k(E_{D*}^O)$ is proportional to the probability of electron transfer at an energy E_{D*}^O (for a
more detailed analysis of the expression for T_{D*} see
(170)).

In a similar manner, it is possible to describe
the transfer of electrons from the valence band on to
an acceptor which is in an excited state (or, which
is the same thing, the transfer of a hole from an
excited acceptor into the valence band)

$$A^* \rightarrow A^- + h^+ \qquad (5.3)$$

A photoexcited acceptor has a mean vacant electron
energy level, E_{A*}^O, whose energy is lower than that of
the vacant level of the ground state by an amount
$\Delta E_A = E_A^O - E_{A*}^O$, where $\Delta E_A > 0$ (see Fig. 5.2). The
current for electron transfer, as in Eq. (4.58), is
written for this case as follows

$$(i_p^c)^* = ec_{A*}T_{A*} \, exp \left\{ -\frac{(E_{v,s} - E_{A*}^O)^2}{4kTE_{R,A*}} \right\} \qquad (5.4)$$

where $E_{R,A*}$ is the reorganization energy for the acceptor
in the photoexcited state and $T_{A*} = (kT/4\pi E_{R,A*})^{\frac{1}{2}} N_v k(E_{A*}^O)$.

When interpreting results with the help of Eqs.
(5.2) and (5.4), it should be borne in mind that the
quantities ΔE_D and ΔE_A may be somewhat smaller than the
energy of the absorbed quantum $\hbar\omega$, due to the rapid
transmission of some part of the energy to the internal

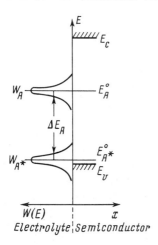

Fig. 5.2. The mutual
disposition of energy
levels for photoex-
citation of acceptors
in solution (photore-
action proceeds via
the valence band).

(e.g., librational) degrees of freedom of the excited
particle. Note that the use of the Franck-Condon principle presupposes that the electron transition may be
regarded as fast. This implies that electron transitions
from excited states are allowed; such transitions may be
forbidden as a result of the symmetry conditions for the
corresponding wave functions (132).

The parameters E_{D*}^O, E_{A*}^O; $E_{R,D*}$; $E_{R,A*}$; T_{D*}, T_{A*} in
Eqs. (5.2) and (5.4) are totally dependent on the inter-
nal properties of the reagents and solvent, as well as
on the electronic structure of the semiconductor, but are
practically independent of the characteristics of the
incident radiation. On the other hand, the concentration
of photoexcited reagents c_{D*}, c_{A*} taking part in the
electrode process is decisively dependent on the incident
radiation.

The photoexcitation frequencies of the donor and
acceptor, ω_{D*} and ω_{A*} respectively, are determined by
the properties of these particles and of the solvent.
In this frequency range the dependences of c_{D*}, c_{A*} on
the frequency of the exciting radiation ω, under sufficiently general assumptions (171), are given by the
relations

$$c_{D*} = c_D \zeta_D \frac{\gamma_D^2}{(\omega-\omega_{D*})^2 + \gamma_D^2} \quad ; \quad c_{A*} = c_A \zeta_A \frac{\gamma_A^2}{(\omega-\omega_{A*})^2 + \gamma_A^2} \quad (5.5)$$

where $\gamma_{D,A}$ are parameters which characterize the lifetime of the photoexcited state (with the assumption that
$\gamma_{D,A} \ll \omega_{D*,A*}$) and have the dimensions of reciprocal

time, while $\zeta_{A,D}$ are dimensionless quantities proportional to the probability of photoexcitation at $\omega = \omega_{D*,A*}$ and to the incident radiation intensity. The frequency distribution of the photoexcitation probabilities is a symmetrical curve with the maximum at the resonance value $\omega = \omega_{D*,A*}$ (a Lorentz distribution), and its width is determined by the parameter $\gamma_{D,A}$.

Equations (5.2), (5.4) and (5.5) describe the dependences of the currents for electrode reactions with the participation of photoexcited reagents on the parameters characterizing the radiation and the physical and chemical properties of the system. The dependences of the currents on the electrode potential are, within the framework of the assumptions, the same as the dependences of the "dark" currents i_n^a, i_p^c (see Fig. 4.6). Thus, a change in the potential drop in the semiconductor leads to an increase in band bending by an amount $e\eta_{sc}$ whilst a change in the potential drop in the Helmholtz layer leads to a shift in the levels $E_{c,s}$ and $E_{v,s}$ with respect to the levels E_D^o* and E_A^o* in the solution by an amount $e\eta_H$.

Let us now briefly examine photoelectrochemical reactions in which excitons take part. Electrode reactions with the participation of excitons are of independent interest, and they could be examined in connection with processes involving the photoexcitation of the semiconductor itself. On the other hand, since exciton energy levels lie within the bandgap (see Fig. 1.8), excitons may be produced by light of the same frequency as that used to excite reagents in solution. Thus photoprocesses of both these types may actually occur simultaneously.

When studying the reactions of photoexcited reagents from the solution, the reactions of excitons present an additional complication. Excitons created within the semiconductor bulk may diffuse to the interface and interact with the ground state reagents in solution. In this context triplet excitons, whose lifetime is much longer than that of singlet excitons, play the most significant part. Equally, there exist mechanisms for electrode reactions of ground state reagents which produce excitons, rather than free carriers, in the bands. In the latter case it is the surface of the semiconductor and not its bulk that serves as the source of excitons.

The transfer of an electron or hole across the interface, as a result of an electrode reaction between

an exciton and a particle in the electrolyte solution
is, in a sense, a chemical reaction between two molecules,
one of which is located in solution and the other in the
semiconductor. For example, the electrode reaction of
an exciton (exc) with a donor may be written as follows

$$exc + D \rightarrow D^+ + e^- \qquad (5.6)$$

In this case, the excitonic hole is captured by the
donor (or, putting it another way, an electron from the
donor passes to the valence band) whereas the excitonic
electron arrives in the conduction band (see Fig. 5.3a).

Let us assume that the exciton level is located
lower than the bottom of the conduction band by an
amount $|E_{exc}|$ (see Section 1.3). Then the proximity of
the energy level $E_{c,s} - |E_{exc}|$ to the energy level of the
donor in solution, E_D^o, serves as the condition for the
electron transfer. The corresponding current, with the
same assumptions as before, is given by (44)

$$i_{exc}^a = ec_D T_{exc,D} \, exp \left\{ - \frac{(E_{c,s} - |E_{exc}| - E_D^o)^2}{4kTE_{R,D}} \right\} \qquad (5.7)$$

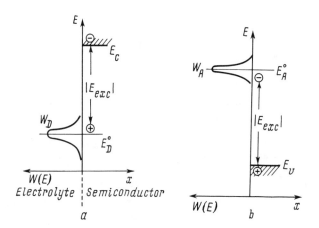

$$
\begin{array}{cc}
\text{Electrolyte} \mid \text{Semiconductor} \\
a \qquad\qquad b
\end{array}
$$

Fig. 5.3. The scheme of electrode reactions of an exciton in a semiconductor
with a reagent in solution: a) with the
participation of a donor (capture of hole);
b) with the participation of an acceptor
(capture of electron).   The figure shows
the edge of only one energy band.

Here the coefficient $T_{exc,D}$ is proportional to the concentration of excitons near the surface.

The electrode reaction between an exciton in the semiconductor crystal and an acceptor in the solution (see Fig. 5.3b) is written as follows

$$exc + A \longrightarrow A^- + h^+ \qquad (5.8)$$

The process described by Eq. (5.8) can be depicted as the capture by the acceptor of an electron, which is part of the exciton, and the transition of the released hole to the valence band of the semiconductor. The corresponding current is given by the relation

$$i_{exc}^c = ec_A T_{exc,A}\, exp \left\{ -\frac{(E_{v,s} + E'_{exc} - E_A^o)^2}{4kTE_{R,A}} \right\} \qquad (5.9)$$

Here $E'_{exc} > 0$, unlike E_{exc} in Eq. (5.7), is the difference between the energy level of the exciton and the top of the valence band. Note that the near-surface concentrations c_D and c_A included in Eqs. (5.7) and (5.9) are those for ground state donors and acceptors respectively.

Ξ 5.2. S e m i c o n d u c t o r   E l e c t r o d e s   a s
a   T o o l   f o r   S t u d y i n g   t h e
R e a c t i o n s   o f   P h o t o e x c i t e d
R e a g e n t s

All the available experimental data supports the conclusion that it is semiconductor electrodes (rather than metal of insulator electrodes)* that are the most convenient for studying the electrochemical reactions of excited states in solutions. This can be explained qualitatively within the framework of theoretical concepts (168, 169). Figure 5.4 shows the band energy diagrams of a metal (a), a semiconductor (b and c), and an insulator (d) in juxtaposition with the energy levels for the ground and photoexcited states of a reagent in solution.

---

*With the help of special contacts injecting free carriers it is possible to deliberately impart sufficient electrical conductivity to an insulator. This method is used to carry out electrochemical research on insulators (172).

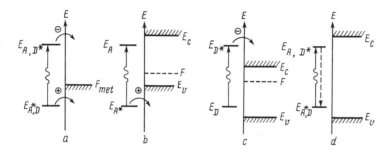

Fig. 5.4. The scheme of reaction of photoexcited reagent on (a) a metal, (b,c) a semiconductor and (d) an insulating electrode (167). In cases a and d the reagent may be both a donor and an acceptor.

In the case of the metal electrode (Fig. 5.4a) the most probable mutual disposition of the energy levels is such that the ground state level of a reagent in solution, taking a donor as a specific example, is located below, and the excited state level above, the Fermi level in the metal, F_{met}. Adjacent to each level (both ground state and excited state) of the donor there is a continuum of energy states in the metal. In the first case these are filled with electrons, whilst in the second case they are empty. This provides an opportunity for the rapid exchange of electrons across the interface. Let us note that, in accordance with the concepts in Section 4.2, transitions of electrons between levels of equal energy are the most probable. This is why the transition of an electron from the excited level E_{D*} of the donor reagent to the metal has a high probability. However, the filling of the ground state level E_D, which became vacant as a result of photoexcitation, by an electron from the conduction band of the metal also has a high probability. The conversion of the energy of the light into the kinetic energy of electrons in the metal, in other words heat, is the net result of the simultaneous occurrence of these two transitions. In other words, the conduction electrons of the metal electrode effectively quench the photoexcitation of the reagent in solution.

The existence of an energy gap in the semiconductor makes a qualitatively different disposition of energy levels quite likely (Fig. 5.4b and c). One of them, either the ground state or the excited state, is adjacent to the bandgap, so that a direct electron transfer with the participation of this level is impossible. This leads to an irreversible photoelectrochemical reaction.

The photoexcited reagent injects an electron into the conduction band of the semiconductor electrode and becomes oxidized (Fig. 5.4c) or, for the other case, the reagent captures an electron from the valence band of the electrode (i.e., injects a hole) and becomes reduced (Fig. 5.4b). The injected carriers are then swept by the electric field of the space charge region into the semiconductor bulk (for simplicity Fig. 5.4 shows the semiconductor at the flat band potential). As a result, in both cases the processes turn out to be irreversible. It must be noted that the reagent reacting at the electrode surface is consumed, and if appropriate measures are not taken, the photoelectrochemical process will gradually grind to a halt.

Finally, for an insulator (or for a very wide-bandgap semiconductor) the opposite extreme to the metal becomes very likely (see Fig. 5.4d). Here the width of the bandgap is so large that it considerably exceeds the energy of the electronic excitation ($E_D* - E_D$ or $E_A - E_A*$). Both energy levels of the reagent are now adjacent to the bandgap, and electron exchange with the electrode proves impossible. Under such conditions, the energy of photoexcitation is transmitted to the molecules of the solvent.

It is most convenient to study photoelectrochemical processes with the participation of photoexited reagents in solution using, as electrodes, semiconductors with relatively large bandgaps although not so large as to lead to the situation shown in Fig. 5.4d. The dark current across the interface, which is usually determined by the rate of minority carrier generation, is small for such electrodes. Thus the photocurrent (i^*) defined as the difference between the current in the light (i_{light}) and in the dark (i_{dark})

$$i^* = i_{light} - i_{dark} \qquad (5.10)$$

is most easily measurable for these materials. It is essential that the photocurrent arises in the radiation frequency range where the semiconductor itself is not photosensitive (that is, below the threshold of the intrinsic absorption of the semiconductor) while the solution is. This phenomenon underlies the so-called photosensitization of semiconductor electrodes to relatively longer wavelength light (in particular, of wide bandgap semiconductors absorbing in the ultraviolet region of the spectrum, to solar radiation).

A vast body of experimental data has now been collected on the photosensitization of a number of semiconductors by substances which readily absorb visible light; these substances are usually dyes (169, 173, 174). Let us investigate the influence exerted by the different factors on the sensitization currents. It is essential that the dye is regenerated once it has reacted. As the dye is consumed the photocurrent drops, and the process stops as soon as the entire near surface (or adsorbed) layer of dye becomes oxidized (during the photoinjection of electrons) or reduced (during the photoinjection of holes). To maintain a steady state photocurrent it is necessary to somehow replenish the stock of dye in the reaction layer. Of course, this can be achieved by maintaining a constant concentration of dye in the solution, insofar as equilibrium is established between the dye in the solution bulk and the dye near the surface (or on the surface). This method, however, is far from always being efficient. For example, in the case of a dye adsorbed at the surface the products of its reduction or oxidation are either not desorbed at all or are desorbed slowly, thus hampering the adsorption of fresh dye. Furthermore, if the concentration of dye is high the solution absorbs a considerable fraction of the incident light. Therefore the addition to the solution of a sufficient amount of a substance called, somewhat inaptly, a supersensitizer is regarded as the most favorable course. A supersensitizer does not absorb light and does not react at the electrode by itself. It also does not interact with the dye-sensitizer, although it is capable of oxidizing the reduced sensitizer or reducing the oxidized one. To achieve this, the redox potential of the supersensitizer should be more positive than that of the couple dye/reduced dye (in the case of photoinjection of holes) or more negative than the potential of the couple dye/oxidized dye (in the case of photoinjection of electrons). The energy scheme for supersensitization is shown in Fig. 5.5. With a sufficiently high supersensitizer concentration and reaction rate constant, a certain steady-state concentration of dye-sensitizer is maintained at the electrode surface and the photocurrent turns out to be stable. The reaction of the non-sensitive substance (the supersensitizer) is then the net result of the photoprocess.

It should be noted that the ions or molecules which are located directly on the surface of the semconductor, for example, those adsorbed on the surface, are the most effective as "photoexcitable reagents." Basically,

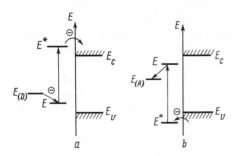

Fig. 5.5. The scheme for supersensitization with the aid of a reagent in solution: a) of donor type, R_D, and (b) of acceptor type, R_A. $E_{(D)}$ and $E_{(A)}$ are the energy levels of supersensitizer-donor and supersensitizer-acceptor respectively.

photoexcited species from the solution bulk can also enter into the electrode reaction. However, species which are not directly near the electrode surface are often deactivated before they can undergo electron exchange with the electrode.

The sensitizing action of the dye increases if it is attached to the electrode surface by a chemical bond. With this aim in view either an ether bond (-O-) is created (here it is convenient to use the "natural" surface OH-groups on oxide semiconductors such as SnO_2 or TiO_2) or an amino-bond (-NH-); in the latter case the surface of the semiconductor is covered by a monolayer of an amino-silane compound). The sensitized photocurrent on such electrodes, usually called "electrodes with chemically modified surfaces," is more stable with time (175-177). In particular, this type of electrode is used in photocells for solar energy conversion (see Section 8.6).

Ξ 5.3. P h o t o s e n s i t i z a t i o n   o f
S e m i c o n d u c t o r   E l e c t r o d e s:
S e l e c t e d   E x p e r i m e n t a l
R e s u l t s

5.3.1. T h e   N a t u r e   o f   t h e   S e m i c o n - d u c t o r   a n d   o f   t h e   D y e - S e n s i t i z e r. It follows from the general picture of photoprocesses with the participation of photoexcited reagents that the mutual arrangement of the ground and excited state levels of the reagent in solution and of the allowed energy bands in the semiconductor is the decisive factor. Basically, one and the same photoexcited reagent, if it can be both reduced and oxidized, can inject both electrons (into an n-type semiconductor) and holes (into a p-type semiconductor). The dye crystal violet is a case in point. Fig. 5.6 shows spectra for the cathodic photocurrent on p-type gallium phosphide and for the anodic

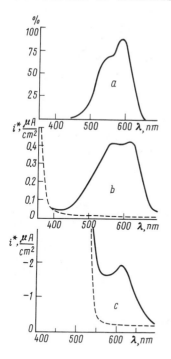

Fig. 5.6. The absorption spectrum of (a) a solution of crystal violet (normalized to the maximum absorption value), the spectra of (b) the anodic photocurrent on an n-type ZnO electrode and (c) of cathodic photocurrent on a p-type GaP electrode in the presence of crystal violet. The photocurrent in the absence of the dye is shown by the dotted lines (168). (Reprinted by permission of the Publisher, The Electrochemical Society, Inc.)

photocurrent on n-type zinc oxide, both in a solution which does not absorb light and in a solution containing crystal violet; Fig. 5.6a also shows, for the sake of comparison, the absorption spectrum of a solution of crystal violet.

In the absence of the dye a photocurrent (the dotted lines in Fig. 5.6b and c) is observed only in the region of intrinsic absorption for the semiconductor and is caused by excitation of its electron-hole ensemble. (This process is examined in the next chapter.) For zinc oxide ($E_g = 3.2$ eV) this occurs for wavelengths shorter than 400 nm, and for the gallium phosphide ($E_g = 2.2$ eV), for wavelengths shorter than 550 nm.

In the presence of the dye a photocurrent (the solid line in Fig. 5.6b and c) on the two electrodes is observed over the whole visible part of the spectrum, that is, where crystal violet absorbs light (indeed, all the three spectra are similar). On zinc oxide the photocurrent is due to the injection of electrons from the excited dye level to the conduction band (cf. Fig. 5.4c); on gallium phosphide it is due to the ejection of electrons from the semiconductor valence band into the vacant dye (ground) level (or, which is the same, to injection of holes into the semiconductor), cf. Fig. 5.4b.

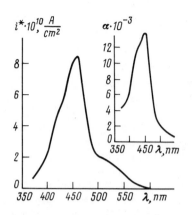

Fig. 5.7. The spectral distribution of the anodic photocurrent on SnO_2 in 0.1 N H_2SO_4 in the presence of $Ru(bipy)_3^{2+}$ (10^{-2} M); the inset shows the absorption spectrum of $Ru(bipy)_3^{2+}$ (178).

Such a dual effect of the sensitizer, however, is not always the case. In particular, bipyridyl complexes of bivalent ruthenium (henceforth denoted as $Ru(bipy)_3^{2+}$) are capable of injecting electrons into n-type tin oxide electrodes (Fig. 5.7) but are incapable of injecting holes into p-type silicon carbide (178). This can be explained, as is seen from a comparison of the energy band diagrams of the two electrodes (Fig. 5.8), by the disposition of the ground and the excited state levels of the $Ru(bipy)_3^{2+}$ complex with respect to the semiconductor band edges. Thus, in the case of silicon carbide the two energy levels are adjacent to the bandgap (the "insulator" case, cf. Fig. 5.4d). At the same time, on the tin oxide electrode the transfer to the conduction band is allowed (179).

In addition to the dyes mentioned above, diethyl-quinocyanine and other cyanine dyes, and also methylene

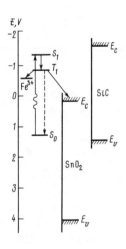

Fig. 5.8. The energy diagram for SnO_2 and SiC electrodes in a solution of $Ru(bipy)_3^{2+}$: S_0 – the ground, S_1 – the excited singlet, and T_1 – the excited triplet states of the dye. Potentials are cited against the Normal Hydrogen Electrode (178).

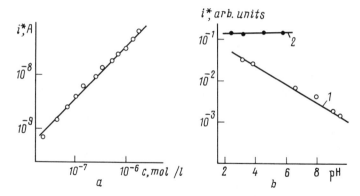

Fig. 5.9. The dependence of the photosensitized
current for a semiconductor electrode on: a) the
dye-sensitizer concentration and (b) on the
solution pH. a) TiO_2 electrode sensitized by
3,3'-diethylthiocarbocyanine iodide, pH 2.2;
b) TiO_2 (1) and CdS (2) electrodes sensitized by
rhodamine B (186).

blue on CdS (174, 180-182), alazarin and rhodamine B on
TiO_2 (183,184), and a number of others (183-185) have
been used as sensitizers for semiconductor electrodes.

5.3.2. T h e   E f f e c t   o f   t h e   S o l u t i o n
C o m p o s i t i o n. The current density due to photoelectrochemical reactions of excited reagents depends
upon the dye concentration (Fig. 5.9a) and also in some
cases upon the solution pH (Fig. 5.9b) (186).   The
latter effect stems from the fact that the pH may determine the flat band potential of the semiconductor (see
Section 3.8.2) and thus change the position of the band
edges with respect to the energy levels (ground and excited) of the dye in solution; these are usually independent of pH.   For example, in alkaline solutions the
flat band potential of TiO_2 and ZnO is much more negative than in acid solutions, and the resulting shift
in the mutual disposition of levels in the system hinders
the photoinjection of electrons from excited molecules
of the dye into these semiconductors.   On the other hand,
in the case of CdS, whose flat band potential does not
depend on pH, the effectiveness of photoinjection is the
same in alkaline and acidic media (cf. the curves for
TiO_2 and CdS in Figs. 5.9b and 3.20).

The composition of the solution may also influence
the processes leading to the deactivation of the excited
species. Let us illustrate this by returning to the

scheme for the SnO_2/Ru(bipy)$_3^{2+}$ system. Usually in the
course of photoexcitation a shortlived singlet state is
initially formed. This rapidly converts (Fig. 5.8) to a
triplet state (for a definition of singlet and triplet
states see footnote on p.25) with a much longer lifetime. This may give sufficient time for the electron
exchange to take place between the excited dye molecule
and the semiconductor: the photoelectrochemical reaction.
There exists, however, one other possibility: the transfer of an electron from the triplet level onto some
vacant level in solution, represented by an oxidizing
species, for example, the Fe^{3+} ion. For this reason the
addition of Fe^{3+} ions to the solution leads to the deactivation of the excited states and the quenching of
the photocurrent. According to (187), small quantities
of metals deposited on the semiconductor surface, which
promote the dissipation of the energy of the excited
molecules, may serve as another reason for quenching of
the photocurrent.

    5.3.3. S u p e r s e n s i t i z e r s. Reducing agents
such as hydroquinone, Cl^-, Br^-, I^-, SCN^- and others
(180,181,188-190) are used as supersensitizers in
reactions leading to the photoinjection of electrons into
semiconductors. Figure 5.10 illustrates the effect of
adding hydroquinone to the CdS/rhodamine B system. It
can be seen that the photocurrent is considerably increased. If the Ru(bipy)$_3^{2+}$ complex is used as the
sensitizer, then the solvent itself (water) can serve
as the supersensitizer, since the oxidized form of the
sensitizer, Ru(bipy)$_3^{3+}$, is capable of oxidizing water
to oxygen

$$Ru(bipy)_3^{2+} \xrightarrow{\hbar\omega} Ru(bipy)_3^{3+} + e^-$$

$$(5.11)$$

$$Ru(bipy)_3^{3+} + H_2O \longrightarrow Ru(bipy)_3^{2+} + O_2$$

although at a very modest rate.

    Special experiments using the rotating ring-disc
electrode technique (see for example, (191)), have shown
directly (180) that the supersensitizer really does
become oxidized in the course of the photoprocess, which
thus boils down to the reduction of a substance which
does not absorb light (for example, hydroquinone) at the
expense of light energy absorbed by the dye, the latter
playing the role of a catalyst of sorts. As a result,

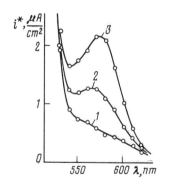

Fig. 5.10. Sensitization and supersensitization of a CdS electrode in 0.2 M Na_2SO_4. Anodic photocurrent: 1) in the absence of sensitizer, 2) in the presence of rhodamine B (2.5×10^{-5} M), 3) in the presence of rhodamine B (2.5×10^{-5} M) and hydroquinone (2.5×10^{-3} M) (189).

a current is generated at the semiconductor electrode (which also itself does not absorb light in this particular region of the spectrum).

It is interesting to note that in the case of chemically modified electrodes one and the same substance may sometimes perform two functions: being attached to the surface, it plays the role of the sensitizer, and being dissolved, the role of the supersensitizer. This may be due to the fact that the reactivity of a substance, generally speaking, depends on its state and may differ for chemisorbed and dissolved species. For example, during the sensitization of silicon electrodes by ferrocene, part of the ferrocene becomes attached to the semiconductor surface, while the remainder stays in solution. The ferrocene attached to the surface is excited by the absorption of light, passes an electron to the silicon conduction band and is oxidized to ferricinium. Remaining attached to the surface, this oxidized form of the sensitizer is able to oxidize the free ferrocene in solution. Thus, the whole process boils down to the photooxidation of dissolved ferrocene (192).

5.3.4. T h e   Q u a n t u m   Y i e l d. The quantum yield for the photocurrent associated with the reactions of photoexcited reagents is usually small. For example, in the case of ZnO photosensitized by crystal violet the quantum yield (electrons per quanta of incident light) is about 0.02 (193); for the TiO_2/rose bengal system it is 0.004 (194). At the same time, the quantum yield related to the number of light quanta absorbed by the dye is tangibly higher. For example, for the SnO_2/rhodamine B system it is 0.09 (175). (The magnitude of the light absorption by the dye layer, which is required to calculate the quantum yield in terms of the absorbed quanta, is measured by making use of the fact that the semiconductor is transparent to the light which is being absorbed by the dye. Therefore optical methods based on the use of optically transparent electrodes can be applied as, for example, in the case of the method of attenuated total reflection (195).)

In all probability, such low efficiencies are explained by the fact that the adsorbed dye layer taking part in the photoprocess is very thin, of the order of a molecular diameter, and therefore it absorbs very little light. This makes the photosensitization by adsorbed dyes a rather inefficient means for making wide bandgap semiconductors sensitive to visible light.

5.3.3. Modelling the Elementary Processes of Photosynthesis. Semiconductor electrodes with a dye-sensitizer make possible the construction of a very simple model for the processes occurring in natural photosynthesis in, for example, green plants. According to the generally accepted notions (see, for example, (196)) the energy transfer and electron transfer processes are combined in the course of photosynthesis. Among the dye molecules (pigments) which absorb the light, in particular chlorophyll, there are molecules of two types: the so-called antennae ensuring the absorption of the light quanta, and the reaction centers in which charge separation (i.e., electron transfer) takes place. It is assumed that the interaction between antennae and reaction centers occurs by means of energy transfer.

Figure 5.11 shows the scheme for a device which models this process (197,198). The semiconductor electrode (SnO_2) is separated from the electrolyte solution by a lipid membrane with a thickness of several nanometers; this prevents immediate electron interaction

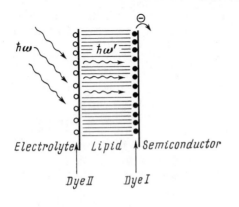

Fig. 5.11. A model of the energy and electron transfer stages in the process of photosynthesis.

between the electrode and the solution. Docosilamine, as well as arachidate, stearic acid, lecithin, etc., are used as materials for such membranes (199-201). The dye I (hydroxycoumarin) is adsorbed on to the outer surface of the membrane. It absorbs the incident light (with a quantum energy of $\hbar\omega$); this process is accompanied by luminescence (the energy of the emitted quantum is $\hbar\omega' < \hbar\omega$). The dye II (cyanine) is adsorbed at the membrane/semiconductor interface. It is insensitive to the light with a quantum energy $\hbar\omega$ but absorbs the light emitted by dye I. Upon photoexcitation dye II injects electrons into the conduction band of the SnO_2.

In other words, there is energy transfer in the membrane from dye I playing the role of antenna to dye II performing, in combination with the semiconductor, the function of the reaction center and carrying out the electron transfer. The spectral characteristics of the system as a whole are determined by dye I which, without direct electron interaction with the semiconductor, nevertheless sensitizes it to visible light.

Despite the approximate character of the model- natural photosynthesizing objects do not apparently contain a macroscopic semiconductor phase - it makes it possible, at least qualitatively, to trace the interconnection of the energy and electron transfer stages in the process of photosynthesis.

Chapter 6

METHODS FOR DESCRIBING ELECTROCHEMICAL PROCESSES
BASED ON PHOTOEXCITATION OF THE SEMICONDUCTOR

This chapter will deal with some electrochemical phenomena occurring on illuminated semiconductor electrodes
under conditions where the absorption of radiation by
reagents in solution is practically negligible. Apart
from being of scientific interest, these phenomena are
important for the practical use of semiconductor electrodes in photoelectrochemical systems.

It is light of a frequency higher than the intrinsic
(fundamental) semiconductor absorption threshold (see
Section 1.3) that has the greatest effect on electrode
processes. In this case the energy of the absorbed
quantum is sufficient for the photogeneration of electron-hole pairs. Redistribution of carriers in the
semiconductor near the solution interface, brought about
by such photogeneration, can radically change not only
the rate, but also the character of the electrochemical
reactions.

Depending on the ratios between the depth of light
penetration in the semiconductor, the diffusion length
and the thickness of the space charge region, as well
as between the rates of electrode reaction and transport
of carriers to the surface, photoelectrochemical processes can occur under very diverse regimes. At the
same time, by solving the equations describing the
kinetics of carrier transport and their transfer across
the interface, it is possible to obtain relatively simple

(but in no way trivial) correlations characterizing the
photoprocess for particular important cases, and to
compare theory and experiment.

There is also a different, quasi-equilibrium,
approach to the description of the photoexcited state of
the semiconductor. This is important for the qualitative
understanding of some of the inherent behavior of photoelectrochemical processes. This approach is dealt with
in the last sections of this chapter.

Ξ 6.1. C a l c u l a t i o n   o f   P h o t o c u r r e n t
        a n d   P h o t o p o t e n t i a l :   A   K i n e t i c
        A p p r o a c h

We shall begin with a qualitative examination of
the processes occurring near the surface of an n-type
semiconductor electrode when carriers are generated by
light. The depth of light penetration, which is of the
order of α^{-1}, where $\alpha(\omega)$ is the light absorption coefficient (see Section 1.3), can vary over a broad range,
depending on the wavelength of the light. This depth is
relatively large for light of a comparatively long wavelength; for short wavelengths the depth of penetration
may become less than the thickness of the space charge
region, L_{SC}. Figure 6.1 shows schematically two extreme
cases: $\alpha^{-1} > L_{SC} + L_p$ (Fig. 6.1a) and $\alpha^{-1} < L_{SC}$ (Fig.
6.1b) where L_p is the diffusion length for holes. The
region where generation of electron-hole pairs takes
place is shown hatched in Fig. 6.1. If a depletion
layer is formed, then holes generated by light in the
region $x < L_{SC}$ are transported by the electric field
to the electrode surface where they take part in the
electrode reaction. Outside the depletion layer, where
$x > L_{SC}$, the photogenerated holes are transported by
diffusion. During their lifetime these nonequilibrium
carriers traverse a distance of the order of the diffusion length. Therefore, holes generated even deeper
(in the case $\alpha^{-1} > L_{SC} + L_p$) into the semiconductor
mainly recombine before they reach the surface and,
consequently, make no contribution to the photocurrent
which is, by definition, (cf. Eq. (5.10))

$$i_{ph} = i_{light} - i_{dark} \qquad (6.1)$$

Let us assume that the rate of the electrode reaction, occurring with the capture of holes at the
surface, is sufficiently large, so that every hole that
reaches the surface is removed by the reaction. Under

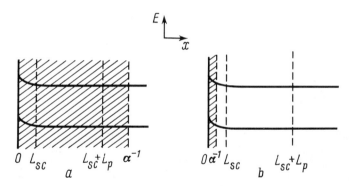

Fig. 6.1. Correlations between the depth of light penetration in the semiconductor α^{-1}, the thickness of the space charge region L_{sc}, and the diffusion length L_p: a) the light is weakly absorbed (the magnitude of α^{-1} is big); b) the light is strongly absorbed (the magnitude of α^{-1} is small).

these conditions, a change in the electrode potential influences the magnitude of the hole photocurrent mainly through a change in the thickness of the depletion layer since L_{sc} depends upon the potential drop in the space charge region ϕ_{sc} (see equation on p.71). In the case of weak light absorption ($\alpha^{-1} > L_{sc} + L_p$, Fig. 6.1a) the thickness of the region from which the surface "gathers" the holes generated by the light changes, and the photocurrent should contain a term which is proportional to L_{sc}. As can be seen from Fig. 6.1a, in the limiting case when $\alpha^{-1} \to \infty$ the photocurrent should be proportional to $L_{sc} + L_p$.

When $\alpha^{-1} < L_{sc}$ (Fig. 6.1b) the region for hole generation is already completely inside the depletion layer. Consequently, a change in its thickness has no effect upon the photoprocess, and the photocurrent should be independent of the potential. Moreover, when $\alpha^{-1} < L_{sc}$, in the case of a fast electrode reaction, obviously all of the photogenerated holes contribute to the photocurrent and, therefore, this potential independent hole photocurrent represents the maximum possible.

An important practical conclusion follows from this purely qualitative picture. As a rule, the magnitude of α^{-1} is small if light absorption occurs as a result of direct interband transitions in the semiconductor (see

Section 1.3). On the other hand for indirect transitions the condition $\alpha^{-1} < L_{sc}$ is rarely met. This is why semiconductor materials characterized by direct transitions can be expected to be more efficient in various photoelectrical and, in particular, photoelectrochemical processes (light energy conversion, photoemission, etc.).

On the basis of the transport equations and those for the electrochemical kinetics examined in Sections 1.5 and 4.1, let us now calculate the dependence of the photoelectrochemical current on the electrode potential, the characteristics of the light flux, and the parameters describing the properties of the physicochemical system in question.

The simplest calculation of the current determined by semiconductor photoexcitation, based on the concepts mentioned above, was first carried out by Gärtner (202). In accordance with his model, the photocurrent i_{ph} (in this particular case, assumed to be entirely a hole current) consists of two contributions

$$i_{ph} = i_{dl} + i_b \tag{6.2}$$

where i_{dl} is the current determined by the generation of holes in the depletion layer, and i_b is the current of holes from the semiconductor bulk. Let us assume that all the holes generated by the light in the space charge region reach the electrode surface and contribute to the photocurrent. For this to be the case, it is necessary, firstly, that the electrode reaction is sufficiently fast (see above), and, secondly, that the time taken for holes to cross the space charge region is less than their recombination time. For the latter condition it is necessary that $L_p \gg L_{sc}$. Using Eq. (1.70) for the hole generation function, we obtain for the value of i_{dl}

$$i_{dl} = - e \int_0^{L_{sc}} g(x)\,dx = eJ_o(e^{-\alpha L_{sc}} - 1) \tag{6.3}$$

According to Eq. (1.68), we get the following equation for the holes in the quasi-neutral bulk ($x \gg L_{sc}$)

$$D_p \frac{d^2p}{dx^2} - \frac{p - p_o}{\tau_p} + \alpha J_o e^{-\alpha x} = 0 \tag{6.4}$$

In the bulk of the semiconductor (i.e, when $x \rightarrow \infty$)
$p = p_0$ serves as a boundary condition for Eq. (6.4).
In (202) $p = 0$ when $x = L_{sc}$ is chosen as the second
boundary condition; this corresponds physically to the
above condition of a very fast electrode reaction, which
"sucks out" practically all of the holes from the space
charge region.   The following expression is the solution
of Eq. (6.4) with these boundary conditions:

$$p(x) = p_0 - (p_0 + Ae^{-\alpha L_{sc}}) e^{(L_{sc} - x)/L_p} + Ae^{-\alpha x}$$

where

$$A \equiv \frac{J_0}{D_p} \frac{\alpha}{\alpha^2 - L_p^{-2}}$$

Calculating the diffusion current of holes

$$i_b = - eD_p dp/dx \big|_{x = L_{sc}}$$

we get

$$i_b = -eJ_0 \frac{\alpha L_p}{1 + \alpha L_p} e^{-\alpha L_{sc}} \tag{6.5}$$

Combining Eqs. (6.3) and (6.5), in accordance with
Eq. (6.2), we finally obtain, for the absolute value of
the photocurrent,[†]

$$i_{ph} = eJ_0 \left(1 - \frac{e^{-\alpha L_{sc}}}{1 + \alpha L_p} \right) \tag{6.6}$$

Equation (6.6) shows how the photocurrent depends on the
characteristics of the radiation (through $\alpha(\omega)$ and J_0),
the kinetic characteristics of the minority carriers
(through L_p), the concentration of majority carriers
and the electrode potential (through L_{sc}).

Note that the magnitude of the photocurrent does
not depend on the equilibrium concentration of minority
carriers p_0. It should be also pointed out that, in

[†]No special distinctions are made below between photocurrent
and its absolute value, the two quantities being designated as i_{ph}.

accord with our model above, if $\alpha^{-1} \gg L_{sc}$, L_p, then, from Eq. (6.6), the photocurrent is proportional to $\alpha(L_{sc} + L_p)$, and if $\alpha^{-1} \ll L_{sc}$, then the photocurrent is independent of L_{sc} (and consequently of ϕ_{sc}), and its absolute value is given by eJ_o, the maximum possible value.

A more general and mathematically more rigorous examination of the problem of the calculation of the photocurrent was carried out in (133). It follows from Eqs. (1.63), (1.64) and (1.70) that in the steady-state the concentrations of electrons and holes in the space charge region $0 \leqslant x \leqslant L_{sc}$ are described by the following equations

$$D_n \frac{d}{dx} \left[\frac{dn}{dx} - n \frac{e}{kT} \frac{d\phi}{dx} \right] + \alpha J_o e^{-\alpha x} = 0 \qquad (6.7)$$

$$D_p \frac{d}{dx} \left[\frac{dp}{dx} + p \frac{e}{kT} \frac{d\phi}{dx} \right] + \alpha J_o e^{-\alpha x} = 0 \qquad (6.8)$$

Here it is again assumed that the time for the transit of holes across the space charge region is less than their lifetime. Hence, recombination in the space charge region, $x < L_{sc}$, can be ignored. Outside the space charge region the concentration of holes is described by Eq. (6.4), and the concentration of electrons is considered to be constant: $n(x) = n_o = N_D$ (see Section 1.5).

If a depletion layer emerges in the semiconductor, then the potential distribution, $\phi(x) < 0$, in the interval

$$\phi_{inv} < \phi_{sc} < -kT/e \qquad (6.9)$$

where ϕ_{inv} is given by Eq. (3.7), may be regarded as fixed and independent of the distribution of the holes (see Eq. (3.10)).

It should be emphasized that Eq. (3.10) is not limited to the potential interval given in Eq. (6.9) in the present case. For example, when a fast electrode reaction involving holes occurs at the surface, their concentration in the space charge region may sharply decrease when compared with the equilibrium concentration. Then, if in addition $\phi_{sc} < \phi_{inv}$, so that an inversion layer would be formed at equilibrium, the

actual charge distribution will be the same as in the
depletion layer because in both cases it is determined
by the charge on the immobile donors. The resultant
space charge layer is called an exhaustion layer.
According to differential capacity measurements and
IR-absorption, an electrode with a "non-equilibrium"
depletion (i.e., exhaustion) layer behaves in many
respects like an electrode with an equilibrium depletion
layer (see, for example, (203, 204)).

It is clear from the principles of physics that the
potential distribution (Eq. (3.10)) created by immobile
charged donors in a semiconductor with a sufficiently
high doping level, as with the distribution created by
excess majority carriers, is practically insensitive to
the passage of a moderately large electron (hole) current
and to the photogeneration of electrons and holes in the
space charge region.

Boundary conditions at the electrode surface for
Eqs. (6.7) and (6.8) are set by the currents i_n and i_p
or by the concentrations n and p of electrons and holes
at x = 0, for example, with the help of relations such
as Eqs. (3.52)-(3.55). Equation (6.4) should satisfy
the condition $p = p_o$ at $x \to \infty$ in the bulk of the semiconductor (the condition $n(x) = n_o$ at $x \to \infty$ is automatically satisfied). In addition $p(x)$, $n(x)$ and $i_p(x)$
should be continuous at $x = L_{sc}$. Thus, there are six
boundary conditions: two at x = 0, one at $x \to \infty$, and
three at $x = L_{sc}$, from which six constants are determined unambiguously in the solution of the three second-
order equations (6.4), (6.7) and (6.8). All three
equations can be solved analytically and the corresponding relations can be obtained in an explicit form
(133).

Using standard methods for the solution of linear
differential equations, after simple, though somewhat
cumbersome calculations, we obtain for the currents of
electrons, i_n, and holes, i_p, at the interface

$$i_n = \frac{eD_n}{\Lambda} (N_D - n_s e^{-Y}) \qquad (6.10)$$

$$i_p = \frac{eD_p}{L_p} (p_s e^Y - p_o) + eJ_o \left(\frac{e^{-\alpha L_{sc}}}{1 + \alpha L_p} - 1 \right) \qquad (6.11)$$

where

$$\Lambda = L_D \sqrt{2} \int_0^{|Y|^{\frac{1}{2}}} exp(t^2) dt$$

and

$$Y = e\phi_{sc}/kT$$

Equations (6.10) and (6.11) provide, in a sufficiently general way, a description of the electron and
hole transport in the semiconductor bulk. To obtain
the final results, it is necessary to link together the
quantities n_s, $i_n|_{x=0}$ and p_s, $i_p|_{x=0}$. This link is
determined by the processes occurring at the semicon-
ductor/electrolyte interface. It is these processes
which underlie the diversity of electrochemical behavior
of the system described by Eqs. (6.10) and (6.11). A
mathematical description of any particular situation
now boils down to algebraic transformations of Eqs.
(6.10) and (6.11) taking into account relations of the
type cited in Section 3.2 and connecting the quantities
n_s and p_s with $i_n|_{x=0}$ and $i_p|_{x=0}$. Without analyzing the
numerous variants in detail, let us examine some im-
portant particular cases at length.

We shall assume that the whole of the change in
potential produced as a result of illumination occurs
in the space charge region of the semiconductor. When
the electrode is illuminated, its potential becomes
equal to

$$E = E_{dark} + \Delta E \qquad (6.12)$$

where E_{dark} is the electrode potential in the dark, and
ΔE is by definition in this case the photopotential E_{ph}
(so $\Delta E = E_{ph}$). In the dark at equilibrium $E_{dark} = E^0$.

We shall further assume that electron-hole recombination through surface states is negligible. Then
for the total current passing through the interface
under photoexcitation of the electrode we have

$$i = i_p(E_{ph}) + i_n(E_{ph}) \qquad (6.13)$$

Equation (6.13) refers to the total current at the illuminated electrode, rather than the photocurrent defined

as the difference between the currents in the light and in the dark (see Eq. (6.1)). From now on we shall not distinguish between the notions of "photocurrent" and "current at the illuminated electrode," since as a rule for photoelectrodes made of wide bandgap semiconductors the dark currents are negligibly small.

Proceeding from Eqs. (6.10)-(6.13) let us calculate the open-circuit photopotential. If the exchange currents i_p^o and i_n^o are sufficiently large, then, from Eq. (4.19), we can expect that $n_s = n_s^o = n_o exp(e\phi_{sc}^o/kT)$ and $p_s = p_s^o = p_o exp(-e\phi_{sc}^o/kT)$. Substituting these expressions into Eqs. (6.10) and (6.11) and assuming that $i = 0$ in Eq. (6.13), we obtain using Eq. (6.12)

$$\frac{D_n N_D}{\Lambda}(1-e^{eE_{ph}/kT}) + \frac{D_p p_o}{L_p}(e^{-eE_{ph}/kT}-1) = J_o\left(1 - \frac{e^{-\alpha L_{sc}}}{1+\alpha L_p}\right) \quad (6.14)$$

In Eq. (6.14) the dependence of Λ on E_{ph} should also be taken into consideration. In accordance with the definition of Λ (see Eq. (6.10)), for $|Y| \gg 1$ we can make the following approximation

$$\Lambda(E_{ph}) = \Lambda^o e^{eE_{ph}/kT} \quad (6.15)$$

where Λ^o is the value of Λ at the equilibrium potential, $E = E^o$.

Substituting Eq. (6.15) in Eq. (6.14) and assuming that $\alpha L_{sc} \ll 1$, we finally obtain

$$E_{ph} = -\frac{kT}{e} ln(1 + bJ_o) \quad (6.16)$$

where

$$b = \frac{\alpha L_p}{1 + \alpha L_p}\left(\frac{D_n N_D}{\Lambda^o} + \frac{D_p p_o}{L_p}\right)^{-1} \quad (6.17)$$

As is seen from Eq. (6.14), for the other limiting case, $\alpha L_{sc} \gg 1$, we obtain an expression of the same form as Eq. (6.16) but with a constant b' in place of the constant b (b' is the limit of b as $\alpha \to \infty$ so that the term before the bracket in Eq. (6.17) is unity). Note that in Eq. (6.17), for an n-type semiconductor the term in N_D is usually much larger than the term in p_o.

Since $b > 0$ and $J_o > 0$, it follows from Eq. (6.16) that $E_{ph} < 0$. Thus the open-circuit photopotential under the given conditions (a depletion layer in the semiconductor and no surface recombination) is always negative. The negative value of the photopotential can be interpreted as follows: when an n-type semiconductor is illuminated, the absolute value of the potential drop in the space charge region $|\phi_{sc}|$ decreases. In other words, under the influence of illumination the bands unbend. Similarly, for a p-type semiconductor under the conditions of depletion layer formation, the photopotential is positive; this again corresponds to unbending of the bands under illumination. It should be noted that a similar effect on band bending under illumination also occurs for an ideally polarizable electrode (see below).

At relatively low intensities of illumination ($bJ_o \ll 1$) Eq. (6.16) predicts a linear increase in the absolute value of the photopotential with increasing J_o, while at high intensities ($bJ_o \gg 1$) the increase is logarithmic (see Fig. 6.2).

Let us now examine the case of an electrode reaction, analyzed above, in which only holes participate. If the exchange current through the conduction band, i_n^o, is negligible, then from Eqs. (4.11) and (6.13)

$$i_n = 0 \quad , \quad i_{ph} = i_p(E_{ph}) \qquad (6.18)$$

Using Eq. (6.11), we obtain from Eq. (6.18)

$$i_{ph} = eJ_o \left(\frac{e^{-\alpha L_{sc}}}{\alpha L_p + 1} - 1 \right) + \frac{eD_p p_o}{L_p} \left(\frac{p_s}{p_o} e^{e\phi_{sc}/kT} - 1 \right) \qquad (6.19)$$

Depending on the intensity of illumination and the numerical values of the parameters characterizing the semiconductor, any of the terms in Eq. (6.19) may dominate.

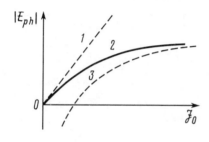

Fig. 6.2. The dependence of the open-circuit photopotential on the illumination intensity (for an n-type semiconductor): 1) linear and 3) logarithmic asymptotes of Eq. (6.16) respectively at small and large J_o; 2) calculated according to Eq. (6.16).

The physical meaning of the first term was examined above. The second term describes the additional current linked with the deviation of the potential drop in the semiconductor from its equilibrium value ϕ_{sc}^o. In semiconductors with sufficiently wide bandgaps the second term is always negligible due to the small value of p_o for all reasonable magnitudes of J_o and ϕ_{sc}. In this case Eq. (6.19) converts into Eq. (6.6).

The effects of surface recombination on the process is examined in (205) where the finite rate of the electrochemical step itself is also taken into account. The expression for the current i_{ph} in (205) is calculated by a method similar to that used in (202) to derive Eq. (6.6). However, instead of the condition $p(L_{sc}) = 0$, a more general correlation is used as the boundary condition

$$D_p \left. \frac{dp}{dx} \right|_{x=L_{sc}} = kp(L_{sc}) \qquad (6.20)$$

which formally reverts to the preceding condition as $k \to \infty$, where k is the parameter related to the hole current at $x = 0$. This current is made up of two simultaneous processes, the electrode reaction (current i_{el}) and the recombination of holes through the surface states (current i_{rec}). These currents can be written as follows

$$i_{el} = ek_{el}p_s \quad , \quad i_{rec} = ek_{rec}p_s \qquad (6.21)$$

where k_{el} and k_{rec} are phenomenological coefficients dependent upon the microscopic characteristics of the surface processes. In the case of $k > D_p/L_p$ (the value of D_p/L_p characterizes the rate of arrival of photogenerated holes in the space charge region from the neutral bulk of the semiconductor), we obtain the resulting photocurrent

$$i_{ph} = eJ_o \frac{k_{el}}{k_{el}+k_{rec}} \left[1 - \frac{e^{-\alpha L_{sc}}}{1+\alpha L_p} - \frac{D_p \alpha}{(1+\alpha L_p)k} e^{-\alpha L_{sc}} \right] \qquad (6.22)$$

and the parameter k proves to be given by

$$k = (k_{el} + k_{rec})e^{e\phi_{sc}/kT}$$

The last term in the brackets is a correction term arising from the fact that the rate constants k_{el} and k_{rec} for the surface processes are finite; it becomes zero when $k \to \infty$. Taking this term into account results in a decrease in i_{ph}. The factor

$$\frac{k_{el}}{k_{el} + k_{rec}}$$

in front of the brackets takes into account the fact that the only contribution to the photocurrent comes from those holes which enter into the electrode reaction; holes recombining on the surface make no contribution to the current i_{ph}. Thus, surface recombination exerts a dual influence on the photocurrent: firstly it changes the relative share of the holes which disappear at the interface that contribute to the photocurrent and secondly it deforms the distribution of holes in the region $x > L_{sc}$, thereby changing their current from the bulk to the surface. For the case $k_{el} \gg k_{rec}$ and $k_{el} \gg D_p/L_p$ Eq. (6.22) converts into Eq. (6.6). The work (205) also contains expressions for the more general case $k \lessapprox D_p/L_p$, these, however, are rather cumbersome. A similar problem is examined in (206) although a somewhat different method is used. In this case with the help of additional suppositions in the model the recombination of carriers in the space charge region is also taken into consideration.

In conclusion, let us deal briefly with the bulk photoelectromotive force generated in an illuminated semiconductor. This is called the Dember potential (207). Physically it originates from the spatial separation of the photogenerated electrons and holes in the quasi-neutral bulk, $x > L_{sc}$, due to the difference in their diffusion coefficients D_n and D_p (or mobilities $u_{n,p} = eD_{n,p}/kT$). The more mobile carriers (usually electrons) generated by the light near the surface move more rapidly into the depth of the semiconductor, charging it with respect to the illuminated surface of the sample. The resultant electric field hinders the transport of the faster carriers and accelerates the slower ones. Thus, under steady-state conditions a constant electromotive force is developed, which compensates for the difference in the diffusion coefficients.

The photocurrent, in accordance with Eqs. (1.61) and (1.62) is, for the one dimensional case

$$i_{ph} = e \left(D_n \frac{dn}{dx} - D_p \frac{dp}{dx} \right) - \frac{e^2}{kT} (D_n n + D_p p) \frac{d\phi}{dx} \qquad (6.23)$$

At open-circuit the full current is equal to zero. Finding the value of $d\phi/dx$ from Eq. (6.23) with due account of the quasi-neutrality condition $dn/dx = dp/dx$, we obtain

$$\frac{d\phi}{dx} = \frac{kT}{e} \frac{D_n - D_p}{D_n n + D_p p} \frac{dn}{dx} \qquad (6.24)$$

Introducing the bulk conductivity of the sample through the relation $\sigma = \frac{e^2}{kT} (D_n n + D_p p)$, we find

$$\frac{d\sigma}{dx} = \frac{e^2}{kT} (D_n + D_p) \frac{dn}{dx} \qquad (6.25)$$

Finally, eliminating dn/dx from Eq. (6.24) with the help of Eq. (6.25), we obtain the relation

$$\frac{d\phi}{dx} = \frac{kT}{e} \frac{D_n - D_p}{D_n + D_p} \frac{d\ln\sigma}{dx} \qquad (6.26)$$

Integrating Eq. (6.26) with respect to x from $x = L_{sc}$ to infinity gives for the Dember potential

$$\Delta\phi_{Demb} = \phi(L_{sc}) - \phi(sc) = \frac{kT}{e} \frac{D_n - D_p}{D_n + D_p} \ln \frac{\sigma(L_{sc})}{\sigma_o} \qquad (6.27)$$

where σ_o is the "dark" conductivity in the sample bulk, when $n = n_o$ and $p = p_o$, while $\sigma(L_{sc})$ is the conductivity at the point $x = L_{sc}$. If $\sigma(L_{sc}) = \sigma_o + \Delta\sigma$ and $\Delta\sigma/\sigma_o \ll 1$, then Eq. (6.27) can be written as follows

$$\Delta\phi_{Demb} = \frac{kT}{e} \frac{D_n - D_p}{D_n + D_p} \frac{\Delta\sigma}{\sigma_o} \qquad (6.28)$$

Equations (6.27) and (6.28) show that the Dember potential arises as a result of the difference in the diffusivity of the carriers; when $D_n = D_p$, the value of $\Delta\phi_{Demb}$ becomes zero.

In the general case $\Delta\phi_{Demb}$ depends upon the geometry of the sample but its order of magnitude is that given by Eqs. (6.27) and (6.28). Under real conditions the Dember potential is usually small and therefore

masked by the considerably larger photoeffects of the
space region.

Ξ 6.2. C o m p a r i s o n   o f   T h e o r y   a n d
        E x p e r i m e n t

    The quantitative comparison between the current-
voltage curve for an illuminated electrode, provided
by Eqs. (6.6) and (6.19), and that measured experimentally was carried out in (205, 208). Measurements were
made of the photocurrent generated in a cell with a
photoanode made from an n-type semiconductor and a
metal auxiliary electrode when the semiconductor was
illuminated by monochromatic light of a frequency
satisfying the condition $\hbar\omega > E_g$.

    On the surface of the semiconductor electrode,
in particular WO_3, under anodic polarization the electrode reaction is the photooxidation of water

$$H_2O + 2h^+ \rightarrow \frac{1}{2} O_2 + 2H^+ \qquad\qquad (6.29)$$

The experiments were designed to study the dependence of the anodic photocurrent on the light intensity
and frequency, the electrode potential and the solution
composition.

    It was shown (208) that over a broad range of
radiation intensity (up to 1 W/cm^2) there appears to be a
linear relation between the photocurrent, i_{ph}, and the
light flux, J_0, and that the corresponding straight
lines i_{ph} - J_0 are very close for electrolytes of diverse composition. Such data qualitatively confirms
the conclusion that the decisive role in the phenomena
discussed above is played by the photoelectric processes in the semiconductor interior, rather than by
surface processes, and that these are single photon
processes. A detailed comparison of the dependences
described by Eq. (6.6) with those found experimentally
was carried out based on the assumption, which is well-
grounded for WO_3 electrodes, that the entire applied
voltage is dropped in the space charge region
(ϕ_H = const, see Chapter 3). Hence, remembering the
difference in the choice of the signs for ϕ_{sc} and E,
we obtain $\phi_{sc} = E_{fb} - E$ where E_{fb} is the flat band
potential (ohmic losses, if they are not small, are
taken into account by a standard procedure).

    If the value of α is small enough so that $\alpha L_p \ll 1$
and $\alpha L_{sc} \ll 1$, then, expanding the exponential in Eq.

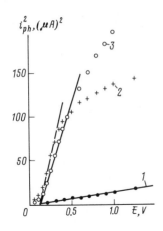

Fig. 6.3. The square of the anodic photocurrent plotted as a function of the electrode potential for WO_3 in 1 M CH_3COONa. Wavelength of the light (nm): 1) 397; 2) 327; 3) 280 (208).

(6.6) as a series, we get

$$E - E_{fb} = \left(\frac{N_D}{2e\varepsilon_{sc}\varepsilon_o} \right) \left(\frac{i_{ph}}{\alpha J_o} \right)^2 \qquad (6.30)$$

Figure 6.3 shows the dependence of i_{ph}^2 on E, which illustrates the functional relation predicted by Eq. (6.30). In this case the linear relationship between i_{ph}^2 and ϕ_{sc} or i_{ph}^2 and E reflects the characteristic dependence of the depletion layer thickness, L_{sc}, on the potential and is analogous to the Mott-Schottky dependence of C_{sc}^{-2} on ϕ_{sc} (see Eq. (3.21)). The straight lines intersect the potential axis at one point which practically coincides, for the materials examined in (205,208), with the flat band potential measured by the differential capacity method. Extrapolation of straight line plots of i_{ph}^2 against E onto the potential axis is widely used as a method for determining the flat band potential. Here the same reservations should be made about possible errors arising from the influence of the Helmholtz capacity and other factors as were made in the discussion of the differential capacity method (see Section 3.7). Using the value found for E_{fb} and the known value of α and taking the diffusion length, L_p, as an adjustment parameter it is possible to obtain agreement between theory and experiment for the entire polarization curve with reasonable values of L_p.

Figure 6.4 shows the potential dependence of the quantum yield, $Y = i_{ph}/eJ_o$, measured and calculated on the basis of Eqs. (6.6) or (6.9). Again note that here the quantum yield for the process is the ratio of the

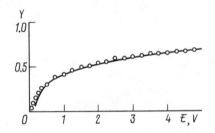

Fig. 6.4. The dependence
of the quantum yield of
anodic photocurrent on the
WO_3 electrode potential.
$N_D=4.10^{15}$ cm^{-3}. Points –
experimental; solid line –
calculated according to
Eq. (6.19) (208).

number of electrons transferred through the external
circuit to the number of photons striking the semiconductor surface.

Another method of checking the theory was used in
(209,210). The spectral distribution of the quantum
yield Y at a fixed value of ϕ_{sc} was calculated from the
absorptio₁ spectrum $\alpha(\omega)$ determined by an independent
method. Figure 6.5 compares the calculated and experimental curves for the anodic oxidation of S^{2-} on CdS electrode: $2S^{2-} + 2h^+ \to S_2^{2-}$. As is seen from Fig. 6.5, good
agreement has been found between theory and experiment.
Thus, the theory of the photoelectrochemical current on
semiconductor electrodes is confirmed both qualitatively
and quantitatively.

It is also possible with photoelectrochemical measurements to study the character of interband transitions
in semiconductors near the fundamental absorption edge.
Still assuming that $\alpha L_{sc} \ll 1$ and $\alpha L_p \ll 1$, we find for the
quantum yield from Eq. (6.6), using Eqs. (1.35) and
(1.36), that

$$Y = A_n \ (\hbar\omega - E_g)^{n/2} \left[L_p + \left(\frac{2\varepsilon_{sc}\varepsilon_0 |\phi_{sc}|}{eN_D} \right)^{\frac{1}{2}} \right] \qquad (6.31)$$

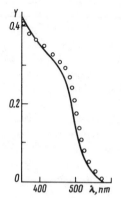

Fig. 6.5. Experimental
(dots) and calculated
(solid line) curves of
the spectral distribution
of the quantum yield of
the anodic photocurrent
on polycrystalline CdS
electrode ($N_D=10^{18}$ cm^{-3}).
(Based on (210).)

It should be borne in mind that in the exponent in Eq.
(6.31) we must have $n = 1$ for direct and $n = 4$ for indirect transitions (for these two cases the values of
A_n are different). The dependences obtained experimentally are given in log-log coordinates in Fig. 6.6
where, with the assumption that $A_n \sim \omega^{-1}$ (see, for
example (18, 20)), straight lines are found in accord
with the theory. Their slope gives $n = 4$, showing that
photoexcitation is occurring as a result of indirect
transitions.

It follows from the data mentioned above that the
comparison of the measured and calculated polarization
curves for the illuminated semiconductor electrode makes
it possible to determine the value of the diffusion
length for the holes by curve fitting. For example, for
WO_3 and TiO_2 in (205, 208) the values of L_p obtained were
5×10^{-5} and 4×10^{-4} cm respectively.

It is convenient to use the following method to determine L_p. When $\alpha L_{sc} \ll 1$, it follows from Eq. (6.6)
that (cf. (6.31))

$$i_{ph} = \frac{eJ_o\alpha}{1 + \alpha L_p} (L_p + \gamma|\phi_{sc}|^{\frac{1}{2}})$$ (6.32)

where $\gamma = (2\varepsilon_{sc}\varepsilon_0/eN_D)^{\frac{1}{2}}$. A plot of i_{ph} against $|\phi_{sc}|^{\frac{1}{2}}$
(or, since $\Delta E = -\Delta\phi_{sc}$, i_{ph} against $E^{\frac{1}{2}}$) gives a straight
line whose slope and intercept on the i_{ph} axis makes it
possible to find the ratio γ/L_p, hence, if γ is known,
it is easy to determine L_p. Such a plot for titanium
dioxide is shown in Fig. 6.7 (which gives $\gamma/L_p = 0.7$
(113)). Note that when $\alpha L_p \simeq 1$ the straight line no
longer extrapolates to the flat band potential (cf. Eq.
(6.30)).

Instead of plotting i_{ph} against $E^{\frac{1}{2}}$ it may prove
more convenient to plot i_{ph} against C^{-1} where C is the
differential capacity. Using this method the diffusion
length of holes in n-type gallium arsenide and phosphide
has been measured (211, 212).

Above it was assumed that in all cases the photocurrent was due solely to the minority carriers. For
example, only holes take part in photoanodic oxygen
evolution on TiO_2 and WO_3. In some cases, however, the
measured photocurrent is greater than the current for
photogenerated holes flowing in the semiconductor to
the surface. Such an effect was observed, in particular,

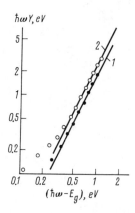

Fig. 6.6. Dependence (in bilogarithmic coordinates) of $\hbar\omega Y$ on $\hbar\omega - E_g$ for the anodic photocurrent on a WO_3 electrode in 1 M CH_3COONa at a potential of 1 V (1) and 5 V (2) (208).

in the photooxidation of certain organic compounds on electrodes made from ZnO (213-216), CdSe (217) and TiO_2 (218, 219). This phenomenon, called photocurrent multiplication, is not physically different in nature from the (dark) anodic current multiplication, investigated in Section 4.5. It can be explained by the participation, in a multi-stage electrode reaction, of electrons from the conduction band as well as photogenerated holes. For example, an organic compound (R) may be oxidized on a photoanode in two stages

$$R + h^+ \rightarrow R^{\cdot+} \tag{6.33a}$$

$$R^{\cdot+} \rightarrow R^{2+} + e^- \tag{6.33b}$$

(cf. Eq. (4.77)). It is evident that such a scheme results in a doubling of the measured photocurrent (therefore this effect is often called "photocurrent doubling" (220)).

Experimentally, however, when the reaction follows the pattern described by Eq. (6.33), the photocurrent is often smaller than simply twice the hole current. This occurs because the intermediate radical product $R^{\cdot+}$

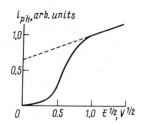

Fig. 6.7. Experimental anodic polarization curve for a TiO_2 electrode plotted as i_{ph} against $E^{\frac{1}{2}}$ (113). The potential E is with respect to the flat band potential.

can enter into other reactions (disproportionation, recombination, etc.) in addition to reaction (6.33b).

To take into account the whole range of these
effects, it is convenient to introduce into Eq. (6.6)
an empirical coefficient M in the same way as was done
for Eq. (4.78). This coefficient reflects the fact
that, due to secondary reactions, one minority carrier
generated by the light may initiate a resultant transfer
of more than one elementary charge across the interface.

≡ 6.3. Calculation of the
       Photocurrent and
       Photopotential:
       A Quasi-thermodynamic
       Approach

In the preceding sections the kinetics of photoelectrochemical reactions were considered in terms of
the solution of the transport equations, taking into
account the spatial inhomogeneity (normal to the surface)
of the photogeneration of the current carriers in the
region near the semiconductor surface. This approach
makes it possible to obtain an accurate solution to the
problem. However, in some cases, for example, with relatively complicated boundary conditions, it turns out
to be rather laborious, and the resultant expressions
are cumbersome. At the same time a quasi-thermodynamic
approach based on the concepts of quasi-Fermi levels
proves more effective than the kinetic approach for the
qualitative understanding and sometimes quantitative
interpretation of the phenomena.

The notion of the quasi-Fermi level, first introduced
by Shockley (17), can be approached in the following
way. Let us assume that, apart from thermal generation,
additional generation of carriers in the semiconductor
takes place under the influence of some external factor,
for example illumination. In the steady-state a dynamic
equilibrium is established between the processes of
generation and recombination of the electron-hole pairs.
As a result, in the irradiated semiconductor some
stationary (but not equilibrium!) concentrations n_o^* and
p_o^* are established. Suppose that the lifetime of the
excited states is sufficiently long. Then it can be
assumed that, as a result of the interaction with the
lattice vibrations (phonons), equilibrium distributions
with a temperature equal to that of the lattice (the
phonon system) are established in the electron and hole
gases which contain the photoexcited carriers. It is

essential that with respect to each other the electron
and hole gases need not be in equilibrium. [†]   Then, each
of the distributions (of electrons and holes) can be
characterized by a chemical and electrochemical potential of its own. We shall denote the electrochemical
potentials of electrons and holes as F_n and F_p respectively. Unlike the case of complete thermodynamic
equilibrium where $F_n = F_p = F$, the values of F_n and F_p,
which are called the quasi-Fermi levels and correspond
to partial equilibrium, are unequal to one another (see
Fig. 6.8).

     Thus, the emergence of non-equilibrium electrons
and holes in bands can be described as a "splitting" of
the initial Fermi level F into two quasi-Fermi levels
F_n and F_p. In the absence of degeneracy, in place of
the equilibrium relation (1.16), we now have a quasi-
equilibrium relation

$$n_o^* p_o^* = n_i^2 e^{(F_n - F_p)/kT} \tag{6.34}$$

where $n_o^* = n_o + \Delta n$, $p_o^* = p_o + \Delta p$. Since the photogeneration of carriers occurs in pairs, outside the space
charge region we have $\Delta n = \Delta p$.

     If the condition $\Delta n \ll n_o$ is met, it follows from
Eq. (1.22a) that for an n-type semiconductor $F_n \simeq F$.
Using Eq. (6.34) and taking into account Eq. (1.16),
for the quantity $F - F_p$ we find

$$F-F_p = kT ln \left(\frac{p_o + \Delta p}{p_o} \right) ; \quad F-F_p \simeq kT ln \frac{\Delta p}{p_o} \quad (\Delta p \gg p_o) \tag{6.35}$$

Similarly, for a p-type semiconductor, if $\Delta p \ll p_o$, then
$F_p \simeq F$ and

$$F_n - F = kT ln \left(\frac{n_o + \Delta n}{n_o} \right) ; \quad F_n - F \simeq kT ln \frac{\Delta n}{n_o} \quad (\Delta n \gg n_o) \tag{6.36}$$

     Thus, the shift in the electrochemical potential
caused by photogeneration is only significant for the

[†] A similar concept of a photoexcited electron gas which is in
thermodynamic equilibrium with the crystal lattice is used in the
Fowler theory of photoemission from metals (see, for example, (38)).

minority carriers. Note that Eqs. (6.35) and (6.36) describe the shift in the electrochemical potential of electrons and holes in a semiconductor due to a change in
their concentration. This is just the same as for the
case of redox species in solution (see relations (2.7)
and (2.15)), and also for semiconductors on the introduction of dopant impurities. In the light of the latter
similarity, the shift in the values of F_n and F_p under
irradiation is sometimes called "illumination doping."

When the intensity of the illumination is not too
high, the magnitudes of $\Delta n, \Delta p$ are proportional to intensity
and therefore the shifts in the quasi-levels ($F - F_p$
or $F_n - F$) are proportional to the logarithm of the intensity; as the latter increases, the further shift in
the quasi-levels become less pronounced due to the
increase in recombination. The edge of the corresponding band represents the limit for the shift in quasi-
level with increasing light intensity within the applicability of Eqs. (6.35) and (6.36). When $\Delta n, \Delta p$ are
extremely high the effects due to the degeneracy of the
carriers should be taken into account and Eqs. (6.35)
and (6.36) cease to be valid.

We will illustrate the use of the quasi-Fermi level
approach by considering the important case of a moderate
steady state photocurrent. Let us assume that on the
semiconductor surface a redox reaction of the following
type takes place:

$$Ox + e^- \rightleftarrows Red \qquad (6.37)$$

and that it is characterized by an equilibrium potential
E^O_{redox}.

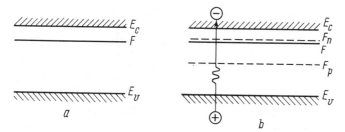

Fig. 6.8. The emergence of the quasi-Fermi
levels for electrons, F_n, and holes, F_p.
n-Type semiconductor: a) in the dark; b)
under illumination (the quantum energy is
bigger than the bandgap).

In the dark at equilibrium (i.e., at the potential E_{redox}^O) the total current (see Eq. (4.12)) is zero. Moreover, the net electron and hole currents taken separately are also zero.

When the semiconductor is illuminated the concentrations of carriers increase as a result of the generation of electron-hole pairs, and this increases the currents for electron and hole transfer to the solution at the given electrode potential. In these conditions, the potentials E_p^O and E_n^O, at which the hole and electron currents respectively become zero, are no longer coincident with each other or with the equilibrium potential E_{redox}^O. In other words, under illumination it is as if the semiconductor electrode has two values of electrode potential simultaneously, one of which determines the rate of reaction with the participation of holes from the valence band, while the other determines the rate of reaction with the participation of electrons from the conduction band.

Let the non-equilibrium distribution of carriers be characterized by the quasi-Fermi levels F_p and F_n. In this particular case to introduce the quantities $F_{p,n}$ it is necessary for equilibrium within the photo-excited hole and electron systems to be established more rapidly than the capture of the individual photo-generated electrons and holes by the surface where the electrode reaction occurs. Below we shall determine when this condition is valid.

Introducing the quasi-Fermi levels and proceeding from general thermodynamic principles, for the deviation of quasi-equilibrium potentials $E_{p,n}^O$ from the equilibrium potential E_{redox}^O we obtain

$$-e(E_{p,n}^O - E_{redox}^O) = F_{p,n} - F \qquad (6.38)$$

As is seen from Eq. (6.38) and the discussion given above, for the case of a sufficiently doped n-type semiconductor, it is only the "hole equilibrium potential" E_p^O that shifts under illumination, whereas for the p-type case only the "electron equilibrium potential" E_n^O shifts.

It is evident that under illumination at the potential E_{redox}^O the total current is not equal to zero. At the same time, there exists a certain quasi-equilibrium potential E_{redox}^*, between E_p^O and E_n^O, at which the total current $i_{ph} = i_p + i_n$ is equal to zero, although the

currents i_n and i_p are not separately zero ($i_p = -i_n$).
The dependence of E^*_{redox} on the parameters of the system
($E^o_{n,p}$; $i^o_{n,p}$) was calculated in (221).

The resultant photocurrent i_{ph} at $E \neq E^*_{redox}$ is
given, according to (221), by a relation which is formally similar to the result in the electrochemical
kinetics of "dark" processes:

$$i_{ph} = i^o_{eff} \left[e^{e(E-E^*_{redox})/kT} - e^{-e(E-E^*_{redox})/kT} \right]. \quad (6.39)$$

(It is assumed here that the whole potential drop occurs
in the semiconductor, that the quantities F_n and F_p are
constant across the space charge region, which is
achieved when the conditions $L_{sc} \ll L_p$, $L_{sc} \ll \alpha^{-1}$ are
met, and that the potential dependence of the "dark"
currents can be neglected because of their extremely
small value.) The role of the effective exchange
current in Eq. (6.39) is played by the quantity

$$i^o_{eff} = (i^o_p i^o_n)^{\frac{1}{2}} exp \{e(E^o_p - E^o_n)/2kT\} \quad (6.40)$$

It is obvious that the effective exchange current i^o_{eff}
under illumination can considerably exceed the "dark"
exchange currents i^o_n and i^o_p .

Equation (6.38) also applies to the particular
case when electron (hole) exchange with the solution
takes place through only one of the semiconductor bands.
For example, let us assume that the electrochemical reaction on an n-type semiconductor occurs entirely
through the valence band. According to the data available (see Section 4.4), such a situation may occur, in
particular, when the electrochemical potential of the
reaction F_{redox} is located much closer to the valence
band edge than to the conduction band edge, and when
$E_g \gtrsim E_R$. Then electron transfer between the level E^o_{ox}
in solution and levels in the conduction band can be
neglected. In such conditions, Eq. (6.38) directly
describes the shift of the potential from the equilibrium value E^o_{redox} to the quasi-equilibrium value
$E^o_p > E^o_{redox}$. The above mentioned regimen when
$i_p = -i_n \neq 0$ and $E = E^*_{redox}$ cannot be realized in
this particular case, due to the condition $i^o_n = 0$.

Since the quantity F_p which characterizes the
minority carriers can change considerably on illumination ($e(F - F_p)/kT \gg 1$), the deviation of the

system from equilibrium will also be considerable. Due
to this fact, a rather noticeable increase in the total
current in the system may occur.

We shall make use of Eq. (4.14) to find the dependence of the electrochemical reaction current on the
potential and (through the quasi-Fermi level) on the intensity of illumination. Suppose that the entire
potential change occurs in the semiconductor (i.e.,
$\eta_H = 0$). Then, taking into account the Boltzmann distribution of holes in the space charge region, we get
from Eqs. (4.14) and (6.38)

$$i = i_p^o [e^{e(E-E_p^o)/kT} - 1] = i_p^o [e^{(eE-eE_{redox}^o - \Delta F)/kT} - 1] \quad (6.41)$$

where $\Delta F = F - F_p$. In other words, in accordance with
Eq. (6.41), it is possible to carry out an electrochemical reaction on the illuminated semiconductor electrode
when the potential is "less" than the dark equilibrium potential by an amount $\Delta E = - \Delta F/e$, i.e., at an "underpotential." Equation (6.41) is also applicable in the
case of a highly irreversible electrode reaction when
$e(E - E_p^o)/kT \gg 1$. It is also easy to obtain the corresponding expression for the general case, when the
overvoltage in the Helmholtz layer cannot be ignored.

Using the concept of quasi-Fermi levels it is
possible, with relative simplicity, to examine a number
of other effects caused by the illumination of the semiconductor surface (and, in particular, of the semicon-
ductor/electrolyte interface). Following (63), let us
show as an example how the change of potential during
the illumination of an ideally polarizable semiconductor
electrode can be found if we assume that recombination
in the space charge region and on the surface can be
neglected.

Since electrons and holes are generated in pairs,
the full space charge per unit surface, Q_{sc}, and consequently the electric field strength ξ_{sc} (see formula
(3.16)) should not change under illumination. The
value of Q_{sc} is a function of the potential drop in the
space charge region ϕ_{sc} (or $Y = e\phi_{sc}/kT$) and of the concentrations of electrons and holes at the boundary of
that region with the quasi-neutral bulk. Following on
from above, under constant illumination these concentrations are equal to n_o^* and p_o^*; illumination
changes the concentrations $n(L_{sc})$ and $p(L_{sc})$ by the

same amount $\Delta = n_o^* - n_o = p_o^* - p_o$, where Δ depends on the illumination intensity. Hence, during illumination the value of ϕ_{sc} (or Y) necessarily changes to compensate for the change in the concentrations n and p at $x = L_{sc}$ and preserve the constancy of Q_{sc}. This condition corresponds to the application of the following equality

$$\xi_{sc}(Y_{dark}) = \xi_{sc}(Y,\Delta) \qquad (6.42)$$

where Y_{dark} is the magnitude of the surface potential in the dark, and Y, its value under illumination.[†] From Eq. (6.42), using Eq. (3.13) with the replacement, during the calculation of $\xi_{sc}(Y,\Delta)$, of $\lambda = p_o/n_i$ by $\lambda + \Delta/n_i$ and $\lambda^{-1} = n_o/n_i$ by $\lambda^{-1} + \Delta/n_i$, we get

$$\frac{\Delta}{p_o} = \frac{F^2(Y_{dark}) - F^2(Y)}{\lambda(e^Y - e^{-Y} - 2)} \qquad (6.43)$$

Equation (6.43) enables us to find Y if we know the values of Y_{dark} and Δ. Assuming that illumination is not too intense, so that $\Delta/p_o \ll 1$, and calculating the derivative $dY/d\Delta$ from Eq. (6.43), we get (63)

$$\frac{dY}{d\Delta} = \frac{1}{p_o} \frac{1 - e^Y}{1 + \lambda^{-2}e^Y} \qquad (6.44)$$

From Eq. (6.44), at strongly anodic ($Y \to -\infty$) or strongly cathodic ($Y \to \infty$) polarization, we have respectively

$$\frac{dY}{d\Delta} = \frac{1}{p_o} \ (Y \to -\infty) \quad \text{and} \quad \frac{dY}{d\Delta} = -\frac{\lambda^2}{p_o} \ (Y \to \infty) \qquad (6.45)$$

If the value of the dimensionless potential $Y_{ph} = Y - Y_{dark}$ is small ($Y_{ph} < 1$), then it is proportional to the expression on the right hand side of Eq. (6.44). For this case the qualitative form of the dependence of the photopotential Y_{ph} on Y is shown in Fig. 6.9.

It follows from Eq. (6.44) that the photopotential Y_{ph} changes sign at $Y = 0$, the flat band potential.

[†] Here we neglect the Dember potential (see Eq. (6.28)) which can be taken into account subsequently.

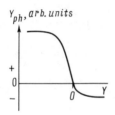

Fig. 6.9. Dependence of
the photopotential on the
surface potential for an
ideally polarizable elec-
trode.

A widely used method for determining the flat band potential of an ideally polarizable electrode is based on this
effect. As an illustration, Fig. 6.10 shows photopotential curves, Y_{ph}, on n- and p-type germanium electrodes
(in the range of ideal polarizability). Note the shift
of the flat band potential when the bulk concentration
of donors (acceptors) in the electrode changes (cf. Fig.
3.17).

Ξ 6.4. <u>The Limits of Applicability</u>
<u>of the Concepts of Quasi-</u>
<u>Fermi Levels in</u>
<u>Electrochemical</u>
<u>Kinetics</u>

The quasi-thermodynamic approach based on the concept of quasi-Fermi levels, set out in the preceding
section, is widespread. In particular, it is broadly
used (see, for example (47)) to describe the kinetics
of electrochemical reactions on illuminated semiconductor electrodes. It is undoubtedly advantageous
because of its descriptive and simple nature. However,
it should be borne in mind that the use of quasi-Fermi
levels is not always justified. To use the concept it
is necessary to observe the following inequality:

$$\tau_c, \ \tau_v < \tau_{cv} \qquad\qquad (6.46)$$

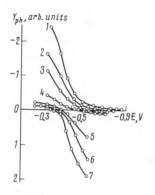

Fig. 6.10. Dependence of the
photopotential on the electrode
potential for a germanium electrode in 1 N NaOH (106). 1)
n-type, 0.04 ohm·cm; 2) n-type,
3 ohm·cm; 3) n-type, 20 ohm·cm;
4) n-type, 40 ohm·cm; 5) p-type,
10 ohm·cm; 6) p-type, 3 ohm·cm;
7) p-type, 0.5 ohm·cm. All
potentials are against the
Normal Hydrogen Electrode.

where τ_c and τ_v are the times taken to establish thermo-dynamic equilibrium in the ensemble of electrons in the conduction band and of holes in the valence band, and τ_{cv} is the time taken to establish equilibrium between the electron and hole ensembles. As has already been mentioned, the interaction of excited electrons and holes with the lattice oscillations (phonons) acts as the usual mechanism in establishing equilibrium in the bands. In-teraction between photoexcited and thermalized carriers in each separate band serves as such a mechanism in some semiconductors. As a rule, the values of τ_c and τ_v are of the order of 10^{-12} - 10^{-13} s. The time τ_{cv} is deter-mined by the interaction of electrons and holes with each other and in the simplest case it coincides with the time for recombination. For the majority of semi-conductor materials this time is significantly larger than 10^{-12} - 10^{-13} s. Thus, the conditions of Eq. (6.46) are met.

At the same time, these conditions alone turn out to be insufficient for the description of electron transfer processes at the interface in terms of the quasi-Fermi level concept. Let us introduce the times $\tau_{n,s}$ and $\tau_{p,s}$ which characterize the interval between photoexcitation of an electron or hole and the capture of the particle on the surface, by, for example, elec-trode reaction. It is obvious that the use of the quasi-Fermi level concept for the description of such processes is possible if these times are bigger than τ_c and τ_v. Otherwise quasi-equilibrium in the bands will not be es-tablished and "hot" particles will participate directly in the electrode process. On the other hand, the times $\tau_{n,s}$ and $\tau_{p,s}$ should be less than τ_{cv}, otherwise full equilibrium will be established in the electron-hole system, and near the surface $F_n = F_p = F$. Hence, the two following systems of inequalities must be observed in order to apply the quasi-Fermi level concept to elec-trons and holes in heterogeneous processes:

$$\tau_c < \tau_{n,s} < \tau_{cv} \quad ; \quad \tau_v < \tau_{p,s} < \tau_{cv} \qquad (6.47)$$

(When only one sort of carrier takes part in the elec-trode reaction, the observance of only one system of inequalities will suffice.)

If the surface quickly captures electrons and holes, the values of $\tau_{n,s}$ and $\tau_{p,s}$ coincide with the time taken for electrons and holes to approach the interface. Suppose that the depth of light penetration in the

semiconductor, α^{-1}, exceeds the thickness of the space
charge region L_{sc}. If L_{sc} is much less than the diffusion length of the minority carriers (L_n or L_p), then,
in order of magnitude terms, $\tau_{n,s} \simeq \alpha^{-2}/D_n$ and
$\tau_{p,s} \simeq \alpha^{-2}/D_p$. Thus, in particular, it follows that if
α^{-1} is bigger than not only L_{sc}, but also L_p (or L_n),
then the condition $\tau_{p,s} < \tau_{cv}$ (or $\tau_{n,s} < \tau_{cv}$) will be
upset for the minority carriers. This is physically in
accord with the evident circumstance for the case when
$\alpha^{-1} \gg L_p$ (or $\alpha^{-1} \gg L_n$) that during the time in which
the carriers approach the interface, equilibrium succeeds
in establishing itself throughout the system.

If $\alpha^{-1} < L_{sc}$, then for the case when a depletion
layer is formed, the time taken for carriers to reach
the interface is determined not, as it was above, by
their diffusion but by migration (drift) in the field
of the space charge region, and $\tau_{p,s}$ ($\tau_{n,s}$) $\simeq \alpha^{-1}/v_d$
where v_d is the drift velocity. Estimates show that
the value of α^{-1}/v_d is 10^{-13} - 10^{-11} s, and that the
inequalities in Eq. (6.47) may be observed.

When the surface process proper cannot be viewed
as indefinitely fast, the times $\tau_{n,s}$ and $\tau_{p,s}$ may be
limited by the characteristic time of that process, for
example, by the time taken for the elementary act of
electrode reaction. From Section 4.2, for reactions
proceeding in accordance with the Franck-Condon prin-
ciple, the conditions $\tau_c < \tau_{n,s}$ and $\tau_v < \tau_{p,s}$ are
observed at any reaction rate, and provided the
reaction is not too slow the conditions $\tau_{n,s}$, $\tau_{p,s} < \tau_{cv}$
are also observed. Thus, since the interval between
the times τ_c and τ_v, on the one hand, and the time τ_{cv},
on the other, is fairly large (from $\simeq 10^{-13}$ to $\simeq 10^{-8}$ s),
the time characteristics of many electrode processes
satisfy the conditions of Eq. (6.47). Consequently the
quasi-Fermi level concept is applicable for a broad
range of systems. At the same time, for a quantitative
description (like that done at the end of Section 6.3
for the calculation of the ideally polarizable electrode
photopotential) the observance of some additional conditions is imperative. It is necessary, among other
things, that the behavior of the quasi-levels in the
space charge region be smooth enough. The latter condition is fulfilled if $\alpha^{-1} \gg L_{sc}$ and recombination in
the space charge region can be neglected. (This means
that formally in the space charge region $\tau_{cv} \to \infty$.
Therefore, even for an ideally polarizable electrode
for which $\tau_{n,s}$ $\tau_{p,s} \to \infty$, the conditions of Eq. (6.47)
are not upset.)

On the other hand, it is obvious that when photogeneration of carriers mainly occurs in close proximity
to a surface on which a fast electrode reaction takes
place, it is quite possible that the photoexcited carriers will be "sucked out" from the semiconductor quicker
than they will come to equilibrium with thermalized carriers in the corresponding band. Under such conditions,
they cannot be characterized by collective quasi-equilibrium levels F_n and F_p. It would be apparently more
correct in this case to deal with the transitions of
carriers from the electrode to the solution as proceeding not through a quasi-equilibrium stage but purely
dynamically in separate independent events (222-224).
An attempt was made in (224) to describe such a dynamic
process quantitatively. According to the model used in
(224) an essential role is played by the quantum energy
levels which occur near the surface when the bands are
heavily bent (see Fig. 3.3). Photoexcited electrons (or
holes) are only partially thermalized and are then localized on these quantum levels. The subsequent thermalization proves to be a rather slow process, and the
electrode reaction occurs first. This reaction is described as the tunnelling of an electron from a quantum
level in the triangular potential well near the surface
through a unidimensional potential barrier into a unidimensional rectangular potential well which models the
reagent in solution (see Fig. 6.11). It is further
supposed (unlike the assumption made in the elementary
act theory expounded in Section 4.2) that after tunnelling, the rearrangement of the solvent around the
reagent occurs rapidly. As a result, the reverse electron transfer on to the electrode proves impossible.

It may be maintained that such a description is
adequate in some cases for electron (hole) transitions

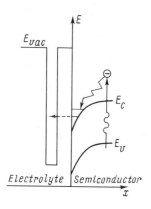

Fig. 6.11. The model used
to describe the tunnelling
of "hot" electrons from a
semiconductor to a reagent
in solution.

on wide bandgap doped semiconductors with carriers of
small effective mass when the band bending is sufficiently strong. At the same time, the model used (in
particular, since it is unidimensional) is too crude
and the quantitative agreement with experiment is hardly
attainable at present.

The option between the two approaches to the description of the kinetics of photoelectrochemical reactions - quasi-thermodynamic or dynamic - may be made
on the basis of experimental data. The first approach
(which applies the concept of quasi-levels) predicts
the existence of some threshold illumination intensity.
Indeed, when the electrode reaction is slow in the dark
(the values of i_n^o and i^o are small), the electrochemical
potential in the semiconductor F and in the solution F_{redox}
are arranged arbitrarily with respect to one another.
Under illumination when the level F is split into F_n and F_p,
the quasi-level F_p, moving away from F with the growth
of light intensity, may reach the level F_{redox} at some
threshold intensity. The onset of photoelectrochemical
reaction corresponds precisely to this intensity. At
lower intensities of illumination the photoelectrochemical reaction does not take place. On the contrary, in
conformity with the dynamic approach, the photoelectrochemical reaction should occur at any illumination intensity. The threshold has not yet been discovered
experimentally. For example, during oxygen photoevolution on TiO_2, within a very broad interval of
light intensities (from 10^{-6} to 4×10^2 W/cm^2), the
photocurrent grows monotonically in proportion to the
intensity (223).

Until now research has been done on semiconductors
with very wide bandgaps and, consequently, with negligibly small concentrations of minority carriers p_o (or
n_o). For such materials the threshold light intensity
determined by the ratio $\Delta p/p_o$ (or $\Delta n/n_o$) may prove to
be too low, and it cannot be fixed experimentally.
Thus, there is no reliable data at present upon which
to base a well-founded opinion in favor of one or another of the two concepts examined above.

Chapter 7

Photocorrosion and Stability

The problem of the stability of semiconductor electrodes with respect to corrosion and, in particular,
with respect to photocorrosion is becoming more prominent
due to the different and ever expanding applications of
semiconductor/electrolyte systems. In the first place,
this problem relates to photoelectrodes for light
energy conversion. This topic is dealt with in Chapters
8 and 9. On the other hand, corrosion-like processes
underlie chemical etching which is widely used for the
processing of semiconductor materials in the production of semiconductor devices.

Basically, the corrosion behavior of semiconductors
may be described within the framework which has been developed for the corrosion behavior of metals (see, for
example, (80, 225, 226)), with due account of the parti-
cularities in the electrochemical behavior of the electrode determined by its semiconductor nature (5, 227).
One of the main differences is the photosensitivity
based on the change, under the influence of light, of
the number of carriers in the bands. This effect underlies a special type of corrosion phenomena which does not
occur in the corrosion of metals, at least not those with
"clean" (non-oxidized) surfaces, that is, photocorrosion.

Ξ 7.1. C o r r o s i o n :   G e n e r a l   C o n c e p t s

The spontaneous decomposition (dissolution, oxidation) of a material as a result of physical and chemical
interaction with its surrounding medium (in particular,

with the electrolyte solution) is called corrosion.
Below we shall examine the corrosion of semiconductor
materials which are homogeneous in their composition.
Principal attention will be paid to the electrochemical
aspects of this process.

Since we analyze decomposition processes without
the net passage of electric current through the sample,
and the solid/solution system on the whole remains elec-
troneutral, oxidation of the solid in the course of
corrosion is bound to be accompanied by the reduction
of other components of the system. For example, the
self-dissolution (oxidation) of semiconductors in aqueous solutions occurs simultaneously with either the
reduction of hydrogen ions and the evolution of molecular hydrogen or with the reduction of dissolved
oxidizing agent. The overall equation of the process
is as follows:

$$\{SC\} + Ox \rightarrow \{SC\}^+ + Red \qquad (7.1)$$

where {SC} denotes the semiconductor material and {SC}$^+$,
its oxidation product (for example, an ion in solution),
Ox is the oxidizing agent, and Red is the reduced form
of Ox.

The reactions of semiconductor dissolution and reduction of oxidant may occur in one act, hence, Eq.
(7.1) reflects the genuine microscopic mechanism of the
process. In this case we mean a "chemical" mechanism
for corrosion.

It happens more often, however, that two partial
electrochemical reactions, one anodic and the other
cathodic, occur at the solid/solution interface
simultaneously and at the same rate,

$$\{SC\} \rightarrow \{SC\}^+ + e^- \qquad (7.2a)$$

$$Ox + e^- \rightarrow Red \qquad (7.2b)$$

which together make up reaction (7.1). It is precisely
this case of "electrochemical" corrosion that is examined in detail below.

Anodic (7.2a) and cathodic (7.2b) reactions on
samples with homogeneous surfaces are localized on the
same areas of the surface, while on heterogeneous samples
they may be spatially separate. Such reactions are
called conjugated. Each of them is characterized by an

equilibrium potential (E^O_{dec} and E^O_{redox} respectively), an exchange current (i^O_{dec} and i^O_{redox}) and a transfer coefficient (see Section 4.1).*

In the electrochemistry of metals the conjugated reactions are regarded as completely independent from each other. This means that it is possible to change the rate of one of them without influencing the other. Being aware of the kinetic parameters of the two reactions, we can determine unambiguously the rate (usually expressed as a current) and the potential of corrosion, making use of the obvious condition

$$i^a = -i^c = |i_{corr}| \qquad (7.3)$$

where i^a and i^c are the current densities for the anodic and cathodic conjugated reactions of Eq. (7.2).

The specific features inherent in the corrosion processes on semiconductors, as compared with metals, originate from the fact that current carriers of both signs take part in the exchange of charges between the solid and the solution: electrons from the conduction band and holes from the valence band. Correspondingly, one condition of the type given in Eq. (7.3) proves insufficient. Both the total balance of electric charges and the balance of each type of carrier in the semiconductor taken separately should be observed in the self-dissolution process as, in the general case, equilibrium between the bands may not be attained.

Let us elucidate this, using a simple extreme case as an example where the partial reaction with the participation of the semiconductor material occurs exclusively through the valence band, while the partial reaction with the participation of the redox couple from solution takes place exclusively through the conduction band (Fig. 7.1a). Although the conjugated reactions occur through different semiconductor bands, they are coupled with each other by the processes of recombination and generation of electrons and holes. However, if the rate of the recombination or generation

---

*It should be stressed that, irrespective of the notations "dec" and "redox", all the reactions in question are related to one and the same class; they are all redox electrochemical reactions (cf. Section 2.2, Eqs. (2.16) and (2.18)).

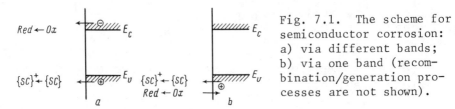

Fig. 7.1. The scheme for semiconductor corrosion: a) via different bands; b) via one band (recombination/generation processes are not shown).

process is small enough when compared with the rates of the conjugated reactions, this connection can be neglected and the reactions in Fig. 7.1a considered as relatively independent.

We will assume that when the electrode is polarized, the potential drop changes only in the space charge region of the semiconductor, so that the change in the potential drop in the Helmholtz layer can be ignored. Fig. 7.2 shows n- and p-type corrosion diagrams (current-voltage curves with the anodic and cathodic branches in semilogarithmic coordinates; cf. Fig. 4.7) for the two conjugated reactions in this case. In the figure the probable diffusion limitation on the supply of minority carriers has been taken into account. It has been assumed as a particular example that:

(i)   the exchange current for the reaction with the participation of the semiconductor material is higher than the exchange current for the reaction of the redox couple in solution, $i_{dec}^{o} > i_{redox}^{o}$;

(ii) the limiting current of minority carriers in the p-type semiconductor is higher than in the n-type semiconductor, $i_n^{lim}(p) > i_p^{lim}(n)$ (here the symbol in parentheses denotes the type of sample conductivity);

(iii) there are no diffusion limitations in the solution; the limiting currents for diffusion of the reagents Ox and Red (not shown here) are bigger than i_{dec}^{o}, i_{redox}^{o} and $i^{lim}(p,n)$.

Using the condition in Eq. (7.3), it is easy to find the corrosion current i_{corr} and the stationary corrosion potential E_{corr} through the intersection of the current-voltage curves for the two conjugated reactions. It is assumed here that the ohmic potential drop (iR) can be neglected (for the case of a finite resistance, R, such graphical analysis can be found on p.265). Note that the mechanism of the process shown in Fig. 7.1a is characterized by the fact that the

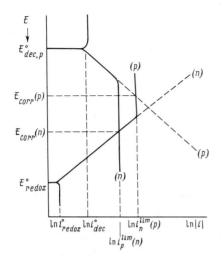

Fig. 7.2. Corrosion diagram. It is assumed that anodic re-reaction proceeds completely via the valence band, and that the cathodic reaction proceeds completely via the conduction band.

corrosion current for any type of semiconductor conductivity is limited by the minority carriers, that is, $i_{corr}(n) = i_p^{lim}(n)$ or $i_{corr}(p) = i_n^{lim}(p)$.

It should be emphasized that in semiconductors with wide bandgaps ($E_g \gtrsim 1$ eV) the equilibrium concentration of minority carriers is very small. This is why the occurrence of electrode reactions with the participation of minority carriers in the absence of any special method for their generation (such as illumination or injection across the interface by another electrode reaction; see below) is barely possible.

Thus, since the anodic partial reaction needs holes and the cathodic reaction needs the conduction band electrons, the rate of corrosion is limited by the rate of supply of the minority carriers to the interface. Therefore the corrosion rate in the dark, irrespective of the conductivity type of the sample, is very low.

A fundamentally different situation occurs when the two conjugated reactions proceed through one and the same band (Fig. 7.1b). For example, the cathodic reaction of oxidant reduction may occur with the consumption of the valence band electrons, rather than with conduction band electrons as was assumed above. As a result of oxidant reduction, holes are then injected into the semiconductor and are further used in the anodic reaction of semiconductor oxidation. Here the cathodic partial reaction "supplies" the anodic reaction with the necessary free carriers. Under these

conditions the kinetics of the anodic reaction on an
n-type semiconductor are no longer limited by the de-
livery of holes from the crystal bulk, since there is
an additional source of holes at the semiconductor/elec-
trolyte interface.

Thus, oxidants which are reduced with the participation of valence band electrons are able to cause intense corrosion of n-type semiconductors, insofar as
holes are required in this process.

The current multiplication effect (see Section 4.5)
introduces additional complications. Under the conditions of current multiplication, holes from the valence
band and electrons from the conduction band take part
simultaneously in each of the conjugated reactions in
the overall corrosion process. The following situations
(for an n-type semiconductor) are possible depending on
the relation between the current multiplication coefficients for the anodic, M^a, and cathodic, M^c, reactions.

When $M^a > M^c$, an intermediate case between the two
extreme cases examined above is observed: the anodic reaction is still limited by the supply of holes to the
interface but the corrosion current is higher than
$i_p^{lim}(n)$ because the cathodic reaction provides an additional flux of holes.

When $M^a = M^c$, all holes supplied by the cathodic
reaction of oxidant reduction are consumed in the course
of anodic dissolution of the semiconductor.

When $M^a < M^c$, surplus holes emerge at the surface
of the corroding semiconductor. The "extra" holes, together with an equal number of electrons injected into
the conduction band in the course of the anodic reaction
(see Eq. (7.2a)), diffuse into the depth of the crystal.
The injection of non-equilibrium carriers into germanium
and silicon in solutions of some oxidants was found by
means of a direct experiment (228-232) using an electrode
device of the "transistor type" described in Section
4.6.2. There the corrosion process acts as a source of
electron-hole pairs at the surface, and in this respect
it resembles the photogeneration of carriers when the
semiconductor is illuminated by "surface-absorbed" light
($\alpha^{-1} \ll L_{sc}$). For example, the corrosion potential of
germanium and silicon in a solution of oxidants injecting
holes depends on the bulk properties of the semiconductor
sample, as in accordance with the same law that determines its photopotential upon illumination in an indifferent electrolyte solution (233, 234).

As has already been mentioned, in the general case the reaction of semiconductor decomposition may be either anodic (cf. Eq. (7.2a)) or cathodic.

Thus, for the binary semiconductor MX (where M is an electropositive and X an electronegative component of the compound) the equation for the partial electrochemical reaction of cathodic decomposition with the participation of electrons from the conduction band can be written as follows:

$$MX + ne^- \rightarrow M + X^{n-} \qquad (7.4a)$$

and the reaction of anodic decomposition with the participation of holes, as follows:

$$MX + nh^+ \rightarrow M^{n+} + X \qquad (7.4b)$$

If, as a specific example, the hydrogen electrode is chosen as the reference electrode, on which the following reaction occurs:

$$H^+ + e^- \rightleftarrows \tfrac{1}{2} H_2$$

then the overall reaction in the cell will occur in accordance with one of the following equations (the notations for the electrochemical potentials for the reactions are in parentheses):

$$MX + \frac{n}{2} H_2 \rightleftarrows M + nH^+ + X^{n-} \qquad (F_{dec,n}) \qquad (7.5a)$$

$$MX + nH^+ \rightleftarrows M^{n+} + X + \frac{n}{2} H_2 \qquad (F_{dec,p}) \qquad (7.5b)$$

For example, for a zinc oxide electrode in hydrochloric acid solution the reactions in Eq. (7.4) are written as follows:

$$ZnO + 2H^+ + 2e^- \rightleftarrows Zn + H_2O \qquad (7.6a)$$

$$ZnO + 2h^+ \rightleftarrows Zn^{2+} + \tfrac{1}{2} O_2 \qquad (7.6b)$$

while Eqs. (7.5) (with the hydrogen reference electrode) are written as

$$ZnO + H_2 \rightleftarrows Zn + H_2O \qquad (7.7a)$$

$$ZnO + 2H^+ \rightleftarrows Zn^{2+} + \tfrac{1}{2} O_2 + H_2 \qquad (7.7b)$$

Having calculated the corresponding changes in free energy for these reactions from the thermodynamic properties of the reagents (see, for example, (52-54)) it is possible to determine the standard equilibrium potentials of the reactions of cathodic semiconductor decomposition $E_{dec,n}^{o}$ and of anodic decomposition $E_{dec,p}^{o}$ (see Section 2.4). For a number of semiconductor materials (in particular, for CdS, CdSe, GaP, GaAs, MoS_2) the values of the decomposition potentials are cited in the literature (55, 235-239).

When discussing the corrosion (and photocorrosion) behavior of semiconductors it is convenient to use Pourbaix diagrams (Section 2.4), that is, the calculated dependence of E_{dec}^{o} on solution pH. These show in which pH region the material is corrosion resistant and which decomposition reaction prevails in the region of corrosion susceptibility. Such diagrams are given for some semiconductor materials in Appendix 2.

The thermodynamic criteria for the occurrence of an electrode reaction, including a partial corrosion reaction, can be written as

$$E < E_{dec,n}^{o}, \quad \text{or} \quad F > F_{dec,n} \qquad (7.8a)$$

for the cathodic reaction with the participation of electrons from the conduction band, and as

$$E > E_{dec,p}^{o}, \quad \text{or} \quad F < F_{dec,p} \qquad (7.8b)$$

for the anodic reaction with the participation of holes from the valence band.

In order to find out in each particular case whether the semiconductor is inclined towards electrolytic (corrosion) decomposition, let us examine the energy band diagram and plot on it the electrochemical potentials for the decomposition reactions (235).

Fig. 7.3 shows schematically the various situations that are possible. The semiconductor is stable with respect to cathodic decomposition if the electrochemical potential for the reaction lies within the conduction band, and it is also stable with respect to anodic decomposition if the electrochemical potential for that reaction lies within the valence band. In both cases if the potential drop is fully concentrated in the semiconductor ("band edges pinned at the surface") this

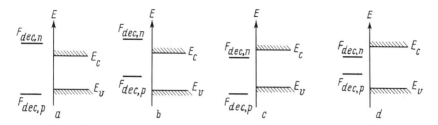

Fig. 7.3. Scheme illustrating the thermodynamic stability
of a semiconductor with respect to decomposition: a) the
semiconductor is absolutely stable; b) stable with respect
to cathodic decomposition; c) stable with respect to anodic
decomposition d) unstable (235).

electrochemical potential proves to be "inaccessible" to
the Fermi level of the semiconductor. The semiconductor
is then absolutely stable, if the levels for both decomposition reactions are beyond the limits of the bandgap
(Fig. 7.3a). More often it is found that the semiconductor is only stable with respect to one of the types
of decomposition: cathodic (Fig. 7.3b) or anodic (Fig.
7.3c). Finally, if the two levels $F_{dec,n}$ and $F_{dec,p}$
are within the bandgap (Fig. 7.3d), then under both
cathodic and anodic polarization the semiconductor is
basically subject to decomposition.

Ξ 7.2. P h o t o c o r r o s i o n

We shall now examine the influence of light on the
corrosion process within the framework of the phenomenological approach developed above. Using diagrams of
the type shown in Fig. 7.2 it is easy to predict the influence exerted by illumination on the corrosion process, provided that the role of the light is restricted
to the excitation of the electron-hole system in the
semiconductor. (Here we consider the photocorrosion
process under the influence of uniform illumination of
the semiconductor surface. Several sections in Chapter
10 will deal with the effects of non-uniform illumination on the process.) The enhancement of minority
carrier concentration will lead to an effective increase
in the current for the reaction occurring through the
minority carrier band. The corresponding curve will
shift, as shown schematically in Fig. 7.4. At the same
time, the light has a much smaller influence on the reaction occurring with the participation of majority
carriers. Therefore, as a result, the corrosion current
will increase and the stationary potential of the

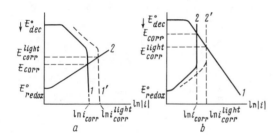

Fig. 7.4. The influence of illumination on the rate of corrosion of an n-type (a) and a p-type (b) semiconductor. Polarization curves for conjugated reactions are shown: 1,2) in the dark; 1',2') in the light.

corroding sample will change, the direction of the shift in E_{corr} depending on the conductivity type of the semiconductor (see Fig. 7.4). Using the relationships for the currents (Chapter 4) and the condition in Eq. (7.3), it is possible to explicitly calculate the potential E_{corr}^{light} and the current i_{corr}^{light} for corrosion under illumination.

Thus, the non-equilibrium electrons and holes generated by the light in the corroding semiconductor are consumed by increasing the rate of the corresponding partial reactions. The simultaneous disappearance of non-equilibrium carriers from the surface of the semiconductor during photocorrosion is formally similar to the process of surface recombination. For this reason the process has been referred to as "electrochemical recombination" (240).

The electrochemical recombination velocity s_{el} can be introduced with the help of a relation which is similar to that used to introduce the notion of the surface recombination velocity (see Section 3.6). Thus, if we denote the rate of removal of the non-equilibrium minority carriers (holes) on the semiconductor surface as a result of the corrosion process as $R_{p,el}$, then the value of s_{el} is given by

$$R_{p,el} = s_{el}[p(L_{sc}) - p_o] \qquad (7.9)$$

From physical considerations we can see that s_{el} is proportional to the rate of corrosion in the dark.

Let us now describe photocorrosion on the basis of the quasi-thermodynamic approach developed by Gerischer (235) and by Bard and Wrighton (236). This approach is based on the notion that the increase in the rate of the semiconductor decomposition reaction on illumination is due to the shift in the quasi-Fermi levels for electrons, F_n, and holes, F_p (see Section 6.3). Consequently

the conditions in Eqs. (7.8) should be modified. For
the cathodic photodecomposition reaction occurring with
the participation of the conduction band electrons the
following condition should be observed:

$$F_n > F_{dec,n} \qquad (7.10a)$$

while for the anodic photodecomposition reaction occurring with the participation of holes:

$$F_p < F_{dec,p} \qquad (7.10b)$$

As soon as one of the conditions in Eqs. (7.10) is
observed, the corresponding reaction becomes thermodynamically favorable (although, of course, kinetic
limitations may slow it down).

From this viewpoint let us now consider the corrosion process on the illuminated semiconductor. Suppose
that in some region of potential, electron transfer at the
semiconductor/solution interface does not take place in
the dark. Then the semiconductor does not corrode and
behaves as an ideally polarizable electrode. This region
of potential is bounded by the potentials for solvent decomposition and/or semiconductor decomposition. The
stationary potential of the electrode in the region of
ideal polarizability is determined by chemisorption processes (in aqueous solutions this is more often than not
the result of the chemisorption of oxygen) or, and this
amounts to the same thing as far as the physics of semiconductor surfaces are concerned, by the charging of slow
surface states. It is then these processes that determine the stationary band bending.

Figure 7.5 shows an energy level diagram of an
n-type semiconductor, chosen as a specific example, in
contact with a solution. Marked on the diagram are
energy levels corresponding to the bottom of the conduction band E_c, the top of the valence band E_v, and
also the electrochemical potentials for the anodic decomposition of the semiconductor and the cathodic reaction of hydrogen evolution. (This is the case when
decomposition of the semiconductor with the participation
of holes is the most probable anodic reaction, whereas
the evolution of hydrogen from water with the participation of conduction band electrons is the most probable
cathodic reaction).

A situation has been chosen in Fig. 7.5 where a
depletion layer is formed in the region near the surface

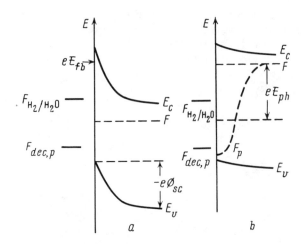

Fig. 7.5. The energy
diagram of a semicon-
ductor/electrolyte
junction: a) in the
dark; b) in the light.

of the semiconductor. Electrons and holes generated by
the light in the electric field of the depletion layer
move in opposite directions: in this case, holes move
toward the interface, and electrons move into the semiconductor bulk. The electric field which emerges as a
result of charge separation partially compensates the
initial field. This is manifested in the fact that the
band bending $e\phi_{sc}$ decreases upon illumination; the bands
unbend (see Section 6.1). This unbending of the bands,
in its turn, leads to a change in the mutual disposition
of the other energy levels in the system. Indeed, let
us assume for simplicity that the potential drop in the
Helmholtz layer is unchanged on illumination (ϕ_H = const)
and, therefore, that the position of the band edges at
the surface is fixed with respect to the reference electrode and, thus, with respect to the energy levels in
solution (the band edges are pinned at the surface, see
Section 3.4). At the same time, the position of the
electrochemical potential F (the Fermi level) in the
semiconductor is fixed with respect to the band edges
in the bulk of the semiconductor (for example, the
value of E_C - F is determined by Eq. (1.22a)). This is
why, under illumination the bands, by unbending, "pull"
the Fermi level, so that a shift in the Fermi level
occurs with respect to its position in the non-illuminated semiconductor. This is shown in Fig. 7.5b. Here
we are referring to the thermodynamic equilibrium Fermi
level F; that is, the electrochemical potential of electrons in the semiconductor as a whole, rather than the
quasi-Fermi levels F_n and F_p. The shift in the value
of F, produced by illumination, can be measured with
the help of a reference electrode. This shift ΔF is
reflected in the photopotential, E_{ph} = $\Delta F/e$.

Let us now investigate the shifts in the quasi-Fermi levels F_n and F_p. For the majority carriers (electrons) the shift in F_n with respect to F is very small (see Section 6.3) and it can be roughly neglected (Fig. 7.5b is plotted with the assumption that $F_n \simeq F$). For the minority carriers (holes), however, the shift is significant, and for a certain intensity of illumination the value of F_p may reach the level of the semiconductor decomposition reaction. In its turn, the quasi-Fermi level for electrons, F_n, reaches the level for the water reduction reaction with hydrogen evolution due to the shift in the Fermi level F. These reactions begin simultaneously and comprise the overall process of photocorrosion.

In order to get some idea of the stability of a semiconductor with respect to photocorrosion, it is again necessary to use Fig. 7.3 and to resort to the same reasoning as was used in the case of corrosion in the dark.

Note that such a mutual arrangement of the levels for the decomposition reactions of the semiconductor and the solvent is possible when both anodic and cathodic decomposition reactions of the semiconductor occur simultaneously, thereby being partial reactions in the photocorrosion process. This situation, which is without analogy in the corrosion of metals, becomes possible due to the fact that the electrons and holes in a semiconductor under illumination behave independently and, in particular, develop quasi-equilibrium electrochemical potentials which differ from one another.

When examining photocorrosion phenomena, one should bear in mind the possibility of other reactions competing with the photodecomposition process. On increasing illumination intensity the cathodic reaction whose equilibrium potential is the most positive will start first, whilst as far as all possible anodic reactions are concerned, the one with the most negative equilibrium potential will start first. Hence, the photocorrosion of the semiconductor may be suppressed by the addition of suitable oxidants or reductants to the solution. In this way, anodic photocorrosion is avoided by using redox systems whose redox potentials are more negative than the potential for semiconductor decomposition: $E^O_{redox} < E^O_{dec,p}$ (i.e., $F^O_{redox} > F^O_{dec,p}$). Such protective systems stabilize semiconductor photoanodes (Fig. 7.6). The condition for stabilization with respect to cathodic photodecomposition is written as $E^O_{redox} > E^O_{dec,n}$.

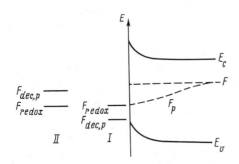

Fig. 7.6. The action of a redox couple introduced into the solution to protect an n-type semiconductor from photocorrosion: I) the semiconductor is stable ($F_{redox} > F_{dec,p}$); II) the semiconductor is unstable with respect to anodic photodecomposition ($F_{redox} < F_{dec,p}$).

These results can also be obtained without recourse to the concept of the quasi-Fermi level (236). Indeed, when $E^O_{redox} > E^O_{dec,p}$, the oxidized form (Ox) of the Ox/Red system is capable of oxidizing the semiconductor material. Therefore the transition of holes generated by the light from the valence band on to the form (Red) in the solution creates the conditions for the destruction of the semiconductor material. On the other hand, when $E^O_{redox} < E^O_{dec,p}$, the oxidized form of the redox couple (Ox) is unable to oxidize the semiconductor. If, then, the reduced form (Red) captures the holes with an efficiency close to 100 per cent, the photocorrosion of the semiconductor is practically averted. Of course, to ensure effective corrosion protection for the semiconductor the concentration of the protecting species near the surface and the rate constant for the corresponding reaction must be sufficiently large. In some cases, the role of the protecting species may be taken by the solvent. Examples of suitable redox species will be given in the next section.

Finally, let us examine the case when, in the dark, equilibrium is established between the semiconductor and the redox couple in solution. In this case the semiconductor assumes a potential $E = E^O_{redox}$. An analysis similar to that made above for the case of an ideally polarizable electrode shows that illumination of the semiconductor results in a shift in both the Fermi level F and the quasi-Fermi level for the holes F_p. Consequently both the forward and the reverse reactions, Eq. (4.1), become faster. In other words, an increase in the exchange current is the first outcome of illumination. However, if the illumination is sufficiently intense, then the values of F_n and F_p are likely to reach the levels for the electrochemical potentials of other reactions (for example, the decomposition of the semiconductor and/or the solvent). When this happens several reactions may occur simultaneously at the

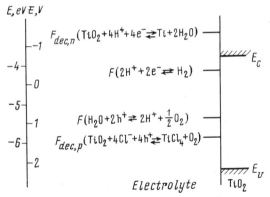

E, eV E, V

$F_{dec,n}$ $(TiO_2 + 4H^+ + 4e^- \rightleftarrows Ti + 2H_2O)$ ———

$F (2H^+ + 2e^- \rightleftarrows H_2)$ ———

$F (H_2O + 2h^+ \rightleftarrows 2H^+ + \frac{1}{2}O_2)$ ———

$F_{dec,p}$ $(TiO_2 + 4Cl^- + 4h^+ \rightleftarrows TiCl_4 + O_2)$ ———

Electrolyte TiO_2

Fig. 7.7. Diagram of the
electrochemical potentials
for reactions on a TiO_2
electrode (pH 7) (235).
In figures 7.7.-7.10. the
energies are read from the
E_{vac} level; all potentials
are against the Normal
Hydrogen Electrode.

semiconductor/electrolyte interface, the total cathodic
current being naturally equal to the total anodic current.

In conclusion let us emphasize that the quasi-thermodynamic approach is equally applicable to all possible
mechanisms of semiconductor decomposition in solution:
spontaneous dissolution (corrosion) in the dark, photocorrosion, and corrosion under the influence of an externally applied polarization.

Ξ 7.3. T h e   P h o t o c o r r o s i o n   B e h a v i o r
of   the   Major   Semiconductor
Materials

To illustrate the propositions of the previous
section, let us examine the energy diagram (235) for a
number of semiconductors in aqueous solutions of electrolytes.

Titanium dioxide, TiO_2, (Fig. 7.7) corresponds to the
case considered above (see Fig. 7.3c). It is resistant
to cathodic photocorrosion, inasmuch as the level $F_{dec,n}$
lies in the conduction band. However, the level $F_{dec,p}$
lies in the bandgap, and TiO_2 would decompose in the
light with the participation of holes if the photoevolution of oxygen from water, whose potential is less
positive, did not compete. Therefore the TiO_2 electrode
is a quite reliable photoanode precisely for the photoelectrolysis of water; here water plays the role of the
species protecting against photocorrosion (241). (Weak
photocorrosion of TiO_2 is sometimes observed. It is
possible to suppress it completely by the addition of
Co^{2+} which is oxidized on the illuminated electrode (242).)

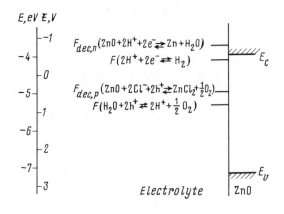

Fig. 7.8. Diagram of the electrochemical potentials for reactions on a ZnO electrode (pH 7) (235).

Zinc oxide, ZnO, (Fig. 7.8) is another example of this case (see Fig. 7.3c). It is relatively resistant to cathodic photocorrosion; in the dark the decomposition of ZnO by reaction (7.6a) occurs only under very strong cathodic polarization when the electrons are degenerate at the surface, and the Fermi level enters the conduction band (107). At the same time, the level for the anodic photodecomposition in chloride solutions lies within the bandgap and is less positive than the level for the decomposition of water with the evolution of oxygen. Therefore under strong illumination in electrolytes containing chloride the photodecomposition of ZnO (reaction (7.8b)) will occur. The situation is different in chloride free alkaline solutions: zinc oxide is resistant to anodic photocorrosion, and the photoevolution of oxygen is the only photoanodic reaction on the electrode (243, 244).

Cadmium sulfide, CdS, in alkaline polysulfide solution (Fig. 7.9) is an interesting case. The photocorrosion can be successfully avoided with the help of the S^{2-}/S_2^{2-} redox couple since the sulfide ions are

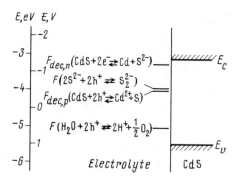

Fig. 7.9. Diagram of the electrochemical potentials for reactions on a CdS electrode ($[S^{2-}]$=1 M) (235).

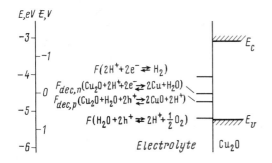

Fig. 7.10. Diagram of the electrochemical potentials for reactions on a Cu_2O electrode (pH 7) (235).

oxidized to polysulfide more readily than the electrode material. This and other similar systems (polyselenides, polytellurides) are widely used in photoelectrochemical cells for solar energy conversion to prevent the photocorrosion of $A^{II}B^{VI}$ and $A^{III}B^{V}$ semiconductor anodes. The potential of the oxygen photoevolution reaction on CdS is much more positive than the photodecomposition potential and cannot be reached; CdS electrodes are unsuitable for water electrolysis without special protection.

Finally, as shown in Fig. 7.10, cuprous oxide, Cu_2O, is highly unstable with respect to both anodic and cathodic decomposition.

That the protecting effect of a redox couple added to the solution is determined by its equilibrium potential can be seen in Fig. 7.11. This figure shows the percentage of the total photocurrent that goes to oxidation of the stabilizing agent (protector) as a function of its redox potential (245-247). (This percentage and also the nature of the products formed in the course of the photoanodic reaction were determined using a rotating ring-disc electrode (191). A semiconductor photoelectrode (CdS or ZnO) was used as the disc electrode and the products of photocorrosion were detected on the basis of the magnitude of the current for their reduction on the

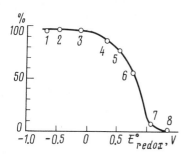

Fig. 7.11. The comparative effectiveness in suppressing the anodic dissolution of CdS of various reducing agents as a function of their redox potentials. All results in $0.2 \underline{M} Na_2SO_4$ with $0.01 \underline{M}$ redox species: 1) SO_3^{2-}; 2) S^{2-}; 3) $S_2O_3^{2-}$; 4) $Fe(CN)_6^{4-}$; 5) I^-; 6) Fe^{2+}; 7) Br^-; 8) Cl^- (247). (Reprinted by permission of the Publisher, The Electrochemical Society, Inc.)

metal ring electrode which surrounds the disc.)   Redox
species with a more positive potential (Cl⁻, Br⁻ and
others) give a relatively weak protective effect; the
share of the overall photocurrent consumed in their
oxidation does not exceed 20-30 per cent, while the rest
of the photocurrent is made up by decomposition of the
semiconductor.   Redox species with less positive potentials, for example polysulfides, are effective protectors
because practically the whole of the photocurrent is the
result of their oxidation.   Similar results were obtained
by the same method for gallium arsenide (248).

Ξ 7.4.   S o m e   M o l e c u l a r   a n d   S t r u c t u r a l
         A s p e c t s   o f   P h o t o c o r r o s i o n

     The quasi-thermodynamic approach cannot, naturally,
provide an exhaustive description of the corrosion processes (in particular photocorrosion).   In a number of
cases essential corrections have to be introduced to
account for kinetic peculiarities of the electrode reactions determined by the crystal structure, the state
of the semiconductor surface, and so on (see for example
(249)).   Thus, reactions of the type given in Eq. (7.5) involving several atoms in the crystal lattice and in the
solution occur, in all probability, in several stages,
with each of the stages being characterized by an activation barrier of its own.   The kinetics of the overall process then depend, in a complicated fashion, on
the kinetic characteristics of the separate stages.

     A detailed description of such multi-stage reactions
runs into great difficulties.   Therefore, at this stage
the kinetic approach is used to reveal the qualitative
behavior of the corrosion processes.   As an example let
us take the microscopic description of the corrosive
destruction of a crystal lattice of a binary compound
MX where, as before, M and X denote the more electropositive and electronegative components respectively.
Fig. 7.12 (237) shows schematically the interface between
an MX crystal and a solution which contains a complexing
agent with ligands L.   The anodic decomposition reaction
may be written as follows (cf. Eq. (7.6b)):

$$2MX + 4h^+ + 2nL \longrightarrow 2[ML_n]^{2+} + X_2 \qquad (7.11)$$

The difference in electronegativities of the components
M and X of the corroding crystal is determined by the
fact (see Fig. 7.12) that the surface X atoms have
superfluous electron pairs on anti-bonding orbitals,
while the M atoms have vacant orbitals with which the

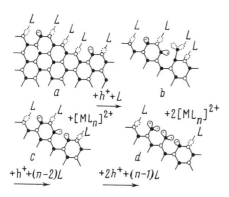

Fig. 7.12. Scheme for the destruction of the crystal lattice in the course of dissolution of a binary semiconductor MX (237) : o — atom M (more electropositive); ● — atom X (more electronegative); L — nucleophilic ligand. Unsaturated electron pairs in orbitals on the X atoms are denoted by dots.

electron-donor ligands, L, react. It is assumed that the localization of a hole on an X atom diminishes the electron density round it (since the states in the upper valence band of crystals such as CdS, ZnO, etc., are usually connected with the electronic orbitals of the electronegative atoms); this weakens the M-X bond and facilitates the interaction of M and electron-donors from the solution (this is the first stage of the reaction). Therefore the reaction begins where the surface atoms have free bonds, in other words, on the "weak" points of the crystal lattice.

The next stage consists of the capture of a second hole at the same place where, after the disruption of the M-X bond, the M atom is held to the crystal lattice by only one bond. Interaction with a second ligand, L, results in a complete loss of the M atom from the lattice and its transfer into the solution as an M^{2+} ion (or complex).

Finally, two more holes are captured at this weakened spot in the lattice by the neighboring M atom, which passes to the solution as an M^{2+} ion. Two X atoms remain on the surface as radicals, and they are ready to recombine with the formation of a molecule X_2 which then also leaves the crystal. This is, for example, how the corrosion of CdS or ZnO probably occurs. In some other cases (GaAs, GaP) the electronegative atoms apparently do not recombine, and the oxidation process continues until the two components of the crystal, both M and X, have been oxidized and passed into the solution in ionic form.

Upon examination the model warrants two principal conclusions. Firstly, the reaction scheme usually includes several consecutive stages, and it may happen

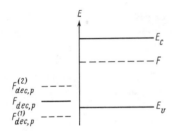

Fig. 7.13. Scheme showing the possible increase in the stability of a semiconductor with respect to photocorrosion when the decomposition reaction occurs in two stages.

that the redox potential of the rate-determining stage differs from the redox potential of the overall reaction (for which, in particular, the diagrams in Figs. 7.7-7.10 were calculated). In this case a similar analysis to that mentioned above should be made precisely for the rate-determining step (237). Fig. 7.13 shows, as an example, the schematic electrochemical potentials which characterize the separate stages of a two-electron reaction of the type in Eq. (7.4b). The level $F^{(1)}_{dec,p}$ for the first stage is chosen to be lower than that of the overall reaction $F_{dec,p}$. In this case, as shown in Fig. 7.13, the semiconductor proves to be more stable than would be expected, on the basis of the electrochemical potential for the overall decomposition reaction.

Secondly, the kinetics of the decomposition of a semiconductor crystal are highly sensitive to the particular electronic structure. This problem has been investigated (250, 251) within the framework of quantum chemistry. For example, the rate of electron transfer at the interface depends on the extent of overlap between the electron orbitals of the atoms. Weak overlap means a smaller probability of electron transfer whenever it is energetically feasible (i.e., a small pre-exponential factor, $k(E)$, in the expression of the type in Eq. (4.29)).

On the basis of these ideas, one would expect a substantial difference in the photocorrosion behavior of "conventional" (or sp-) semiconductors as compared with d-semiconductors. Indeed, it is known that in materials such as CdS, CdSe, ZnO, etc., the interatomic bonds are mainly ionic in character. The conduction band and the valence band are formed, in the main, from orbitals of the electropositive and electronegative components respectively. On illumination holes are generated as a result of interband transitions between the valence band (p-band) and the conduction band (s-band). The localization of holes on interatomic bonds, which is tantamount to the weakening of those bonds, can be regarded (see above) as the *sine qua non* for both the

anodic dissolution of the semiconductor and its photocorrosion (7).

In some semiconductor materials, however, another
type of electronic excitation is possible, namely a
transition between two d-bands. Tributsch (250, 251)
examined a class of dichalcogenides of the transition
metals MoS_2, $MoSe_2$, WS_2, and others. When light is
absorbed, electronic transitions occur for these materials between d-bands corresponding to the orbitals of
the metal (in particular, Mo). Since d-electrons are
much more localized on the corresponding atoms than
either s- or p-electrons, their excitation has only a
very small influence on the atomic interaction. It was
therefore expected that such transitions would not
weaken the bonds in the crystal lattice to any considerable extent, and that, consequently, these materials
would be more stable with respect to photocorrosion. At
the same time, the holes generated by the light prove to
be capable of oxidizing dissolved species. This situation can be basically used, for example, to create
stable photoanodes for photoelectrochemical cells.

However, experiments showed that MoS_2 and similar
electrode materials (probably with the exception of PtS_2
and RuS_2 (252)) undergo tangible anodic photocorrosion.
To explain this fact, the assumption was made (250, 251)
that the photoholes in MoS_2 or $MoSe_2$ oxidize water to
$OH^•$ radicals which further interact chemically with the
S and Se atoms of the semiconductor surface oxidizing
them to sulfates and selenates. For this reason, in
order to stabilize d-semiconductors in, for example,
photoelectrochemical cells for light energy conversion,
it is still necessary, as it was for the sp-semiconduc-
tors, to introduce redox couples into the solution.
This has diminished the initial optimism, although electrodes made from the dichalcogenides of the transition
metals need to be investigated further before it is
possible to draw any final conclusions concerning their
photocorrosion behavior.

The phenomenon of silicon photopassivation and
photoactivation (253, 254) is an interesting example of
kinetic effects in the corrosion of semiconductors.
Silicon is a highly electronegative element, and it
should spontaneously and rapidly dissolve in water with
the evolution of hydrogen. However, the surface of
silicon is, in the majority of aqueous electrolyte solutions, covered with a passivating oxide film which
protects it from corrosion.

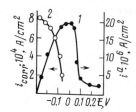

Fig. 7.14. The anodic polarization curve (1) and the dependence of the etching rate i_{corr} (2) on the potential for silicon in 10 M KOH (30°C). The potentials in Figs. 7.14 and 7.15 are against the Hydrogen Electrode in the same solution (253).

According to (253, 254), on anodic polarization the rate of self-dissolution of silicon does not increase as might be expected, but, on the contrary, rapidly decreases: the silicon is passivated. The anodic polarization curve (Fig. 7.14) has a shape typical of passivating metals (see, for example, (226)): with increasing anodic potential the current initially increases then passes through a maximum and drops sharply to a value which is then practically independent of potential. To carry out the anodic passivation of silicon it is sufficient to apply current which is 1.5-2 orders less than the rate (expressed as a current) of silicon self-dissolution at the stationary potential.

The ability of the silicon electrode to go over reversibly from the passive to the active state under the influence of light, and, *vice versa*, from the active to the passive state, is a non-trivial feature of its behavior in aqueous solutions of alkalis. For example, if actively dissolving silicon (the rising branch of the curve in Fig. 7.14) is the initial state, then upon illumination its potential spontaneously and sharply shifts (in the galvanostatic regimen) towards more positive values, that is, towards the "passive" region (Fig. 7.15a). The dissolution then discontinues completely, and photopassivation takes place. On the other hand, if the silicon has already been anodically passivated (Fig. 7.15b), then upon illumination its potential shifts towards less positive values. The point on the curve in Fig. 7.14 which characterizes the state of the system moves from the descending to the ascending branch, and active self-dissolution begins, i.e., photoactivation occurs. As shown in Fig. 7.15, the behavior of n- and p-type samples is basically the same.

Although the detailed picture of the processes occurring under photopassivation and photoactivation is, as yet, not quite clear, they can apparently be related (254) to the acceleration upon illumination of one of the two conjugated reactions which make up the electrochemical component of silicon corrosion. Either the

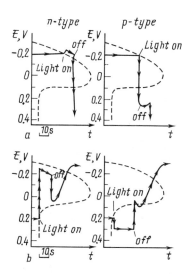

Fig. 7.15. The change of
potential with time during
photopassivation (a) and
photoactivation (b) of
silicon in 10 M KOH (20°C).
The anodic polarization
curve (cf. Fig. 7.14) is
shown by the dotted line
(253).

anodic (on the active surface) or the cathodic (on the
passive surface) partial reaction accelerates depending
upon the initial state of the silicon (active or passive).
This causes a potential shift of a corresponding sign,
and the system "jumps over" the hump in the polarization
curve from one stable state to the other.

It can be assumed that this phenomenon will also
occur in the case of other "non-noble" materials which are
covered with a continuous passivating film in the course
of corrosion.

Chapter 8

THE CONVERSION OF LIGHT ENERGY INTO ELECTRICAL POWER

The task of utilizing solar energy is becoming more and more urgent as a result of the gradual exhaustion of fossil fuel resources and the increasing problem of environmental pollution. The technological waste from existing energy installations and the "thermal" pollution they produce presents a threat to the ecology. The sun, in contrast, is a practically inexhaustible source of ecologically pure energy which brings about no change in the thermal balance of the Earth. The overall quantity of solar energy striking the Earth is immense. Outside the atmosphere, near the equator, the incident power per m^2 of the Earth surface is about 1,350 W.

Whatever the method of solar energy conversion chosen, it is necessary to overcome the same inherent problems: firstly, solar energy is diffuse, the light flux has a relatively low power density; secondly, there are considerable fluctuations in the power density over the daily and annual cycles. This introduces the problem of solar energy accumulation and storage, in particular in the form of the energy rich products of photoelectrochemical reactions.

§ 8.1. Photogalvanic and Photovoltaic Cells

In recent years solar energy conversion by means of the photoelectrochemical reactions of cells which may include semiconductor electrodes has become the most prominent practical application of photoelectrochemistry.

Photoelectrochemical devices for the conversion of solar
energy are subdivided into two groups depending on where
the light absorption occurs, and, consequently, on the
nature of primary photoprocess. In one case the light
is absorbed in the solution (these are the so-called
photogalvanic cells), while in the other case the light
is absorbed in the electrode (as with the solid-state
photocells such cells are called photovoltaic).

In photogalvanic cells the absorption of light in
the solution triggers a chain of chemical reactions
which culminates in the formation of products possessing
a large amount of chemical energy. These products reach
the surfaces of the electrodes and there enter into electrochemical reactions with the result that a current
flows in the external circuit. A cell containing a solution of the dye thionine and ferrous ions is a classical example. When thionine is excited by light it
undergoes electron transfer quenching by the ferrous ion
to produce ferric ion and reduced thionine (this reaction does not occur in the dark). The reduced form
of the thionine is then oxidized at one of the electrodes in the cell while the ferric ions are reduced on
the other. The products of these electrode reactions -
thionine and ferrous ion - return to the solution. As a
result, the chemical composition of the solution as a
whole is kept constant and the conversion process results
solely in the generation of an electric current in the
external circuit. The system is sensitive to visible
light. It is stable and relatively cheap. The elim-
ination of back reactions (for example, the reaction of
the free radicals, produced as the intermediate product
of the electron transfer quenching, and ferric ion) is
the major problem which remains to be satisfactorily
solved. These reactions lead to premature loss of the
energetic products. As a result, the quantum efficiency is low (about 0.1 per cent). Analysis shows (see,
for example, (255)) that the maximum efficiency for a
photogalvanic cell cannot exceed several per cent. Thus,
in spite of their relative cheapness and convenience
their practical application is doubtful.

It is clear from this discussion that electrochemical, as well as photochemical, energy conversion occurs
in at least two stages; namely, an initial photoexcitation followed by the transfer of the energy stored
in the primary products of excitation to other species.
In the long run, these are then transformed into more
stable final products rich in energy.

A photovoltaic cell consisting of a semiconductor in contact with a transparent electrolyte is a very convenient system for light energy conversion. The photosensitivity of such a system is mainly determined by the interband transitions of the semiconductor. Electrons and holes are therefore the primary excited products. Their energy is used to carry out electrochemical reactions which, for some reason, do not occur in the dark. In order to prevent the deactivation of the excited particles (recombination of the electron/hole pairs) and then to achieve efficient transmission of the stored energy to the external circuit, it is most convenient to use the electric field of the double layer at the semiconductor/electrolyte interface to spatially separate the electrons and holes.

Photoelectrochemical cells with semiconductor electrodes possess a significant advantage when compared with solid state solar cells; this has already been demonstrated by the initial research in this area. For the reasons examined below, a much lower degree of perfection of the semiconductor crystal structure is required. Therefore cheap and readily available materials can be used as electrodes. This is a decisively significant factor for large scale solar energy conversion, and it is the main reason for the broad development of work aimed at the creation of photoelectrochemical cells with semiconductor electrodes. The mode of operation of such cells was first examined by Gerischer (47, 256). A vast body of information concerning particular systems is reviewed in (223, 257-280).

Electrochemical reactions occur at the anode and cathode of the photocell, one of these reactions (or both) being stimulated by the absorption of light in the semiconductor. If the same electrode reaction takes place at both electrodes (in the forward and reverse directions), then the overall composition of the electrolyte in the cell does not change and the net result of the photoprocess is the flow of current in the external circuit. Such a photocell, which is a complete analogue of the solid-state solar cell, is called a "liquid-junction solar cell", or regenerative type photoelectrochemical cell. If two different reactions occur on the electrodes of the photocell, the composition of the solution changes and the overall photoprocess is the storage of energy in the form of the chemical energy of the electrolysis products. The operation of photoelements for photoelectrolysis is based on precisely this principle. These two kinds of light energy conversion,

into electrical power and into chemical energy, are discussed in this and the following chapter. A comparative
analysis and evaluation of the prospects for the different
methods of energy conversion can be found in Section 9.5.

8.2. The Efficiency of Solar Energy Conversion in Photoelectro-chemical Cells

Before getting down to the detailed analysis of
semiconductor photoelectrochemical cells, let us examine
the general principles which determine the maximum possible efficiency for light energy conversion.

The spectrum of radiation reaching the Earth beyond
the boundaries of the atmosphere is close to that of an
ideal black body with a temperature of about 6,000 K.
As it passes through the atmosphere the overall quantity of incident energy decreases and its spectral distribution changes, the spectral maximum shifting toward
longer wavelengths (Fig. 8.1). The magnitude of this
shift depends upon the altitude of the sun over the
horizon due to the change in "air mass" (AM) penetrated
by the sun rays.

From general thermodynamic principles, the maximum
possible efficiency of energy conversion is equal to

$$\eta_{therm}^{max} = \frac{T_S - T_E}{T_S} \tag{8.1}$$

where T_S is the temperature of the energy source (the
sun), T_E is the temperature of the energy receiver (the
surface of the Earth). For $T_S = 6,000$ K and $T_E = 300$ K,
we obtain $\eta_{therm}^{max} \simeq 0.95$. However, a more accurate
account of the "energy" and "entropy" components of the
solar radiation leads to the conclusion that the amount

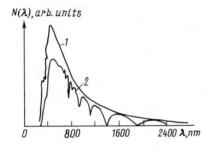

Fig. 8.1. The solar
spectrum above the
Earth's atmosphere
(1) and at sea level
(2).

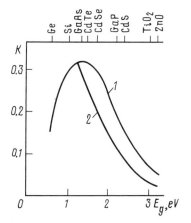

Fig. 8.2. Theoretical
dependence of the solar
energy conversion effi-
ciency on the semiconduc-
tor bandgap: 1) $K=K_{thresh}$;
2) $K=K_{thresh} \cdot K^{chem}_{stor}$ for
water photoelectrolysis
(47).

of diffuse light energy that can be converted into useful work is substantially lower. On this basis a thermodynamic efficiency, η_{therm}, of approximately 0.7 is
obtained (see, for example, (281)). Since the conversion
of light energy is actually a threshold process which
only occurs at quantum energies $\hbar\omega$ greater than a certain
minimum value $\hbar\omega_{min}$, the efficiency that can be obtained
is heavily restricted. For semiconductor systems the
minimum threshold energy, $\hbar\omega_{min}$, is equal to the width
of the bandgap E_g; in the case of direct electron transitions it is somewhat higher (see Section 1.7). Light
with lower quantum energies does not result in interband
transitions. On the other hand, when $\hbar\omega > E_g$, the quantum energy is not fully utilized: the surplus energy,
$\hbar\omega - E_g$, is wasted in the excitation of vibrational
degrees of freedom in the solid, in other words in
heating the semiconductor. The threshold nature of the
energy conversion process can be taken into account (see,
for example, (282)) by the introduction of a dimensionless factor

$$K_{thresh} = \frac{E_g \int\limits_{E_g}^{\infty} N(E)(1-R)\,dE}{\int\limits_{0}^{\infty} EN(E)\,dE} \qquad (8.2)$$

where $N(E)$ is the number of light quanta with energy
$E = \hbar\omega$ striking the surface of the photodetector in unit
time, and R is the light reflection coefficient for the
surface, which is itself dependent on E. Figure 8.2
(curve 1) shows the dependence of K_{thresh} on E_g for the

solar spectrum (where we have assumed that R << 1). The
upper scale gives the values of E_g for a number of semiconductors (cf. Appendix 1). It is clear that the
optimum bandgap, from this particular viewpoint, lies
within the interval 1.1 - 1.4 eV. In particular, silicon,
gallium arsenide, indium phosphide and cadmium telluride
have bandgaps which are close to this optimum; of these
the first two are also widely used in the manufacture of
solid-state solar cells.

However, Eq. (8.2) only gives an upper limit for the
conversion efficiency. The real efficiency (9, 282) is
the product of K_{thresh} and several other factors, each
of which takes into consideration a certain form of
energy loss in the overall conversion process. These
additional factors are listed below:

(i) The efficiency of energy storage in the elementary act of conversion is determined by the change in
electrochemical potential of the minority carriers in
the semiconductor for the given illumination. This quantity makes itself manifest in solid-state solar cells,
as well as in liquid-junction solar cells, in the form
of the "open-circuit photovoltage (photopotential),"
E_{ph}^{oc}. In cells where light energy is converted into
chemical energy, the storage efficiency is determined
by the change in free energy ΔG in the course of the
overall chemical reaction occurring in the illuminated
cell. The corresponding factor is equal to

$$K_{stor}^{electr} = \frac{eE_{ph}^{oc}}{E_g} \quad , \quad \text{or} \quad K_{stor}^{chem} = \frac{\Delta G}{E_g} \qquad (8.3)$$

(ii) The efficiency of the mutual separation of
non-equilibrium carriers, produced by the light, makes
itself manifest in the value of the short-circuit photocurrent for the cell $i_{sh.c}$. The factor characterizing
the efficiency for the conversion of the incoming photon
flux into electrical current, the quantum yield (per absorbed quanta) is

$$Y = \frac{i_{sh.c}/e}{\displaystyle\int_{E_g}^{\infty} N(E)(1-R)\,dE} \qquad (8.4)$$

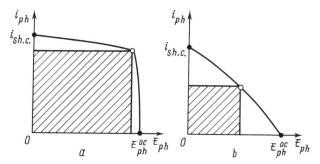

Fig. 8.3. Working characteristics of
a photocell for the conversion of
light energy into electricity: a) high
fill factor, b) low fill factor. o –
working point of the photocell corresponding to the maximum output.

The value of Y depends on the various "quenching" processes for the excited states in the system.

(iii) Finally, the effect of ohmic losses in the
cell (which are made up of voltage drops in the semiconductor and solution bulk) can be taken into account
with the help of the fill factor

$$f = \frac{(i_{ph} \cdot E_{ph})_{max}}{i_{sh.c} \cdot E_{ph}^{oc}} \qquad (8.5)$$

The numerator is the maximum value of the product of the
photocurrent and photovoltage which can be obtained from
the working characteristic (the photocurrent/photovoltage
curve) for the cell (Fig. 8.3). The fill factor, f, is at
its largest and approaches unity when the working characteristic is rectangular (Fig. 8.3a). When there is a
considerable ohmic resistance in the cell, the characteristic is approximately linear (Fig. 8.3b), and the output
power comprises only a fraction of the maximum possible
($i_{sh.c} \cdot E_{ph}^{OC}$). This fraction is given by f.

Thus, the real efficiency is given by the product

$$\eta = \eta_{therm} \cdot K_{thresh} \cdot K_{stor} \cdot Y \cdot f \qquad (8.6)$$

where each of the cofactors, as can be seen from Eqs.
(8.2)-(8.5), is less than unity.

≡ 8.3. L i q u i d - J u n c t i o n   S o l a r   C e l l s :
        P r i n c i p l e s   o f   O p e r a t i o n   a n d
        E n e r g e t i c s   o f   C o n v e r s i o n

Non-equilibrium electrons in the conduction band
and holes in the valence band produced by steady-state
illumination are effectively spatially separated by the
electric field in the region near the surface of the
semiconductor. The main difference between photoelectrochemical cells and solid-state solar cells is that the
photogenerated carriers at the interphase boundary, in
the course of electrochemical reactions, pass the stored
energy from light to chemical species in solution.

  Regenerative photocells (liquid-junction solar cells)
function as follows. Let us assume that the sign of
semiconductor surface potential facilitates the formation
of a depletion layer. In this layer the minority carriers generated by the light are transferred by the
electric field to the semiconductor/solution interface
where they enter into the electrode reaction. The
majority carriers are transferred to the semiconductor
bulk and thus, via the external circuit, to the auxiliary
electrode (counter electrode) of the photoelectrochemical
cell where the same reaction is carried out in the
opposite direction.

The simplest regenerative photocell consists of two
electrodes (one of them a semiconductor, the other a
metal) in an electrolyte solution containing a redox
couple Ox/Red. Both electrodes should be inert, in
other words, they should serve only to transfer charge
between the species in solution and the external electrical circuit; their material takes no direct part in
the electrode reactions.

The open-circuit photovoltage $\mathcal{E}_{ph}^{oc}$, short-circuit
photocurrent $i_{sh.c}$, fill factor f, and efficiency η are
the main characteristics of the photocell.

Let us examine the operation of such a photocell,
taking as an example the following system: n-type CdS
photoanode/alkaline solution of the couple S^{2-}/S_2^{2-}/metal
cathode (Fig. 8.4). Equilibrium is established throughout the cell in the dark:

$$S_2^{2-} + 2e^- \overset{\rightarrow}{\underset{\leftarrow}{\rightleftharpoons}} 2S^{2-} \tag{8.7}$$

The two electrodes assume the equilibrium potential of
this redox couple, and the Fermi levels of the metal and

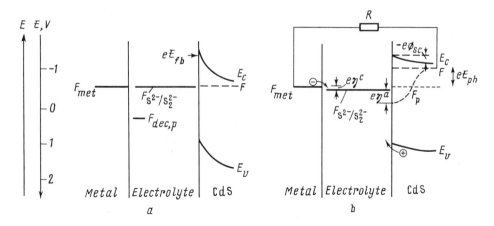

Fig. 8.4. The energy diagram for a regenerative type
photocell with a CdS photoelectrode in a solution containing the S^{2-}/S_2^{2-} couple: a) in the dark; b) in the
light. Potentials are cited against the Normal Hydrogen Electrode.

CdS electrodes and the solution level $F_{redox} = F_{S^{2-}/S_2^{2-}}$
all become equal (see Fig. 8.4a). To achieve a good separation of the electrons and holes generated by the
light it is necessary, as has already been mentioned,
for there to be a depletion layer in the semiconductor.
This condition can be achieved if the equilibrium potential of the redox couple is more positive than the
flat band potential of the semiconductor ($E_{redox}^{o} > E_{fb}$).
When the electrode is illuminated, the bands unbend (see
Fig. 8.4b) and the Fermi level of the semiconductor, F,
shifts, which leads to a change in the electrode potential. As can be seen in Fig. 8.4, this change in the
electrode potential E_{ph} (the photopotential) is equal to
the shift of the Fermi level for the illuminated electrode, with respect to the non-illuminated case, divided
by the charge on the electron. In the simplest case
(which is often realized experimentally) the potential
drop in the Helmholtz layer, ϕ_H, remains constant when
the electrode is illuminated. Furthermore, it does not
depend on the value of E_{redox}^{o}. In other words, the
band edges are pinned at the surface (see Section 3.4).
Under these conditions the maximum value of E_{ph} (which
is the open-circuit photopotential, E_{ph}^{OC}) is equal to
the initial potential drop in the space charge layer in
the dark: the difference $|E_{redox} - E_{fb}|$. Thus it is
self-evident that the combination of semiconductor and
redox couple should be chosen so that this difference
is as great as possible.

In some cases, however, the potential drop in the Helmholtz layer does not remain constant either under illumination or when E^O_{redox} changes. As a result, the photopotential is less than $|E^O_{redox} - E_{fb}|$. For example, for silicon and gallium arsenide electrodes the open-circuit photopotential was found to be practically constant and in the range 0.4 - 0.5 V for a series of redox couples despite the fact that their equilibrium potentials, E^O_{redox}, varied by more than 1.5 V (283-285). Similar behavior has been observed for WSe$_2$ and MoSe$_2$ electrodes. In these cases the change in ϕ_H for different redox couples considerably exceeds the change in ϕ_{sc}. This type of redistribution of potential at the interface seriously worsens the current-voltage characteristics of the photocells. The reasons for it were discussed in Section 3.8; it arises either as a result of a high density of surface states or as a result of chemical interaction between the semiconductor and the solution species. For example, the oxidation of the silicon surface occurs in solutions of redox couples with high E^O_{redox}; this results in a shift in the flat band potential E_{fb}. That is why the value of E_{fb} is no longer independent of E^O_{redox}, as was assumed above (286).

Let us now consider the effects of the formation of quasi-Fermi levels. As a result of the photogeneration of electron/hole pairs, the quasi-Fermi levels for minority and majority carriers F_p and F_n shift, which is shown in Fig. 8.4b; for the majority carriers (electrons) we can assume that $F_n \approx F$. Under illumination both the anodic and cathodic partial reactions at the CdS electrode are accelerated in the sulfide/polysulfide solution since $F_p < F_{S^{2-}/S_2^{2-}}$ and $F_n > F_{S^{2-}/S_2^{2-}}$. This causes "electrochemical recombination" of the nonequilibrium carriers produced under illumination and, as a result, the energy absorbed in the photocell under open-circuit conditions is converted into heat. When the circuit containing the external load R is completed the anodic and cathodic reactions are spatially separated: the holes are transferred from the semiconductor photoanode to the solution, so that S^{2-} ions are oxidized to S_2^{2-}, while the electrons travel, via the external circuit, to the metal auxiliary electrode (cathode) where they are consumed in the reduction of S_2^{2-} to S^{2-}. The potential drop across the external load is equal to $i_{ph}R$, where i_{ph} is the photocurrent and depends on both the light intensity and the load resistance R. The value of R and, hence, the working point on the $i_{ph} - E_{ph}$ characteristic (Fig. 8.3) are selected in such a way that the electrical power output, $i^2_{ph}R$, of the cell is a maximum.

As can be seen from Fig. 8.4b, we can write for a regenerative photocell

$$F_n - F_p = e(i_{ph}R + \eta^a + \eta^c) \qquad (8.8)$$

where η^a and η^c are the overvoltages for the reactions on the photoelectrode and auxiliary electrode. For the semiconductor electrode this refers to that part of the overvoltage which is localized in the Helmholtz layer; $\eta = \eta_H$. The value of R, apart from the useful load, includes the internal resistance of the cell (the ohmic resistance of the electrodes and solution); this can usually be made insignificant. Photocells with p-type semiconductor photocathodes function in a similar way. In this case the condition $E^o_{redox} < E_{fb}$ should be observed.

Regenerative photocells may contain two, rather than one, photosensitive electrodes (one of them being n-type and the other p-type). The total photopotential is then made up of photopotentials produced at the two electrodes. The maximum value of the photopotential E^{oc}_{ph} is then equal to the difference in the flat band potentials of the two semiconductor electrodes; see Fig. 8.5.

Preventing the photodissolution of the semiconductor electrode is a very important problem for the practical realization of efficient photocells. In order to overcome

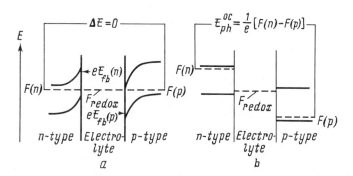

Fig. 8.5. The energy diagram for a regenerative type photocell with two photoelectrodes: a) in the dark; b) in the light. $E_{fb}(n)$ and $E_{fb}(p)$ are the flat band potentials of the n-type and p-type semiconductor electrodes respectively.

this problem, the equilibrium potential of the redox couple is selected to be less positive than the potential for the photodecomposition of the n-type semiconductor with the participation of holes: $E^O_{redox} < E^O_{dec,p}$, where $E^O_{redox} = E_{S^{2-}/S_2^{2-}}$ (cf. Figs. 8.4 and 7.9); for a p-type semiconductor photoelectrode the condition is $E^O_{redox} > E^O_{dec,n}$. It is significant that these conditions also restrict the value of the initial band bending and, consequently, the conversion efficiency. (To maximize the efficiency, the value of E^O_{redox} for an n-type electrode should be made as positive as possible, while for a p-type electrode it should be made as negative as possible.) Thus, the stability of regenerative photocells is, to a certain extent, achieved, at the expense of their efficiency (236, 262).

On the basis of our analysis above we can formulate the requirements for the semiconductor material and the redox couple to be used in a photocell in order to achieve effective solar energy conversion.

Requirements for the photoelectrode material:

(i) the bandgap should be at the optimum value (see Fig. 8.2);

(ii) there should be a suitable doping impurity concentration (N_D or N_A) so that $L_{sc} > \alpha^{-1}$ (where α is the light absorption coefficient); this is necessary for a good separation of photocarriers and high quantum yield;

(iii) there should be effective light absorption within the semiconductor (a large value of α); this is usually found in materials with direct interband transitions.

Requirements for the redox couple:

(i) it should fulfil the condition $E^O_{redox} > E_{fb}$ for an n-type or $E^O_{redox} < E_{fb}$ for a p-type electrode and have a large value of $|E^O_{redox} - E_{fb}|$;

(ii) it should fulfil the condition $E^O_{redox} < E^O_{dec,p}$ for an n-type or $E^O_{redox} > E^O_{dec,n}$ for a p-type electrode;

(iii) the reactions on both electrodes of the cell should be perfectly reversible (the overvoltages η^a and η^c should be small);

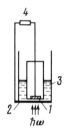

Fig. 8.6. Scheme for a thin-layer regenerative type photocell: 1) photo-electrode; 2) transparent auxiliary electrode; 3) electrolyte solution; 4) external load.

(iv) the solution should be of adequate transparency:

(v) there should be a low ohmic resistance.

Furthermore, to ensure the effective operation of the cell, there should be good mass transfer in the solution between the regions near the surfaces of the anode and cathode. In particular this is achieved by using a thin-layer structure for the photocell (Fig. 8.6). An optically transparent electroconductive film (e.g., SnO_2) serves as the auxiliary electrode. This is deposited on the glass window in the photocell wall through which the photoelectrode is illuminated. Due to the very thin solution layer used, diffusion and natural convection ensure adequate transport of the oxidized and reduced components of the redox couple between the elec-trodes. Thus concentration polarization in the electro-lyte is kept to a minimum when the cell is functioning.

The type of photocell described can also be basi-cally used for energy storage. To achieve this mass transport between the electrodes is now, however, hin-dered by the use of a separating diaphragm. The products of the reactions then accumulate near the corresponding electrodes, and a "dark" electromotive force emerges in the cell; its sign being the opposite to that of the photoelectromotive force of the cell under operation. This electromotive force can be calculated with the help of Eq. (2.16). In the dark such a cell discharges through the external load, releasing the energy accu-mulated in the light.

In conclusion, let us stress that in photoelectro-chemical cells the "active region" in which the light absorption and separation of photogenerated carriers occurs (that is, the space charge layer with a thickness of the order of 10^{-6} cm) is located at the illuminated semiconductor surface. In this respect photocells of this type are like solid-state cells based on semicon-ductor/metal contacts (the so-called "Schottky diodes")

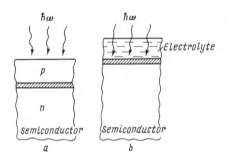

Fig. 8.7. Scheme for a solid-state solar cell (a) and a photoelectrochemical cell (b). The space charge region is shown hatched.

but differ from solar cells with p-n junctions (Fig. 8.7). For the p-n junction the "active region" is located at the depth of the order of 10^{-4}-10^{-5} cm below the semiconductor surface. Consequently, the degree of perfection of the crystal lattice demanded by semiconductor photoelectrodes is less than that demanded of the material for solid-state solar cells made from the same semiconductor. Indeed, while for the electrodes of photoelectrochemical cells the size of crystallites need not exceed 10^{-5}-10^{-6} cm (the thickness of the space charge region), in the case of solid-state cells their size should be more than 10^{-4} cm; otherwise recombination of photocarriers at grain boundaries will considerably decrease the photocurrent and conversion efficiency of the solid state cell. This is the major advantage of electrochemical photocells, and has resulted in the interest in their further study.

8.4. P h o t o c e l l s   w i t h   P o l y c h a l c o - g e n i d e   E l e c t r o l y t e s

Binary semiconductor compounds $A^{II}B^{VI}$ and $A^{III}B^{V}$, with close to optimal photoelectric characteristics and therefore widely used in solid-state solar cells, have also proved to be excellent electrodes in regenerative-type photocells (liquid-junction solar cells). Research work on a broad front (see, for example, (287, 288)) has made it possible to choose the most favorable combination of electrode and electrolyte solution. For photoanodes made from $A^{II}B^{VI}$ semiconductors with the common formula CdX, and also for some $A^{III}B^{V}$ semiconductors, alkaline solutions containing X^{2-}/X_2^{2-} redox couples where X = S, Se or Te have proved to be most suitable. Today the efficiencies achieved in the best photocells of this kind (10-12 per cent) are close to the mean efficiency values for standard solid-state cells. (It is essential that it has proved possible to obtain these high efficiencies for cells with polycrystalline semiconductor electrodes.)

In these photocells the redox couple performs two functions: it determines the initial band bending in the semiconductor (and, at the same time, the photovoltage E_{ph}), and it also protects the photoelectrode material from photocorrosion. This is because for these combinations of semiconductor and electrolyte the conditions $E^O_{redox} > E_{fb}$ and $E^O_{redox} < E^O_{dec,p}$ are observed simultaneously.

The relatively low transparency of the polychalcogenide is a complicating factor in their application: polytelluride is highly colored and absorbs visible light (its absorption maximum is at 512 nm) whilst polysulfide scatters the light considerably. When the electrode is illuminated through the solution this leads to substantial losses in light energy (up to 50 per cent for a solution of thickness 1-2 mm).

The auxiliary electrode should possess good electrocatalytic activity to ensure the reversibility of the "dark" electrode reaction. Active electrodes, either made from Teflon-bonded carbon with cobalt or nickel added as catalysts or made from copper and cobalt sulfides, have been specially developed for the polychalcogenide electrolytes (289, 290). The cathodic overvoltage (η^c) on such electrodes is less than 25 mV for typical solar cell current densities (10 mA/cm^2).

Various methods have been used to produce the semiconductor photoelectrodes. Single crystal electrodes are used as a standard for photoelectrochemical behavior since they, as expected, possess the best characteristics.

To illustrate several CdX - X^2/X_2^{2-} combinations, Table 2 (291-293) contains the values of the quantum yield for the photocurrent Y, the maximum open-circuit photovoltage E^{OC}_{ph}, the conversion efficiency η, and also the wavelength of monochromatic light used to measure these characteristics.

At present, CdSe$_{0.65}$ Te$_{0.35}$ layers obtained by smearing a paste of CdSe + CdTe on to a titanium substrate, followed by drying and sintering, have nearly the highest conversion efficiency among the polycrystalline chalcogenide electrodes (294-296). Tablets made by pressing CdSe powder followed by sintering at

TABLE   2.   Characteristics of Photocells with Chalcogenide
Photoelectrodes

Semiconductor (single crystal)	E_g, eV	Redox couple	λ, nm	Y	E_{ph}^{oc}, V [a]	η, per cent
CdTe	1.5	Te^{2-}/Te_2^{2-}	633	0.6	0.7	11
CdTe	1.5	Se^{2-}/Se_2^{2-}	633	0.6	0.7	8-10
CdSe	1.7	Se^{2-}/Se_2^{2-}	633	0.6	0.6	9.2
CdS	2.4	S^{2-}/S_2^{2-}	500	0.6	0.8	6.8

[a] The value of E_{ph}^{oc} here and below corresponds to the usual level of insulation.

925-1,100°C and then doping in cadmium vapor at 500-700°C (297) are also quite efficient. In cells with alkaline polysulfide electrolyte the solar-to-electrical conversion efficiency is 5.1 per cent (as against 7.5 per cent for single crystal electrodes under the same conditions), while the short-circuit photocurrent is 12 mA/cm^2 and the fill factor is 0.45. High efficiencies (9 per cent for monochromatic light with λ = 577 nm) were obtained with electrodes pressed from CdS and CdSe powder mixtures when the component ratio was 9:1 (298).

Thin-film polycrystalline CdS and CdSe electrodes can be obtained by spraying a solution of $CdCl_2$ and thiourea (or selenourea respectively) on to a heated substrate. For these electrodes the efficiencies reach 7.8 per cent (monochromatic light, λ = 640 nm (299-302)). Polycrystalline film electrodes have also been produced by the electrolytic co-deposition of Cd and Se from acidic $CdSO_4/SeO_2$ solutions on to titanium substrates with subsequent heating. A short-circuit current of 12 mA/cm^2 (303, 304) was obtained in this case in polysulfide electrolyte under solar irradiation.

Other methods for producing cheap polycrystalline electrodes have not as yet yielded satisfactory results. For example, it is possible to obtain anodic sulfide films (CdS and Bi_2S_3) on cadmium (210, 297, 305) and bismuth (305-307) in Na_2S solutions. However, the conversion efficiency for cells with such electrodes is small.

Amongst the $A^{III}B^{V}$ semiconductor compounds gallium arsenide displays the best characteristics in chalcogenide electrolytes (308-311). Cells with single crystal n-type gallium arsenide photoanodes, carbon cathodes and 0.8 M K_2Se + 0.1 M K_2Se_2 + 1 M KOH electrolyte solution have a solar energy conversion efficiency of 12 per cent, a quantum yield of 0.6, and an open-circuit photovoltage of 0.4 V (310).

Polycrystalline GaAs films (with a grain size of around 3 μm) deposited on a graphite substrate by chemical vapor deposition also possess better characteristics among the polycrystalline photoelectrodes. An efficiency for solar energy conversion of 7.8 per cent (312, 313) has been attained in solutions of Se^{2-} and Se_2^{2-}.

The combination of two photosensitive electrodes made from n- and p-type semiconductors, for example n-type CdSe (photoanode) and p-type CdTe (photocathode) with alkaline polysulfide electrolyte (314) has been used in liquid-junction solar cells. This method, however, did not yield high efficiencies. The main stumbling block in the practical realization of photocells with two photoelectrodes is that the combination of the two semiconductor materials and the redox couple must simultaneously fulfil four conditions: two conditions relating to the values of E_{redox}^{o} and E_{fb} and two conditions relating to the values of E_{redox}^{o} and E_{dec}^{o}. The selection of such a system is a difficult task. If, however, it is successfully solved, one may expect solar energy converters of even higher efficiency.

Let us now deal in greater detail with some of the specific points inherent to the operation of liquid-junction solar cells. Many of the cells mentioned above have a relatively long operating lifetime. For example, the most effective photocells with gallium arsenide and cadmium selenide anodes are capable of functioning for two months under solar irradiation without any noticeable decrease in the photocurrent or efficiency. During this time a total charge of $3.5 \times 10^4 C/cm^2$ (287) is passed by the electrode. This is more than sufficient to dissolve the electrode many times over. Thus, the photocorrosion of the semiconductor is almost completely suppressed. This in itself is not enough, however, to enable us to say that liquid-junction solar cells can be used in practice. Although corrosion can be practically excluded, there are other, more refined, effects which degrade the semiconductor electrode, thus restricting

its operating lifetime. Let us examine these effects
and the various methods for overcoming them.

(i) Defects at the surface and in the layer near
the surface of the semiconductor contribute to the recombination of carriers generated by the light, and
therefore decrease the quantum efficiency. A great
number of point defects, dislocations, etc. emerge as a
result of the mechanical processing of the semiconductor
(cutting, grinding, polishing). Etching single crystal
electrodes removes the damaged layer from the surface.
As a result of etching the efficiency sometimes increases
by a factor of 10 to 15 (315-317). In addition, special
"anti-reflection" etching of GaAs in the mixture of
$H_2O_2 + H_2SO_4$ diminishes the reflection of light from the
photoelectrode surface (310).

(ii) Surface states located in the semiconductor
bandgap may also cause a decrease in efficiency. For
example, majority carriers may tunnel from the semiconductor bulk via surface states to the solution redox
couple (cf. Fig. 4.8). This side reaction competes with
the main reaction involving the minority carriers and
diminishes the conversion efficiency. This is apparently
the case in the system

$$GaAs - Se^{2-}/Se_2^{2-}$$

The adsorption of additives, for example, Ru(III) and
Pb ions on to the semiconductor surface (and in the case
of polycrystalline electrodes into the grain boundaries),
restructures the system of surface states, thus making
this tunnelling impossible. The efficiencies for GaAs
solar cells mentioned above (p.249) were obtained after
treatment of the electrodes with Ru(III) salts (310, 312,
313, 318).

(iii) The exchange of chalcogenide ions between
the solution and the semiconductor surface layer is a
significant process in the deterioration of the characteristics of semiconductor electrodes in polychalcogenide
electrolytes. The exchange occurs when the semiconductor compound and the redox couple in solution have different anions (for example, a CdSe electrode in a polysulfide electrolyte or CdS in a polyselenide electrolyte).
As a result the semiconductor surface layer is replaced
by a compound of the same cation but with another anion.
This exchange can occur in either direction: a layer of
selenide develops on cadmium sulfide photoanodes in
polyselenide solution, and, *vice versa*, in the operation

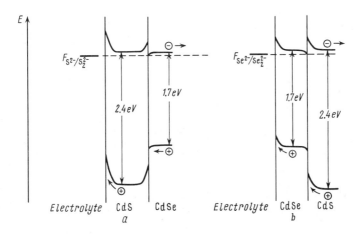

Fig. 8.8. The influence of S/Se substitution on the performance of the photoelectrode: a) CdSe electrode with a CdS layer on the surface; b) CdS electrode with a CdSe layer on the surface (322). (Reprinted by permission of the Publisher, The Electrochemical Society, Inc.).

of cadmium selenide electrodes in polysulfide solution the selenium is replaced by sulfur. These phenomena occur up to a depth of the order of 10^{-6} cm (319-323).

As a result of this replacement of one chalcogenide by another a potential barrier to the motion of free carriers emerges at the contact between the "foreign" surface layer and the electrode bulk. This results in a drop in the photocurrent and the efficiency. Figure 8.8 shows the band schemes for a CdSe electrode with a CdS layer on its surface in a solution of S^{2-}/S_2^{2-} redox couple and for a CdS electrode with a CdSe surface layer in a solution of Se^{2-}/Se_2^{2-}. Due to the large difference ($\simeq 0.7$ eV) in the width of the bandgap for CdS and CdSe a barrier occurs between them at the contact which, in the former case, blocks the motion of the photogenerated holes from CdSe to the electrode surface (Fig. 8.8a). Only the holes generated in the "outer" CdS layer can therefore take part in the photoelectrochemical reaction, while the deeper electrode interior is "switched off." In the latter case (Fig. 8.8b) a barrier to the holes does not arise, and for this reason the CdS electrode does not significantly degrade in selenide solutions. An additional harmful effect of the surface ion replacement is the increased recombination velocity (see point (i) above) which arises because of the many crystallographic defects which occur in the surface layer due to the mismatching of the two crystal lattices.

It is possible to prevent the deterioration of the
characteristics of CdSe electrodes in polysulfide solution by adding small amounts of selenide to the sulfide
solution (0.02 M Se^{2-} in 1 M S^{2-} solution). Even such
small additions of selenide are enough to slow down the
ion replacement process or, at least, to considerably
decrease the amount of sulfur in the electrode surface
layer.

These complicating factors are now the most formi-
dable obstacles to the practical use of these photocells.
Despite the high values of conversion efficiency, the
enhancement of their stability and service life is the
most pressing problem today.

Ξ 8.5. P h o t o c e l l s   w i t h   T r a n s i t i o n
       M e t a l   D i c h a l c o g e n i d e
       E l e c t r o d e s

The highly efficient systems examined in the preceding section appear the most promising at present.
At the same time, there are other classes of photoelectrodes (as well as of redox electrolytes) which have
not yet been studied in detail and whose potential remains hidden. These systems are being intensively
examined at present as a possible alternative to the
cadmium chalcogenide electrodes. Firstly, we should
mention the dichalcogenides of molybdenum and other
transition metals (MoS_2, $MoSe_2$, WSe_2, etc.). These have
relatively narrow bandgaps (E_g = 1 - 1.8 eV) and are
therefore highly promising for solar energy conversion:
for these materials K_{thresh} is about 20-25 per cent (cf.
Fig. 8.2).

In aqueous solutions these materials, as has already
been mentioned, are susceptible to photocorrosion. They
can be protected from photocorrosion in the same way as
the $A^{II}B^{VI}$ and $A^{III}B^{V}$ semiconductors, that is, by the
choice of suitable redox couples in solution. Among
these the most efficient are I^-/I_3^- and Fe^{2+}/Fe^{3+} (131,
324-328). These couples also give a large value of
$|E^{O}_{redox} - E_{fb}|$ necessary for the attainment of high
efficiency.

As shown in Section 8.3, in the simplest case the
magnitude of the open-circuit photopotential is
$E^{OC}_{ph} = |E^{O}_{redox} - E_{fb}|$. It is therefore a linear function
of the equilibrium potential of the redox couple, E_{redox},
if the flat band potential, E_{fb}, remains constant irrespective of the presence of the redox couple in solution.

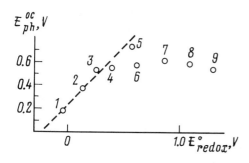

Fig. 8.9. Open-circuit photo-voltage for a cell with an n-type WSe_2 photoanode and the redox couples: 1) Ti^{3+}/Ti^{4+}; 2) SO_3^{2-}/SO_4^{2-}; 3) hydroquinone/quinone; 4) $Fe(CN)_6^{4-}/Fe(CN)_6^{3-}$; 5) I^-/I_3^-; 6) Fe^{2+}/Fe^{3+}; 7) Ru^{3+}/Ru^{4+}; 8) Br^-/Br_3^-; 9) Ce^{3+}/Ce^{4+} (327). The dotted line is drawn with a unit slope.

Experiments have shown (327, 329-331) that in the case of $MoSe_2$ and WSe_2 (as in the case of Si and GaAs, see above) such a dependence is only observed over a limited interval of E_{redox}^o (Fig. 8.9). It was also found that electrolytes with very high redox potentials (for example, Ce^{4+}/Ce^{3+}) proved less effective than those of more moderate oxidizing ability. Thus we see a typical case of the unpinning of the band edges at the surface. It is possible to explain these results if we assume (329) that these compounds are not completely stable with respect to strong oxidizing agents. Apparently, they are oxidized and the change in the surface dipole that thus occurs (see Section 3.7) brings about an undesirable shift of E_{fb} towards more positive values, diminishing the effective magnitude of $|E_{redox}^o - E_{fb}|$. The combination of dichalcogenide photoelectrodes with some redox couples (for example, Cl_2/Cl^-) in non-aqueous solvents, for example ethanol and acetonitrile, where such a shift in E_{fb} evidently does not occur (288, 332) (see also (333)), may prove to be a highly promising approach from this viewpoint.

Table 3 lists the characteristics of regenerative photocells with photoelectrodes of this type (327): these are the fill factor f, the open circuit photo-voltage E_{ph}^{oc} and the efficiency of solar energy conversion η.

The best results in terms of the efficiency and fill factor were obtained for cells with $MoSe_2$ photoanodes. Figure 8.10 shows the working characteristic of such a cell. It has an almost rectangular form (cf. Fig. 8.3a), and this ensures a high value for the fill factor. The $MoSe_2$ electrode is capable of functioning under very high levels of illumination when the short circuit current reaches 120 mA/cm^2. Thus, it is basically suitable for the conversion of concentrated (that is, focussed) solar

TABLE   3.   Characteristics of Photocells with Photoelectrodes
of Transition Metal Dichalcogenides

Semiconductor	E_g, eV [a]	f	E_{ph}^{oc}, V	η, per cent
MoS_2(n-type)	1.75	–	0.42	–
$MoSe_2$(n-type)	1.4	0.67	0.56	4-5
$MoTe_2$(n-type)	1.0	–	0.35	–
WSe_2(p-type)	1.57	0.23	0.71	2

[a] Values of E_g were assessed on the basis of the threshold quantum energy of anodic photocurrent.

light.   The operating life of cells with $MoSe_2$ photo-anodes in iodide solution is 9 months at a photocurrent density of ca. 10 mA/cm^2 (327, 329).   The total number of coulombs passed through the electrode in that period of time would be enough to dissolve it 5,000 times over, were the photocorrosion not suppressed.

Using n-type WSe_2 crystals with especially active surfaces (prepared by cleavage of the crystal along the cleavage plane), it is possible (334, 335) to obtain rather high conversion efficiencies: 14 per cent for monochromatic light at λ = 590 nm, with a fill factor of 0.46.

Thus, dichalcogenides of molybdenum and other transition metals constitute a highly promising class of semiconductor material for regenerative-type photocells.

E 8.6.   P h o t o e l e c t r o d e s   w i t h   C h e m i c a l l y   M o d i f i e d   S u r f a c e s

The reversible organic redox couple ferrocene/ferricinium may be used to protect silicon and other semiconductor electrodes from photocorrosion.   Under illumination, photooxidation of ferrocene to ferricinium occurs, and the ferricinium is then reduced back to ferrocene on the auxiliary electrode.   The conversion of visible and infrared light (the bandgap of silicon is E_g = 1.11 eV) into electricity occurs with an efficiency of 1-2 per cent (336).

The conversion efficiency is increased if the ferrocene is attached to the silicon surface with the help of

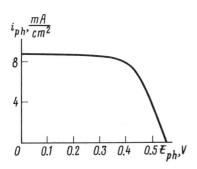

Fig. 8.10. The working characteristic for a regenerative type photo-cell with n-type MoSe$_2$ and Pt electrodes in I$^-$/I$_3^-$ electrolyte under solar irradiation (90 mW/cm^2) (325). (Reprinted by permission of the Publisher, The Electrochemical Society, Inc.)

intermediate -O- or -NH- groups: when the surface of the electrode is chemically modified or derivatized (see Section 5.2). Such electrodes are much more efficient than the same electrodes merely submerged in a solution of the redox couple (337, 338). Thus under illumination, ferrocene attached to the surface is oxidized to ferri-cinium which, whilst remaining attached to the electrode surface, is capable, in its turn, of oxidizing the dis-solved ferrocene. The solution species plays the role of a sort of supersensitizer, regenerating the reagent bound to the photoelectrode surface.

Under strong illumination approximately half of the surface attached ferrocene is permanently in the oxidized form (as ferricinium). This means that the "surface" ferrocene is more readily oxidized (by roughly 0.2 V) than the dissolved form (evidently for some kinetic reason). This effect enhances the conversion efficiency. For example, for a gallium arsenide photo-anode the efficiency is 3 per cent when the electrode is simply immersed in a solution containing a mixture of ferrocene and ferricinium, but it reaches 5 per cent if the surface of the same electrode is modified by chemically attached ferrocene (338).

Surface derivatization has given positive results in aqueous and non-aqueous solutions for n- and p-type silicon (337, 339-341), gallium arsenide (338) and germanium electrodes (342) (the latter, however, has poor photoelectrochemical characteristics in regenerative photocells).

Amorphous silicon presents an interesting, although little studied, material for photoelectrodes. Though the efficiencies and stabilities of the regenerative photo-cells with amorphous silicon photoanodes and ferrocene/

ferricinium solutions are comparatively low (343) the good photoelectric characteristics of this semiconductor, which have already been applied in the production of solid-state solar cells (see, for example, (344, 345)), give every ground to hope that the use of cheap amorphous silicon in photoelectrochemical solar energy conversion may be highly promising.

Chapter 9

CONVERSION OF LIGHT ENERGY INTO CHEMICAL ENERGY

In optical-to-chemical energy conversion, the energy of the light (specifically, solar energy) is used to carry out a chemical reaction which results in "energy-rich" products. The splitting of water into hydrogen and oxygen is a most attractive reaction, since hydrogen is both a convenient fuel and technological raw material, and water is a readily available and cheap reagent (346).

Direct photolysis of water requires ultraviolet light with an energy of about 6 eV; this is practically non-existant in the solar spectrum. For this reason, scientists have spent many years searching for an effective photochemical process using light of lower quantum energy to achieve this reaction. A number of reviews of photochemical solar energy conversion have been published; see for example (281, 347-354); certain aspects of the photochemical approach are examined in (355-359).

Let us deal in greater detail with photoelectrochemical methods of energy conversion. Photoelectrochemical reactions are subdivided into photocatalytic and photoelectrolytic, depending upon the sign of the free energy change, ΔG, in the course of the reaction (267, 269). Photocatalytic reactions ($\Delta G < 0$) can also occur in the dark. However, they are slow for kinetic reasons. In these cases the light energy helps to overcome the corresponding activation barrier with which a certain overvoltage is associated. (It should be pointed out that the term "photocatalysis" is not strictly accurate.

The concept of catalysis implies an acceleration of a chemical reaction due to a lowering of the activation energy, rather than the supply of energy from outside.) Those photoreactions incapable of taking place spontaneously without the input of energy are historically called photoelectrolytic ($\Delta G > 0$). In these cases the energy of the light is used to drive an uphill chemical process.

The first electrochemical cell with a semiconductor electrode to convert light energy into the chemical energy of hydrogen and oxygen by the photoelectrolysis of water was described by Fujishima and Honda in 1972 (8). Their electrochemical cell used a titanium dioxide photoanode and a platinum cathode separated by a porous diaphragm. The anode and cathode compartments of the cell were fitted with burettes to collect the gases evolved on the electrodes. Upon illumination of the photoanode, oxygen was evolved on it, whilst hydrogen was evolved on the cathode. Some time later, at the peak of the energy crisis, this work attracted world wide attention, although the basic principles used had been known for several years. (In particular, Boddy had observed oxygen photoevolution on titanium dioxide electrodes in 1968 (360).) This work (8) stimulated the further development of research into water photoelectrolysis and, in a broader sense, in photoelectrochemical solar energy conversion.

Ξ 9.1. Cells for Water Photoelectrolysis: the Principles of Operation and the Energetics of Conversion

Consider a semiconductor electrode in contact with a solution in the dark. Suppose that the solution does not contain a redox couple which is reversible at the electrode, so that electronic equilibrium is not established between the semiconductor and the solution. Over a certain potential region the electrode will be ideally polarizable. Its stationary potential and band bending are then usually determined by chemisorption processes.

Figure 9.1a shows the energy diagram for an ideally polarizable semiconductor electrode, in this case n-type, in contact with an aqueous solution. The initial band bending is such that a depletion layer is formed near the surface of the semiconductor. For simplicity we will assume that the band edges are pinned at the surface (see Section 3.4) and, consequently, that their

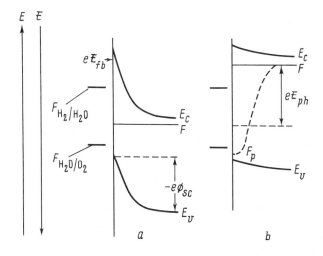

Fig. 9.1. The energy
diagram for a semicon-
ductor/aqueous elec-
trolyte junction (a)
in the dark and (b) in
the light.

position is fixed with respect to the energy levels in
the solution both in the dark and under illumination of
the electrode. The electrochemical potentials for water
oxidation, with the release of oxygen, and reduction,
with the release of hydrogen, are shown in Fig. 9.1.
The difference between these levels, in other words the
free energy change $\Delta G > 0$ for the reaction

$$2H_2O \rightarrow 2H_2 + O_2 \tag{9.1}$$

is equal to 1.23 eV per electron.

Under illumination the band bending, $e\phi_{sc}$, decreases
(the bands unbend). As a result, the Fermi level, F,
shifts with respect to its position in the non-illuminated electrode as shown in Fig. 9.1b. This shift in the
Fermi level is accompanied by a change in the electrode
potential, which is, by definition, the photopotential,
$\mathcal{E}_{ph}$ (cf. Sections 7.2 and 8.3). The generation of non-
equilibrium electrons and holes leads to the formation of
quasi-Fermi levels. It should be borne in mind (see
Section 6.3) that for the majority carriers (in this
case electrons) the shift in the quasi-Fermi level, F_n,
with respect to F is very small while for the minority
carriers (holes) it is considerable. At a certain illumination intensity the quasi-Fermi level for the holes,
F_p, may reach the level for the evolution of oxygen from
water, while the quasi-Fermi level for the electrons
(which roughly coincides with the Fermi level F) may
reach the level for hydrogen evolution (Fig. 9.1b). (In
aqueous solution the region of ideal polarizability of
the electrode is always restricted by these potentials.)

The following inequalities serve as the thermodynamic conditions for the occurrence of the anodic and
cathodic reactions on the illuminated semiconductor,[†]
as discussed in Section 2.3,

$$F_p < F_{H_2O/O_2} \quad \text{and} \quad F_n > F_{H_2/H_2O} \tag{9.2}$$

If these conditions are met, then the simultaneous
evolution of oxygen and hydrogen on the electrode
becomes possible:

$$2H_2O + 4h^+ \rightarrow O_2 + 4H^+ \tag{9.3a}$$

$$4H^+ + 4e^- \rightarrow 2H_2 \tag{9.3b}$$

and the photoelectrolysis of water takes place (cf. Eq.
(9.1)). Indeed, when aqueous suspensions of some semiconductor materials are illuminated, the simultaneous
release of oxygen and hydrogen is observed (see, for
example, (361)); however, the rate of the process is low.

The effectiveness of photoelectrolysis is tangibly
enhanced if the reaction occurring with the participation
of the majority carriers is transposed on to a separate
(auxiliary) electrode. Indeed, the cathodic partial reaction is, in itself, practically unaffected by illumination (since $F_n \simeq F$), and the entire "driving force"
of the photoelectrode is concentrated on the anodic partial reaction occurring with the participation of holes.
It is therefore convenient to carry out the cathodic reaction on a metallic electrode possessing good electrocatalytic properties for hydrogen evolution, for example,
on a platinum electrode. The energy level diagram for
this cell is shown in Fig. 9.2. The electrodes are
separated by a porous diaphragm (not shown in the figure).
In the dark (Fig. 9.2a) the electrochemical potential for
electrons is constant throughout the system, since the
reversible hydrogen potential is established on the
platinum electrode (under an atmosphere of H_2), and in
the short-circuited cell the two electrodes are in

[†]It should be stressed that, in general, the value of the
photopotential for the semiconductor electrode, E_{ph}, does not, in
itself, allow one to judge the feasibility of any photoelectrochemical reaction because this value (in this case equal to $\Delta\phi_{sc}$) is
not directly related to ΔF_p.

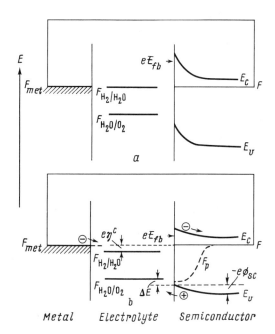

Fig. 9.2. The energy
diagram for a photocell
for water photoelectrol-
ysis: a) in the dark;
b) in the light. The
cell shown in Figs.
9.2 - 9.4 is short-cir-
cuited.

equilibrium. Under illumination (Fig. 9.2b), photogenerated holes move towards the interface and there
enter into the oxidation of water, while electrons are
transferred to the bulk of the semiconductor electrode and
from there, through the external circuit, to the auxiliary electrode where the reduction of water takes place.
Since a current, i, flows in the circuit, the Fermi level
of the metal electrode no longer coincides with the
level F_{H_2/H_2O} in the solution but now exceeds it by $e\eta^c$,
where $\eta^c(i)$ is the cathodic overvoltage. (Strictly
speaking, the value of the overvoltage η^c (Fig. 9.2b)
also includes the ohmic voltage drop due to the internal
resistance of the cell, made up of the electrode and
solution resistances.)

Photoelectrodes made from p-type semiconductors act
in a similar way, the only difference being that the
quasi-Fermi level for electrons shifts considerably
under illumination (accelerating the cathodic partial
reaction), while the quasi-Fermi level for the holes remains close to the Fermi level of the semiconductor
$(F_p \simeq F)$.

The effective occurrence of water photoelectrolysis
requires the exclusion of all side reactions, in particular photocorrosion of the semiconductor electrode. This
requires, as has already been mentioned, the observance
of the following condition:

$$E^o_{H_2O/O_2} < E^o_{dec,p}$$

(for n-type semiconductors) and

$$E^o_{H_2/H_2O} > E^o_{dec,n}$$

(for p-type semiconductors). Thus, for the photoelectrolysis of water we require that:

(i) The width of the bandgap should exceed the difference between the electrochemical potentials for the hydrogen and oxygen electrode reactions in water (1.23 eV), thus

$$E_g > (F_{H_2/H_2O} - F_{H_2O/O_2})$$

(ii) The quantum energy of the light should exceed the width of the bandgap, $\hbar\omega > E_g$.

(iii) The flat band potential for an n-type semiconductor should be more negative than the hydrogen electrode potential, or for a p-type semiconductor should be more positive than the oxygen electrode potential.

The last condition is essential if the quasi-Fermi level for electrons (holes) is to achieve the electrochemical potential for hydrogen (oxygen) evolution (see Figs. 9.1 and 9.2). Otherwise, photoelectrolysis will only be possible if some external voltage is applied to the cell.

It should be emphasized that the existence of a depletion layer in the semiconductor, despite the assertion which is often met in the literature, is not a *sine qua non* for the occurrence of a photoelectrochemical reaction, or in particular for the photooxidation of water. As shown above, the true condition for such a reaction is given by the inequalities in Eq. (9.2). However, the presence of a depletion layer assists in the separation of the electrons and holes, generated by light, in the space charge region. In the case of "surface-absorbed" light ($\alpha^{-1} < L_{sc}$) this prevents recombination which, in turn, leads to a high quantum yield, and, in the final analysis, the effective conversion of the photon flux into electrical current (or chemical energy). Thus, significant photocurrents are usually found if $E > E_{fb}$ (n-type) or $E < E_{fb}$ (p-type).

There is a certain qualitative similarity between

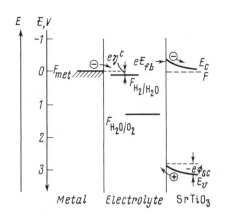

Fig. 9.3. The energy diagram for a photocell for water photoelectrolysis with a SrTiO₃ photoanode and a metal cathode in the light. Potentials are cited against the Hydrogen Electrode in the same solution.

the role played, on the one hand, by the flat band potential in the kinetics of photoelectrochemical reactions on semiconductors, and, on the other hand, by the potential of zero (free) charge in the kinetics of electrochemical reactions on metals in dilute electrolyte solutions (the so-called ψ' correction). For both cases these are special points on the potential scales. Any deviation from them influences not only the energetics of the processes, but also the surface concentrations of charged reagents (in the former case the electrons and holes; in the latter case the ions), and in some cases the processes bringing these reagents to the interface at which the reaction takes place.

As examples of real cells for the photoelectrolysis of water we shall consider the schematic energy diagrams for cells with photoelectrodes made from strontium titanate (SrTiO₃) and titanium dioxide (TiO₂). The method for plotting such diagrams is described in Section 3.8. The first of them (Fig. 9.3) resembles the scheme in Fig. 9.2. The bandgap of strontium titanate is 3.2 eV. This means that the material only absorbs in the ultraviolet region. Its flat band potential is approximately -0.2 V with respect to the hydrogen electrode. The cell SrTiO₃-Pt can, under illumination, split water into hydrogen and oxygen. For the titanium dioxide electrode the situation is qualitatively different (Fig. 9.4), even though as far as the bandgap is concerned (3 eV) this material is practically the same. However, its flat band potential is somewhat more positive than that of the hydrogen electrode (by 0.05 V). Thus, even under very strong illumination and at open-circuit, when the bands unbend completely, the potential for hydrogen evolution is not achieved. Furthermore, the flat band potential of TiO₂ changes with pH in exactly the same manner as the potential of the hydrogen and oxygen

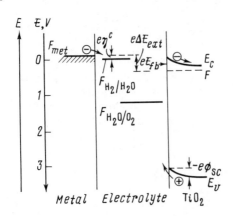

Fig. 9.4. The energy diagram for a photocell for water photoelectrolysis with a TiO_2 photoanode and a metal cathode in the light. Potentials are cited against the Hydrogen Electrode in the same solution.

electrodes (see Fig. 3.20). Hence, when the acidity of the solution changes, all the levels in Fig. 9.4 shift on the energy/potential scale so that it is impossible to shift the flat band potential of TiO_2 with respect to the water/hydrogen reduction potential by changing the pH of the solution in the cell as a whole. This is why voltage from an external source, ΔE_{ext}, needs to be applied to the TiO_2-metal electrode cell to shift the potential of the cathode to more negative values than that of the hydrogen electrode and, at the same time, to maintain a depletion layer in the TiO_2 to separate the photocarriers. This process is called photoassisted electrolysis: in contrast to photoelectrolysis proper, examined above, in which the decomposition of water is achieved with only the energy supplied by the light and without the application of an external voltage.

It should be pointed out that the consumption of energy from an external voltage source should be taken into account when calculating the conversion efficiency. For photoassisted electrolysis the value of K_{stor}^{chem} (Eq. (8.3)) should be written in the following way

$$K_{stor}^{chem} = \frac{\Delta G - e\Delta E_{ext}}{E_g} \qquad (9.4)$$

For the photoelectrolysis of water ($\Delta G = 1.23$ eV) the dependence of $K_{thresh} \cdot K_{stor}^{chem}$ on E_g can be found in Fig. 8.2. It is easy to calculate the magnitude of the current i in the short-circuited cell, as well as the potentials of the anode and the cathode, if we know their polarization curves. Let us use the following conditions:

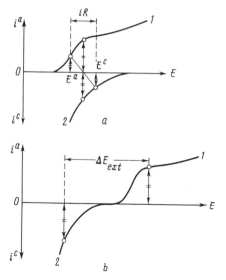

Fig. 9.5. The scheme for determining the current in a photoelectrochemical cell a) in the absence and b) in the presence of an external voltage: 1 and 2 - the polarization curves of the anode and cathode respectively. The horizontal lines are pairs of segments of equal length.

$$i = i^a = -i^c$$

$$E^a - E^c = iR + \Delta E_{ext} \qquad (9.5)$$

where i^a and i^c are the anodic and cathodic currents at the corresponding potentials, E^a and E^c, and R is the internal ohmic resistance of the photocell. Figure 9.5 shows schematically the polarization curves for the anode ($i^a - E$) and the cathode ($i^c - E$), and shows how to calculate E^a and E^c at arbitrary (zero or finite) values of R. To graphically determine the values of the potentials E^a and E^c it is necessary to plot, in Fig. 9.5a, a straight line with a slope determined by the resistance R for the cell (where R is the total resistance of the semiconductor and the solution) in the i - E coordinate system. For small values of R the straight line is practically vertical, and the potentials E^a and E^c are equal to one another. If the value of the resistance R is large, then E^a and E^c differ by the magnitude of iR. In the simplest case when the Ohmic voltage drop can be neglected (this is the case that is examined below) and no external voltage is applied, $E^a = E^c$ (cf. Fig. 9.3). It should be borne in mind that the polarization curve for the photoelectrode changes depending upon the illumination intensity; simultaneously, the photocurrent, i, and the electrode potentials E^a and E^c also change.

To give a quantitative assessment of the influence of the various factors on the efficiency of water photoelectrolysis, let us write down the energy balance for

the process. This can be done in either of two ways in
accord with the two approaches to the description of the
kinetics of photoelectrochemical reactions given in
Chapter 6. Using the concept of quasi-Fermi levels, we
obtain (for n-type semiconductors) (47)

$$F_n - F_p = \Delta G + e(\eta^c + \eta^a) \qquad (9.6)$$

where η^c and η^a are the anodic and cathodic overvoltages
in the photoelectrochemical cell (again, see Section 8.3;
for the semiconductor electrode we believe that $\eta = \eta_H$),
and ΔG is the change in free energy characterizing the
reaction proceeding in the photocell.

Comparing Eq. (9.6) with the corresponding Eq. (8.8)
for a photocell converting light energy into electrical
energy, we can see that ΔG in Eq. (9.6) has the same
meaning as $ei_{ph}R$ in Eq. (8.8): this is effectively the
assimilated part of the energy of the absorbed quantum.
However, the "make-up" of these two quantities is different: $ei_{ph}R = eE_{ph}$ is determined by the potential of
the redox couple E^0_{redox} with respect to the flat band
potential E_{fb}, whereas ΔG is determined by the free
energy of the reaction proceeding in the photoelectrolysis cell.

On the other hand, considering, in accordance with
(257), the transfer of electrons and holes to the solution as proceeding from the edges of the corresponding
bands (see Fig. 9.2-9.4), we can write down another form
of the energy balance:

$$E_g = \Delta G + e\phi_{sc} + (E_c - F) + e\eta^c + \Delta E \qquad (9.7)$$

where ΔE is the energy difference between the top of
the valence band at the surface, $E_{v,s}$, and the redox
level F_{H_2O/O_2}.

Using Eq. (9.7), let us examine the contribution
of the various terms to the full energy balance. The
cathodic overvoltage η^c may be lowered (< 0.1 V) by the
choice of a suitable cathode. At the same time, the
anodic overvoltage included in the quantity ΔE, will,
in all probability, not be less than ≈ 0.5 V due to
the low electrocatalytic activity of ordinary (in other
words, not activated by special catalysts) semiconductor
electrodes. The magnitude of the band bending, $e\phi_{sc}$,
in the light for effective separation of carriers should
not be lower than ≈ 0.2 eV. The value of $E_c - F$ is of
the same order. Any decrease in the latter quantity,

for example by increasing the concentration of doping
impurity in the semiconductor, is undesirable since it
will diminish the thickness of the space charge region,
L_{sc}, and violate the condition $L_{sc} > \alpha^{-1}$ (α is the
light absorption coefficient) necessary to ensure the
most advantageous use of the absorbed photons.   Thus,
in total these losses comprise at least 1 eV. Hence,
the width of the bandgap should exceed the free energy
change for the reaction of water decomposition
($\Delta G = 1.23$ eV) by at least 1 eV, requiring a bandgap of
~ 2.2 eV.   This is much higher than the optimal value
for the bandgap of 1.1 - 1.4 eV (see Section 8.2).   For
$E_g = 2.2$ eV the value of K_{thresh} is only about 10 per
cent (Fig. 8.2).

Hence, the main difficulty on the road towards the
realization of water photoelectrolysis using solar energy
is inherent to the very nature of the problem: the change
in free energy for carrying out this reaction is too
big when compared with the energy of the quanta which
comprise the bulk of the solar spectrum, though the
application of an external voltage (photoassisted electrolysis) to compensate for the energy deficit could
improve the situation significantly.   To solve this
problem we need to find different, roundabout approaches,
such as the use of multistage, multi-quanta processes.
Some of these will be examined in Section 9.3.3.   At the
same time, the numerous studies of water photoelectrolysis, carried out within the framework of the traditional one-quantum approach, although they have brought no
final solution to the problem, have greatly advanced the
scientific and methodological basis of many diverse
aspects of photoelectrolysis.   The results of these
studies are given below.

It should be noted that for harder radiation (ionizing radiation), although the above restriction on the
width of the bandgap is insignificant (362), other more
considerable losses occur (for details see Section 11.4).

Let us return to the anodic overvoltage η^a.   For
photoelectrodes made from wide-bandgap semiconductors
(e.g., $SrTiO_3$, TiO_2) the quantity $\Delta E = F_{H_2O/O_2} - E_v + e\phi_{sc}$,
which includes η^a (see Eqs. (9.6) and (9.7)), amounts to
at least 1 eV; see Fig. 9.3 and 9.4.   In other words,
the energy of the photogenerated holes exceeds the
equilibrium value for the reaction (9.3a) by 1 eV.   As
a result, the anodic reaction proceeds under a regime
of a limiting hole current (all the photogenerated holes
which reach the semiconductor/electrolyte interface are
immediately captured).

Thus, it is as if the anodic overvoltage were disguised by this "surplus" energy of the photogenerated holes. Even at very high illumination intensities when the photocurrent is 6 A/cm^2, no kinetic limitations from the electrode reaction have been discovered; the photocurrent still remains fully limited by the rate of photogeneration of holes (363). It would be possible to measure η^a by selecting a series of semiconductors in which the value of E_g gradually decreased until the activation energy required for the oxygen evolution reaction became important. During such an experiment it is imperative to maintain all the energy levels in the semiconductor, with the exception of the top of the valence band, constant with respect to the energy levels in the solution. However, this proves to be impossible because, as the width of the bandgap decreases for oxide semiconductors (for this experiment oxide semiconductors are chosen because these materials are chemically resistant to the conditions for anodic oxygen evolution) the flat band potential shifts towards more positive values (Fig. 9.6), and $\Delta E_g \simeq e|\Delta E_{fb}|$.

The reason for such behavior for oxide semiconductors in aqueous solution is not, as yet, totally clear. It has been postulated (364) that this is due to an interaction between the oxygen atoms of the oxide and those of the water molecules, which stabilizes the position of the valence band with respect to the levels in the solution. (It should be borne in mind that the valence band is formed by the orbitals of the more electronegative component of the semiconductor compound – oxygen.) This apparently predetermines the values of $E_{v,s}$ at the surface of various semiconducting oxides in aqueous solutions, despite other differences in their characteristics.

Consequently, as the width of the bandgap for oxide semiconductors decreases, the energy of the top of the valence band with respect to F_{H_2O/H_2} remains at the same level, while the energy of the bottom of the conduction band decreases (Fig. 9.7). This leads to applying even greater voltages, ΔE_{ext}, to carry out photoassisted electrolysis (364, 365). For this reason, the use of relatively narrow bandgap oxide photoanodes, such as WO_3 and Fe_2O_3, did not lead to the expected increase in efficiency: such materials are good absorbers of visible light, but this advantage is brought to naught by the loss in energy due to the need to apply an external voltage to the cell.

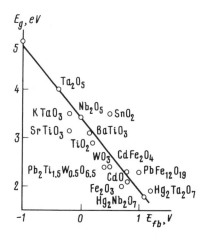

Fig. 9.6. The connection between the flat band potential (against the Hydrogen Electrode at pH 13.3) and the bandgap for oxide semiconductors (364).

Note that the disposition of the electrochemical potential F_{H_2O/O_2} with respect to the bandgap of the photoelectrode, for example, $SrTiO_3$ or TiO_2, imposes certain restrictions on the mechanism of the oxygen evolution reaction. The transfer of photogenerated holes directly from the valence band to the "oxygen" level is associated with the dissipation of a large amount of energy, $\Delta E \simeq 1$ eV, and therefore seems scarcely probable. This has brought some authors to the conclusion (see, for example, (241, 366-368)) that the holes from the valence band first transfer on to a surface level (or set of levels) located in the bandgap in the vicinity of E_{red}^0 (with the energy evolved imparted to the semiconductor lattice) and then transfer to the solution (cf. Fig. 4.9).

§ 9.2. E l e c t r o d e s   f o r   W a t e r
        P h o t o e l e c t r o l y s i s

Several dozen semiconductor materials have already been used as photoelectrodes for the photoelectrolysis of water. The greater part of these materials have been n-type semiconductors. None of them possesses the full set of characteristics needed for effective photoelectrolysis in solar light. However, these materials can be used for the conversion of light of higher energy. Below we examine chiefly those materials which have proved acceptable as far as some of their characteristics (stability, flat band potential, etc.) are concerned.

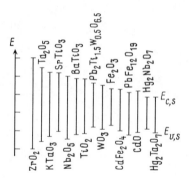

Fig. 9.7. The relative position of the band edges $E_{c,s}$ and $E_{v,s}$ at the surface of oxide semiconductors in aqueous solution. In arranging the semiconductors the value of $E_{c,s}$ is chosen to decrease from left to right (364).

Ξ 9.2.1. T i O $_2$ - a   M o d e l   P h o t o a n o d e   f o r   t h e   S t u d y   o f   P h o t o e l e c t r o l y s i s. The titanium dioxide electrode was the first one to be used in a cell for the photoelectrolysis of water (8). Although it soon became obvious that it cannot be used for the conversion of solar radiation (the bandgap is too wide and the flat band potential insufficiently negative), it still remains, due to its chemical resistance and relative simplicity of manufacture, the main model electrode for the development of the theory and methodology of photoelectrolysis.

Crystals of TiO_2 with the rutile structure, and sometimes with the anatase structure, are used for the manufacture of electrodes. The surface of the single crystal rutile electrodes is polished and etched to remove the damaged layer produced in the course of processing. This etching is done in molten sodium hydroxide or hot (about 250°C) sulfuric acid (369).

The thermal oxidation of titanium is the principal method for the production of polycrystalline electrodes. This is carried out either in oxygen at atmospheric pressure (or sometimes at reduced pressures of the order of 10^{-3} Torr (370)) or in air (371, 372) at a temperature of 800-1,000°C. To produce large area electrodes, a thin layer (0.2-0.3 μm) of titanium metal is sputtered onto a flexible polymer film, and the layer partially oxidized (373). Other methods include TiO_2 plasma sputtering (374, 375), and hydrolysis or pyrolysis of titanium-containing organic compounds on a conducting substrate (376). Finally, another widespread method for making polycrystalline rutile electrodes is to press pellets of TiO_2 powder and then to sinter these at a temperature of about 1,000°C.

In practically all cases these methods produce insulating titanium dioxide of stoichiometric composition. To sufficiently increase the conductivity of the material for use as an electrode, electron donors are introduced. More often than not, these are oxygen vacancies, Ti^{3+} ions, or sometimes hydrogen atoms. The following methods are used: heating under vacuum (about 10^{-5} Torr) at about $800°C$ for 1-2 hours, or partial reduction in a hydrogen atmosphere (at $600-800°C$). Sometimes cathodic polarization of the electrode at the potential for hydrogen evolution (377-381) or irradiation by a beam of high-energy electrons (382) is used. The concentration of donors in samples produced by the different methods is usually of the order of $10^{18} - 10^{20} cm^{-3}$.

When comparing these methods for the introduction of donors into TiO_2, the authors (383) came to the conclusion that heating under vacuum is accompanied by a partial disordering of the TiO_2 crystal structure, leading to the emergence of oxygen vacancies and interstitial Ti^{3+} ions. When TiO_2 is reduced in hydrogen gas, oxygen does not leave the crystal lattice but is converted to OH-groups, with Ti^{3+} ions transferring to the lattice sites; consequently the resulting degree of disorder introduced into the crystal is insignificant. As a result, the quantum yield for the anodic photocurrent in the latter case is somewhat higher than that in the former (384). Cathodic polarization, accompanied by the penetration of hydrogen atoms into interstitial sites in the TiO_2, does not give stable results since the hydrogen atoms can enter and leave the TiO_2 lattice with equal ease (see below).

The anodic oxidation of titanium in alkaline solutions (385-389) is a simple method for producing titanium dioxide electrodes. The anodic film of TiO_2 thus produced already has the appropriate electrical conductivity.

It should be pointed out that in the majority of the work the influence of the conditions for electrode production on the photoelectrochemical behavior of the electrodes has not been examined in detail. Neither have there been a sufficient number of attempts to optimize these conditions. For this reason the temperatures given above for processing the electrodes are only tentative.

The quantitative comparison of the properties of electrodes produced by the different methods entails

certain difficulties because in the majority of cases
the magnitude of the anodic photocurrent (i.e., the relative characteristic of photosensitivity) rather than
the quantum yield (i.e., the absolute characteristic)
was measured. Nevertheless, there is no doubt that
single-crystal rutile electrodes give the highest sensitivity (243, 371, 384), the quantum yield being close
to unity. However, polycrystalline electrodes produced
by the thermal oxidation of titanium foil or by pressing
and sintering TiO_2 powder, in spite of the simplicity
and cheapness of production, may also have rather good
characteristics (quantum yield = 0.7 (390)). Anodic oxide
films are inferior to electrodes produced by the other
methods: their quantum yield (379, 391) and the "breakdown" potential at which the dark anodic current increases sharply are much lower. It is significant that
the quality of the electrode produced anodically can be
substantially improved (almost to the level for single
crystal electrodes) by heating in an inert atmosphere
at $600-700^{\circ}C$ for one hour. This treatment, according
to X-ray studies, causes the amorphous modification of
anodic TiO_2 to crystallize with the rutile structure
which is characterized by large photocurrents (389).

The shape of the anodic polarization curve for the
titanium dioxide electrode is shown in Fig. 9.8. According to (209), the onset potential for the anodic photocurrent is close to the flat band potential E_{fb}. Since
the latter for TiO_2 is a linear function of the solution
pH (see Fig. 3.20), the polarization curve shifts along
the potential axis (measured against a reference electrode whose potential does not depend on pH, for example
the calomel electrode). In solutions of different pH
the shape of the curve may be somewhat different but the
value of the photocurrent at a given value of $E - E_{fb}$
is practically independent of pH (392). The current
efficiency for oxygen evolution is usually close to 100
per cent (243, 366).

The spectral dependence of the photocurrent is
shown in Fig. 9.9. The photocurrent usually appears at

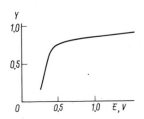

Fig. 9.8. The dependence of
the quantum yield for the
photooxidation of water on a
TiO_2 electrode ($N_D=6\times10^{17}$ cm^{-3})
on the potential (209). Poten-
tials are cited with respect to
the flat band potential.

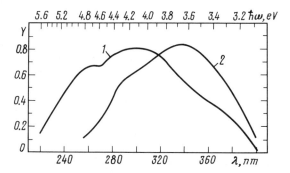

Fig. 9.9. The spectral distribution of the quantum yield of anodic photo-current on titanium dioxide electrodes produced by the oxidation of titanium foil under different conditions. The thickness of the TiO_2 layer: 1) 2 μm, 2) 7 μm (393).

wavelengths less than about 400 nm, and this corresponds to the width of the bandgap, $E_g = 3$ eV. The shape of the curve is largely determined by the method used to make the electrode (393-395). The photocurrent is proportional to the incident light intensity up to very high values ($J_{inc} = 380$ W/cm^2) (363).

The titanium dioxide photoanode has a high chemical stability. The potential for its anodic photodecomposition is more positive than that for the oxidation of water (see Fig. 7.7), and this reliably protects it from photocorrosion. For this reason photodecomposition in the direct sense of the word is not observed, with the possible exception of strongly acidic solutions (396, 397). Nevertheless, in the course of prolonged electrolysis, changes do occur in the electrode, and as a result the photocurrent decreases (398, 399). According to ESCA studies (400), the Ti^{3+} ions in the TiO_2 lattice, which serve as electron donors, are oxidized to Ti^{4+} in the course of oxygen photoevolution, while the oxygen vacancies are filled by oxygen; consequently the titanium dioxide recovers its stoichiometry. (The same thing happens with other titanium-containing materials: $SrTiO_3$, $BaTiO_3$ (400), and apparently $TmTi_2O_7$ (401).) As a result, the surface layer of the semiconductor becomes insulating. For a photocurrent density of 10-15 mA/cm^2 this process takes 8-10 hours. Since the donors (Ti^{3+} ions) which determine the electrical conductivity of TiO_2 are usually located on interstitial sites, rather than lattice sites, they are comparatively highly mobile. In

the electric field of the depletion layer they are effectively transported to the surface where they are oxidixed.
A proposal was made (402) that in order to remove this undesirable process the TiO_2 should be doped with substitutional impurities, rather than interstitial ones: substitutional impurities migrate much more slowly in an
electric field and so the resultant electrodes should be
more stable. With this aim in view it is possible, for
example, to replace part of the oxygen with fluorine
(403), instead of using oxygen vacancies and Ti^{3+} ions
as the source of the electrical conductivity.

The effective operation of a water photoelectrolysis
cell with a titanium dioxide photoanode and a metal
cathode is strongly dependent upon the operating conditions. As has already been mentioned, such a cell is
unable to decompose water without the application of an
external voltage. This is because the flat band potential of TiO_2 is insufficiently negative when compared
with the potential of the hydrogen electrode. As a result, the hydrogen evolution potential is not achieved
at the cathode. If an external voltage of 0.3 - 1 V is
applied, it is possible to obtain a photoelectrolysis
efficiency of up to 4-8 per cent for monochromatic ultraviolet light with a wavelength of 330 nm (363, 404, 405).
For sunlight in which the proportion of ultraviolet light
is small, the efficiency is usually no more than one per
cent (394). Nevertheless, industrial cells for the electrolysis of water using TiO_2 photoanodes have been paten-
ted (406-411). For the present, however, they have not
found practical application. Some attempts to improve
the spectral characteristics of titanium dioxide electrodes will be examined below.

In a number of studies, beginning with the first
paper by Fujishima and Honda (8), instead of applying an
external voltage, a cell with separated anodic and
cathodic compartments containing electrolyte solutions
of different pH was used. For example, the titanium dioxide photoanode was in 1 M NaOH solution, while the
platinum cathode was in 0.5 M H_2SO_4 solution. Due to the
pH difference, the electrochemical potential for the
hydrogen electrode reaction differed in two solutions by
approximately 0.8 eV, and this made it possible to carry
out photoelectrolysis. (When calculating the efficiency
for this type of photocell, the quantity ΔE_{ext} in Eq.
(9.4) is replaced by the value of the electromotive force
for the corresponding concentration cell.) As an example,
let us mention a cell that was constructed for solar
energy conversion using a TiO_2 photoanode with a total

area of 0.17 m^2. In Tokyo (latitude 36° N) this produced a daily average of 1.1 dm^3 of hydrogen gas (corresponding to an efficiency of about 0.4 per cent) (391).

Ξ 9.2.2. P h o t o e l e c t r o l y s i s   W i t h o u t   a n E x t e r n a l   V o l t a g e. We now consider semiconductor photoanodes which, unlike TiO_2, are capable of splitting water into oxygen and hydrogen in cells without an externally applied voltage. First among these are strontium titanate ($SrTiO_3$), barium titanate ($BaTiO_3$), and also potassium tantalate ($KTaO_3$). These are n-type semiconductors having the perovskite structure.

Both single crystal and ceramic electrodes have been investigated. In the latter case, judging by the results obtained in (185, 412), the efficiency for energy conversion is much lower. Single crystal $SrTiO_3$ electrodes were preheated under vacuum or in a hydrogen atmosphere to make them sufficiently conducting (this was described in Section 9.2.1 in connection with the production of TiO_2 electrodes).

Of the materials mentioned above, strontium titanate is the most efficient. Its flat band potential, as already mentioned, is more negative than the hydrogen electrode potential and this is why the photoelectrolysis of water is possible without the application of an external voltage. This is also true of potassium tantalate. It is somewhat unexpected that barium titanate is almost as good as an anode material for the photoelectrolysis of water as strontium titanate. Indeed, according to (413, 414), the flat band potential of $BaTiO_3$ is more positive than the potential of the hydrogen electrode. These authors failed to explain why there was a significant photocurrent at a potential which is about 0.3 V more negative than the flat band potential. It should be noted that although a photocurrent may be observed when the short-circuited cell is illuminated, when a small external voltage is applied this increases significantly along with the conversion efficiency. The maximum efficiency achieved for ultraviolet light is almost 20 per cent (415-417). Naturally in sunlight the efficiency is much smaller.

Table 4 lists the characteristics for these electrode materials: the width of the bandgap E_g; the maximum reported quantum yield, Y, for the anodic photocurrent with an indication of the energy $\hbar\omega$ of the light quanta used for the measurement; the flat band potential E_{fb} with respect to the hydrogen electrode; and the maximum

TABLE   4.   Characteristics of Photoanodes which Require no
External Voltage

Semiconductor	E_g, eV	Y	$\hbar\omega$, eV	E_{fb}, V	η, per cent
$SrTiO_3$	3.2	1[a] (415,416)	3.8	-0.2 (417,418)	20 (415)
$BaTiO_3$	3.3	0.3[b] (413)	3.9	0.3 (413,414)	--
$KTaO_3$	3.5	0.4 (418,419)	4-4.5	-0.2 (109)	4-6 (418,419)

a) The value of Y was measured at low light intensities; at higher
intensities it falls (363).

b) As the temperature increases, Y increases on average by 0.11 per
cent per °C (414).

efficiency for water photoelectrolysis η. All the electrodes are characterized by high stability over a broad
pH range. In alkaline solutions all of the anodic current
goes to oxygen evolution; in acidic solutions some hydrogen peroxide is also formed (415, 417).

The common defect of these materials is that their
bandgaps are too wide and, therefore, they are only sensitive to ultraviolet light. This makes them almost
useless for the conversion of sunlight. Attempts have
been made (420, 421) to reduce the width of the bandgap
of $BaTiO_3$ and $SrTiO_3$ by partial replacement of Ti by Fe
or Cr, O by F, and Sr and Ba by La, thereby sensitizing
the materials to longer wavelength light. Indeed, the
compounds thus obtained (for example, $BaTi_{1-x}Fe_xO_{3-x}F_x$
where the value of x varies from 0.02 to 0.1) are sensitive to longer wavelength light than the original
$BaTiO_3$ and $SrTiO_3$ but they are unable to split water
under illumination without an external voltage.

Among the other perovskites that have been investigated,
potassium tantaloniobate ($KTa_{0.77}Nb_{0.23}O_3$) resembles
$KTaO_3$ in its photoelectrochemical properties. The maximum efficiency achieved for water photoelectrolysis is
4-6 per cent (for ultraviolet light) (413, 419). The
perovskites $Pb_{0.86}La_{0.14}(Zr_{0.1}Ti_{0.9})_{0.965}O_3$ and
$Pb_{0.92}La_{0.08}Ti_{0.98}O_3$ turned out to be sensitive to light
of wavelengths shorter than 430 nm and 410 nm respectively. The effectiveness of these materials as photoanodes is, however, low due to their large resistances
(413).

9.2.3. Oxide Photoanodes Sensitive to Solar Light. Tungsten and iron trioxides, as well as some other n-type oxide semiconductors, have relatively narrow bandgaps and are, therefore, basically capable of splitting water using visible light. As mentioned in Section 9.1, any reduction in the bandgap is accompanied by a shift in the flat band potential towards more positive values. Thus these materials require the application of external voltages ($\gtrsim$ 0.5 V) in order to carry out electrolysis.

Tungsten trioxide electrodes have been examined in the form of single crystals (422) and also as polycrystalline films obtained by the oxidation of tungsten foil (423), thermal decomposition of ammonium tungstate on a glass substrate (424), WO_3 sputtering on to a tungsten substrate (425), and by other methods. As is the case of the oxides mentioned above, heating the WO_3 under vacuum leads to an increase in the donor concentration and, consequently, in the electrical conductivity.

Polycrystalline Fe_2O_3 electrodes have been produced by oxidizing the metal in the flame of a gas burner (426, 427), sintering Fe_2O_3 powder (428), and chemical vapor deposition of the oxide on a platinum or titanium substrate with subsequent heating (429, 430). Single crystal samples have also been examined (431).

The anodic polarization curves for WO_3 (Fig. 6.4) and Fe_2O_3 electrodes are well-described by Eq. (6.6). The analysis of curves like that in Section 6.2 made it possible to evaluate the diffusion length for holes, L_p, in polycrystalline samples of Fe_2O_3; a value of about 10^{-6} cm was found (432).

Table 5 gives the photoelectrochemical characteristics of WO_3 and Fe_2O_3 as well as those for the iron titanates which are sensitive to the visible region of the spectrum. The efficiency of energy conversion in a cell with an Fe_2O_3 electrode is about 1 per cent (illumination by a Ne-lamp) (438). The $YFeO_3$ electrode is characterized by a lower photosensitivity as compared to Fe_2O_3 (432).

The semiconductors investigated in this section are somewhat less stable than such materials as, for example, TiO_2 and $SrTiO_3$. Nevertheless, they are relatively stable, WO_3 in acid and neutral solution, Fe_2O_3 and the other iron compounds in neutral and alkaline solution, at least under anodic polarization. Under cathodic

TABLE   5.   Characteristics of Photoanodes Sensitive to Visible
Light

Semiconductor	E_g, eV	Y	$\hbar\omega$, eV	E_{fb}, V(vs. hydrogen electrode)
WO_3	2.7	1 (422)	4.6	0.6 (422)
Fe_2O_3	2.2	0.3-0.35 (431,433,434)	3.6	0.3-0.5[a] (431,435,436)
$FeTiO_3$ Fe_2TiO_4 Fe_2TiO_5	2.2	0.15 (437)	3.5	--

a) The complex form of the $C^{-2} - E$ curves makes it possible only to assess E_{fb} approximately.

polarization WO_3 decomposes (422). Partial replacement of O by F increases the stability of WO_3 (439).

Ξ 9.2.4. Z i n c   O x i d e,   T i n   O x i d e   a n d
O t h e r   O x i d e s. In aqueous solutions the zinc oxide electrode is not highly corrosion resistant either in the dark (107) or in the light. Indeed, its anodic decomposition potential in neutral chloride solution is less positive than the potential for water oxidation (Fig. 7.8). Thus, zinc oxide is not thermodynamically protected from corrosion. Nevertheless, under certain conditions (in strongly alkaline solutions) ZnO is a sufficiently stable photoanode (243, 244, 386), probably for kinetic reasons. In this case, the decomposition of water, accompanied by oxygen evolution (and also by hydrogen peroxide production (440)) is the main anodic reaction. However, for water splitting the zinc oxide electrode is ineffective. This is due both to the broad bandgap (3.2 eV) and the insufficiently negative flat band potential ($E_{fb} \simeq 0$ V with respect to the hydrogen electrode (107)) of the material; as a result, an externally applied voltage is necessary for photoelectrolysis.

Tin oxide, SnO_2, is rather stable as a photoanode. The threshold wavelength for anodic photocurrent is 350 nm and the quantum yield is 0.27 (for monochromatic ultraviolet light, λ = 254 nm). The photoelectrolysis of water occurs in cells with SnO_2 photoanodes and platinum cathodes with an externally applied voltage (about 0.5 V); apart from O_2, H_2O_2 is also produced at the anode in acidic and weakly alkaline solutions. On the whole, the properties of SnO_2 as a photoelectrode are somewhat inferior to those of TiO_2 (441).

A number of other oxide electrodes have been investigated (364, 429) (see also the various reviews (223, 258-260)). For different reasons, all of these materials turned out to be unfit as photoanodes for solar energy conversion: for example ZrO_2, Ta_2O_5 and Nb_2O_5 are quite stable but their bandgaps range from 3.5 to 5 eV, and, consequently, they are only sensitive to ultraviolet light; Cu_2O, Bi_2O_3, and V_2O_5, amongst others decompose under anodic polarization; the flat band potentials of CdO and of some mixed oxides, all of which are relatively stable and photosensitive to the visible region of the spectrum, are too positive.

Ξ 9.2.5. P h o t o c a t h o d e s. Attempts have been made to use the $A^{III}B^V$ and $A^{II}B^{VI}$ semiconductor compounds, which have relatively narrow bandgaps and have therefore been successfully applied both in solid-state and liquid-junction solar cells, in the water photoelectrolysis cells. However, in the absence of specially introduced redox couples, they are subject to photodecomposition under anodic polarization (see Section 7.3). Although a number of methods have been proposed to prevent photodecomposition (see Section 9.3.2), satisfactory results have, as yet, not been achieved. This led to the idea of transferring the "photosensitive element" in the cell from the anode to the cathode and changing to the p-type form of these materials, since at negative potentials they can be expected to be sufficiently stable. The corresponding photoelectrochemical cell consists of a p-type semiconductor photocathode, on which hydrogen is evolved under illumination, and a metal anode, on which oxygen is evolved; reactions (9.3) occur.

The bandgap of gallium phosphide, GaP, is 2.2 eV, and thus it absorbs a considerable part of the solar spectrum. At the same time, to carry out water photoelectrolysis in the absence of an external voltage it is necessary (see below) for the flat band potential of the photocathode to be more positive than that of the oxygen electrode. Unfortunately, in the case of gallium phosphide, and of other materials that have been examined, this condition is not met. Thus, it is necessary to apply a considerable voltage from an external source to the GaP photocathode to carry out water photoelectrolysis. This considerably reduces the efficiency of

energy conversion: it is only 0.1 per cent (442-444).
The stability of the GaP photocathode under conditions
of prolonged operation is also slightly unsatisfactory
(445, 446).

Some other materials (GaAs, Ga(AlAs), Te) were
found to be ineffective as photocathodes. All of them
require an even higher external voltage than GaP and,
therefore, the efficiency of light energy conversion is
insignificant (443, 447-449).

The perovskites based on Rh_2O_3 represent an interesting group of p-type photocathode materials. They
have a common formula $MRhO_3$ (where M is Lu, Dy, La or
some other rare-earth element) and are produced by
sintering a mixture of powders of Rh_2O_3 and the oxide of
the corresponding metal at $1,350^{\circ}C$ and 3000 atm. Water
photoelectrolysis was observed in a photocell with a
$LuRhO_3$ cathode and a nickel anode with an external
applied voltage of 0.6 V (450).

Finally, let us mention two attempts (451, 452) to
increase the efficiency of hydrogen production on photocathodes by using a "carrier" to circumvent the kinetic
difficulties. Methylviologen (MV^{2+}) was used as the
"carrier." The reduction of methylviologen occurs on
the photocathode (for example, p-type gallium arsenide)

$$MV^{2+} + e^- \rightarrow MV^{\overset{\cdot}{+}}$$

and is followed by the catalytic reduction of water by
the resulting radical cation with the formation of
hydrogen and regeneration of the methylviologen

$$MV^{\overset{\cdot}{+}} + H_2O \xrightarrow{\text{catalyst}} MV^{2+} + 1/2\ H_2 + OH^-$$

Small particles of PtO_2 and of platinized alumina,
Al_2O_3, suspended in solution were used as a catalyst.
However, so far, this method has not brought any substantial increase in the efficiency of light energy
conversion because the reactions occurring in the cell
are still accompanied by considerable overvoltage losses.

9.3. Methods for Enhancing the Efficiency of Water Photoelectrolysis

The photoelectrochemical method of converting solar
energy will only be applied on a large scale if the cost

of the hydrogen so produced does not exceed the cost of
hydrogen produced by other means. According to economic
estimates (453), it can be assumed that photoelectrolysis
of water will become economically profitable when the
efficiency of the process reaches at least 10-15 per
cent. Until now the achievement of this sort of efficiency has been hampered by two factors. The first of
these is a fundamental problem; the solar spectrum is
insufficiently "energy rich" to allow effective water
photoelectrolysis in a single quantum process without
the application of some external energy source. The
second of these is of a subsidiary character, the problem of the deficiencies in the available electrode
materials (unfavorable spectral sensitivity, low corrosion resistance, and so on).

   However, attempts are being made to improve photoelectrolysis cells. Research in this field is along
two major lines:

   (i) the improvement of the photoelectrode characteristics of specific semiconductor electrodes by, for example,
photosensitizing wide bandgap semiconductors to visible
light; the use of materials with narrower bandgaps, provided
that their corrosion resistance is enhanced, and so on;

   (ii) the improvement of the characteristics of photoelectrochemical cells as a whole by creating new types
and designs of electrodes and cells, in particular by
combining two photoelectrodes in one cell or by connecting
two cells in series.

   The results of this work are briefly discussed below.

Ξ 9.3.1. S e n s i t i z a t i o n   o f   W i d e - B a n d g a p
S e m i c o n d u c t o r s   t o   V i s i b l e   L i g h t.
Attempts have been made to solve the problem of semiconductor sensitization, firstly, by doping semiconductors
with special impurities and, secondly, by adsorbing
species on the surface of the electrode which are able
to absorb visible light.

   Under illumination, the atoms of deep-lying donor
impurities in the semiconductor may become ionized, giving
electrons in the conduction band, while those of acceptor
impurities may capture electrons from the valence band,
thereby creating holes. As shown in Fig. 1.6, these
transitions require quanta of an energy which is less
than the bandgap, E_g, in contrast to the interband transitions examined above. These processes are reflected

in a maxima in the semiconductor absorption spectrum
located at longer wavelengths than the intrinsic
absorption edge. For the process to proceed in the
steady state, it is essential that the vacant levels
produced by the photoionization of the donor impurities
are effectively refilled by electrons from the valence
band, whereas acceptor impurities which have captured
photoelectrons must pass them on to the conduction band.
These electron transitions take place as a result of
thermal excitation. Thus, in addition to the light
energy, it is also necessary to supply heat to an electrode functioning in the impurity absorption mode (this,
however, partially occurs by itself since some part of
light absorbed by the semiconductor is inevitably converted into heat; see Section 8.2). However, the impurity light absorption coefficient is considerably
lower, sometimes by several orders of magnitude, than
the intrinsic absorption coefficient; consequently the
associated photocurrent is also lower.

The experimental data on the influence of doping
on photoelectrode properties are scarce and often con-
tradictory. All of the data refer to the titanium
dioxide photoanode. The doping of titanium dioxide by
admixtures of metals has been carried out by a variety
of methods: pre-doping of the titanium metal followed
by oxidation (454, 455); pressing of a mixture of the
powder-like titanium and doping metal oxides with subsequent sintering (390); the production of a TiO_2 film,
containing the necessary impurity, on the substrate by
chemical vapor deposition (456); pyrolysis of a solution
of a mixture of titanium and the doping metal salts on
a titanium substrate (457). The amount of impurities
varied from 0.01 to several per cent. It was found that
some impurities, for example, aluminium, niobium, yttrium
and boron (454, 455, 458), can increase the value of the
photocurrent and prevent long term changes in the TiO_2
(see Section 9.2.1) without changing the spectral sensitivity of the material. According to (458), these
impurities create a stable defect structure in the
rutile lattice, thereby lending it a high conductivity
which does not diminish in the course of photoelectrolysis. Other impurities expand the spectral range of
the anodic photocurrent, making titanium dioxide sensitive to visible light. Among these are chromium (390,
458-460), manganese, iron (461), cobalt (462) and beryl-
lium (457). However, the photocurrents in the visible
region of the spectrum are very small, as is to be expected.

   Another method for the photosensitization of semiconductor electrodes - the adsorption of dyes on their
surface - is based on the application of electrode processes with the participation of photoexcited chemical
species in solution (see Chapter 5). The dye-sensitizer
is built up on the surface of the semiconductor by
spraying under vacuum or by "chemical modification" of
the semiconductor surface (see Section 5.2). Thus, the
titanium dioxide electrodes in cells for the photoelectrolysis of water are sensitized to sunlight by cyanin
(260), rhodamine B$_2$ (463), bipyridyl complexes of ruthenium (II) [Ru(bipy)$_3^{2+}$] (464, 465) and by other dyes.
Examples of the sensitization of other semiconductor
materials with wide bandgaps to visible light were given
in Section 5.3.

   This method of sensitization is characterized by
two principal drawbacks. Firstly, the amount of dye adsorbed on the electrode surface is very small (of the
order of a monolayer). Light absorption by such layers
is insignificant and, as a result, the electrolysis
quantum yield is low ($\lesssim 10^{-2}$). When a thicker layer of
dye is built up, the light absorption increases but at
the same time the electrical resistance of the layer
increases, inasmuch as the organic dye is, in itself,
an insulator. The use of porous photoelectrodes im-
pregnated with sensitizer has been proposed. For example,
an electrode produced by pressing and sintering ZnO
powder was kept in a solution of the dye rose bengal.
A monolayer of dye was adsorbed on the internal surfaces of
the pores. Since zinc oxide is transparent to visible
light, the entire internal surface of the pores "works"
in such an electrode, and the effective quantum yield
reaches 0.15 (466-468). However, it is still not quite
clear whether it is possible to ensure effective mass
transfer at such an electrode in the course of prolonged
photoelectrolysis. The second difficulty stems from the
fact that the adsorbed dye is oxidized (reduced) under
illumination and thus consumed. In order to carry out
the steady state photoelectrolysis of water it is necessary that the oxidized form of a dye, for example the
bipyridyl complex of Ru(III), in its turn, oxidizes
water to O_2:

$$Ru(bipy)_3^{2+} \xrightarrow{\hbar\omega} Ru(bipy)_3^{3+} + e^-$$

$$2Ru(bipy)_3^{3+} + 2OH^- \longrightarrow 2Ru(bipy)_3^{2+} + \tfrac{1}{2} O_2 + H_2O$$

Thus water plays the role of a supersensitizer ensuring
the regeneration of the dye. Unfortunately, it is impossible to implement this process with any significant
efficiency (179, 465, 469-472). The use of other substances as supersensitizers for the regeneration of the
dye, instead of water, is tantamount to abandoning the
whole idea of water photodecomposition.

Thus, all attempts to sensitize wide bandgap semiconductor electrodes to visible light for the photoelectrolysis of water have so far been unsuccessful.

Ξ 9.3.2. P r o t e c t i v e   C o a t i n g s   t o   A v o i d
P h o t o c o r r o s i o n. As was shown in Sections 9.2.1-
9.2.3, it is possible to create stable photoelectrodes
from oxide semiconductors but their characteristics are
far from optimal. In particular, it has been impossible
to combine, in one and the same oxide, a sufficiently
narrow bandgap with a relatively negative flat band potential. This problem could be solved by using $A^{III}B^{V}$
and $A^{II}B^{VI}$ semiconductors but these are rather unstable
in aqueous solutions. To cope with this problem it has
been suggested that the surface of the unstable semiconductor be coated with a thin film of metal or with some
other photocorrosion-resistant semiconductor. Such a film,
being transparent to light, should at the same time,
according to these authors (473, 474), protect the electrode from decomposition.

Thin films of a noble metal (for example gold or
palladium) are usually vacuum-deposited on to the surface
of the semiconductor (GaP, Si and others) single crystal.
At a thickness of about 10-20 nm this does not significantly reduce the intensity of the light passing through
it to the semiconductor. Two totally different cases can
be identified depending on the regime of metal deposition.
When the process is carried out slowly, the metal forms
a continuous layer on the semiconductor. When such
an electrode is immersed in an electrolyte solution, it
behaves practically as a metal electrode. At the same
time, the overvoltage for electrochemical reactions is
somewhat decreased when the electrode is illuminated.
This occurs because a semiconductor electrode covered
by a continuous metal film is nothing other than a metal
electrode connected in series with a photosensitive
metal/semiconductor junction (a Schottky diode). The
photopotential developed at this junction acts in the
same way as an external voltage, polarizing the metal
electrode. As yet, the potential for oxygen evolution
has not been attained on such an electrode (475, 476).

On the other hand, when the film is deposited at a rapid rate it becomes porous. Thus, direct contact between the semiconductor and the solution is not entirely excluded. However, (475, 477) porous metal films provide the semiconductor no protection from corrosion.

Although covering semiconductor electrodes with metal films did not solve the problem of protecting these photoanodes from corrosion, a useful effect was discovered in the course of this research: the electrocatalytic action of metal atoms on semiconductor surfaces. This effect is especially marked in the case of photocathodes, as opposed to photoanodes. Photocathodes function in a reducing atmosphere, and the problem of corrosion is not so acute. A relatively small - of the order of a monolayer - amount of silver, gold or palladium on the surface of a p-type gallium phosphide considerably enhances the photoevolution of hydrogen, or in other words decreases η^C (478-481).

The protection of unstable semiconductors towards corrosion with the aid of a thin film of a more stable semiconductor (482-486) has the following potential advantage: such a "two-layer" electrode will absorb more solar light energy than an ordinary one. To achieve this, the outer layer should be made from a semiconductor with a wider bandgap than the substrate, as is done in heterojunction solid-state solar cells. The short wavelength part of the spectrum is then absorbed in the outer layer, while the longer wavelength part penetrates the over layer and is absorbed in the substrate. Such an electrode will function effectively if the minority carriers generated by light in the interior of the electrode are free to penetrate the outer layer and reach the electrolyte interface where, along with carriers generated in the outer layer, they can enter into the electrode reaction.

In practice, this scheme has not been fully realized (433, 487, 488). For example, a continuous film of TiO_2 protects semiconductor materials such as GaAs, GaP and Si from photocorrosion. However, the photoelectrochemical behavior of the electrode is entirely determined by this film. The protected semiconductor, as in the case of the continuous metal film, proves to be switched off. In other words, the two-layer electrode has no advantages when compared with an ordinary titanium dioxide electrode. This is because of the potential barrier to the photogenerated minority carriers from the narrow bandgap semiconductor which

emerges at the junction between the two semiconductors.
This barrier excludes practically all of the minority
carriers photogenerated in the narrow bandgap material
from the electrode process. Figure 8.8a showed schematically such a barrier for the contact between two n-type
semiconductors with different bandgaps E_g. Indeed, while
the majority carriers (electrons) generated by light in
the narrower bandgap semiconductor substrate move, without hindrance, into the electrode bulk, the minority carriers (holes) from the substrate are unable to overcome
the barrier at the contact between the two semiconductors.
Only those holes generated in the outer layer take part
in the oxygen evolution reaction at the electrode. The
presence of the substrate may slightly shift the electrode potential at the expense of the photoelectromotive
force produced at the heterojunction (489).

The two methods for protecting semiconductors from
photocorrosion examined in this section have not given
satisfactory results. For this reason $A^{III}B^V$ and $A^{II}B^{VI}$
materials, which are promising as regards their photoelectric properties, are not used as anodes for water
photoelectrolysis. This highlights the goal of finding
semiconductors with relatively narrow bandgaps (1.5-2 eV)
which are stable with respect to photocorrosion.

≡ 9.3.3. C o m b i n a t i o n s   o f   T w o
P h o t o e l e c t r o d e s   i n   O n e   C e l l
a n d   o f   T w o   P h o t o c e l l s. As has already been stated, the insufficiently high energy of
the quanta predominating in the solar spectrum, when
compared with the change in free energy for the reaction,
still presents a major obstacle towards the effective
conversion of solar energy by water photoelectrolysis.
Moreover, the experimental results warrant the conclusion that water splitting is hardly possible using the
simplest one-quantum process, at least not without the
use of additional voltage from an external source. It
should be emphasized that this conclusion not only relates to photoelectrochemical converters but also to a
wide class of photochemical processes for splitting
water into hydrogen and oxygen. Thus, the quest for a
suitable process which would make it possible in photoelectrolysis to use the energy of more than one quantum
of light seems promising. Photoelectrolysis in a cell
with two photosensitive electrodes is one such process.

The use of a cell with two semiconductor electrodes
instead of one should basically increase the efficiency.
Consider an electrolysis cell with electrodes made from

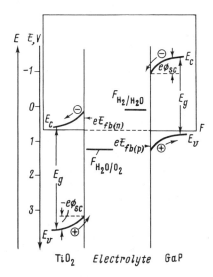

Fig. 9.10. The energy diagram for a cell for water photoelectrolysis with two semiconductor photoelectrodes: n-type TiO_2/p-type GaP. Potentials are cited against the Normal Hydrogen Electrode.

n- and p-type semiconductors, both of which are illuminated. This type of cell, proposed by Nozik (490), makes it possible to carry out a two-quantum photoelectrochemical process, and to thus gain in the overall energy balance. The two electrodes may be made from different semiconductors or from the same semiconductor but one of them should be doped with a donor and the other with an acceptor impurity. The energy scheme for such a cell is shown in Fig. 9.10, where the pair n-type TiO_2/p-type GaP serves as an example. The minority carriers take part in the reactions at each electrode, namely, oxygen evolution on the TiO_2 with the participation of holes, and hydrogen evolution on the GaP with the participation of the conduction band electrons.

For cells with electrodes made of the same semiconductor, if the band edges are pinned at the surface, the top of the valence band at the surface $E_{v,s}$ takes the same position for both electrodes as does the bottom of the conduction band $E_{c,s}$ (Fig. 9.11). Consequently, an analogous relation to Eq. (9.7) can be written for the energy balance in the conversion process,

$$E_g = \Delta G + \Delta E(n) + \Delta E(p) \qquad (9.8)$$

It should be pointed out that ΔE denotes the difference in energies between the electrochemical potential for the hydrogen (oxygen) electrode reaction in solution and the edge of the corresponding band at the semiconductor surface. As can be seen from Eq. (9.8), the entire width

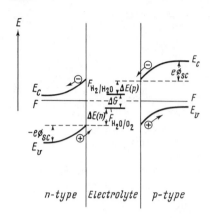

Fig. 9.11. The energy diagram for a cell for water photoelectrolysis with two photoelectrodes made from the same semiconductor but of different conductivity type. The symbols in brackets denote the conductivity type.

of the semiconductor bandgap is used effectively in the cell described (there are no losses caused by band bending, $e\phi_{sc}$). Moreover, there is no need to apply an external voltage to such cell even if, for example, the flat band potential of the photoanode is more positive than F_{H_2/H_2O}. Thus, the TiO_2 photoanode in combination with a metal electrode does not give water photoelectrolysis without an external voltage, whereas the cell n-type TiO_2/p-type GaP requires no external power supply (490).

In practice it is not easy to realize the self-evident advantages of the cell with two semiconductor photoelectrodes. It should be borne in mind that replacing the metal auxiliary electrode with a semiconductor electrode entails a certain loss in electrocatalytic activity. Since both electrodes of the cell now have low exchange currents, the overvoltages η^a and η^c become considerable. In addition, an efficient cell with two photoelectrodes has to meet a greater number of constraints characterizing the individual semiconductor electrode materials and their electrolyte interfaces (E_g, E_{fb} etc.).

In order to ascertain the potential of this approach experiments were conducted with cells using n-type semiconductors as photoanodes and p-type semiconductors as photocathodes (445, 490–495). The surface areas (or illumination intensities) of the two electrodes were selected in such a way that photocurrents at the anode and cathode were equal in their absolute value. The results (494) are summarized in Table 6. From Table 6 it is clear that, in comparison to the TiO_2-metal system cited on p. 274, the replacement of the metal auxiliary electrode with a second semiconductor photoelectrode (494) does not

TABLE 6. Efficiencies of Cells with a Combination of both
p- and n-Type Photoelectrodes

Combination of electrodes	Efficiency in per cent (for sunlight)
$TiO_2(n)$ - $GaP(p)$	0.1
$TiO_2(n)$ - $CdTe(p)$	0.04
$SrTiO_3(n)$ - $GaP(p)$	0.67
$SrTiO_3(n)$ - $CdTe(p)$	0.18

lead to a substantial increase in the efficiency of photoelectrolysis.

We should also mention an original electrode structure proposed by Nozik (491, 492, 496) which he calls a "photochemical diode." Two pieces of semiconductor of the different conductivity types, or of semiconductor and metal, serve as the separate electrodes in the diode. The pieces are pressed together and connected by an ohmic electrical contact. The size of a diode may vary from fractions of a micrometer to several millimeters. The thickness of the semiconductor pieces must always exceed both the thickness of the space charge region and the depth of light penetration. A multitude of such diodes are then suspended in an electrolyte solution which is mixed and illuminated. The photochemical diode functions as a galvanic couple; the electrode reactions occur on the working surfaces of cathode and anode with equal rates, and the resultant mixture of products escapes into the solution. If the overall process is water splitting, as is the case with n-type TiO_2/p-type GaP diodes in aqueous solution (492), then a mixture of gaseous products - oxygen and hydrogen - is formed, and it has to be separated into its components. (Sometimes H_2 and H_2O_2 are the products of photoelectrolysis, and their separation presents no difficulties.) It is quite probable that the proposed structure will be applied in some special applications of solar energy conversion (for example, in water desalination (497-499)).

Another possible way to implement water photoelectrolysis with the use of relatively low-energy quanta is to subdivide the full process into two stages, with each of them being carried out in a separate photocell (Fig. 9.12). These photocells are not electrically connected but are connected through the solution. In the first of these photocells water is reduced to hydrogen on the

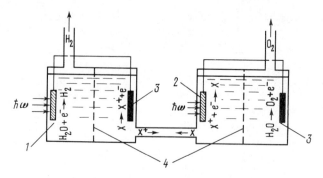

Fig. 9.12. The scheme for two-stage
water photoelectrolysis: 1) photocathode;
2) photoanode; 3) auxiliary electrode;
4) diaphragm.

cathode, while a reagent X is oxidized on the anode.
This reaction proceeds more readily than the oxidation
of water to oxygen. The resultant oxidized form of X,
X^+, is transported to the second photocell where it
undergoes reduction on the cathode (regenerating the
initial reagent X), while water is oxidized to oxygen
on the anode. Thus, the production of hydrogen and
oxygen is the net result of the overall process. The
reaction occurs in accord with Eq. (9.1) but now using
two quanta of light, rather than one, for each electron
transferred in the overall electrolysis reaction. The
reagent X performs the function of a "carrier". It is
neither included in the final products, nor is it consumed.
It is not impossible to find a carrier such that the
change in free energy for the reaction in each of the
two cells will be less than the 1.23 eV necessary for
the single stage splitting of water. This will naturally
make it possible to use longer wavelength light, to drive
the reactions in the photocells, than can be used for the
direct single quantum process. Such a combination of two
photoelectrochemical cells connected in series models,
to a certain extent, the combination of the two photo-
systems involved in photosynthesis in green plants (see,
for example, (196)). No system of this type has yet been
realized but this appears to be a quite promising avenue
for future research.

Ξ 9.4. P h o t o e l e c t r o c h e m i c a l
        S y n t h e s i s

The overwhelming majority of studies of photoelectrochemical conversion of solar energy into chemical
products have still been concentrated on the problem of

water splitting as a method for obtaining hydrogen.
This has largely distracted the attention of researchers
from the study of other photoelectrochemical reactions
and from working out methods for the production of chemical products other than hydrogen. At the same time, many
photoelectrochemical processes have the advantage, when
compared with water photoelectrolysis, that their change in
free energy, ΔG, is less than the 1.23 eV required for
water splitting. Therefore, such reactions can be carried out with higher efficiencies using sunlight.

As already mentioned, one should distinguish, among
photoelectrochemical reactions, between photoelectrolysis
and photocatalysis. As an example, let us examine the photodeposition reactions of copper on to particles of titanium dioxide suspended in solutions of copper salts in the
presence and absence of the acetate ion (500). Suspensions of particles function as microgalvanic couples
(Fig. 9.13). The illuminated part of the particles's
surface serves as the anode where the photooxidation of
water to oxygen (reaction 9.3a) or of acetate ions (the
so-called photo-Kolbe reaction) takes place

$$CH_3COO^- + h^+ \longrightarrow \tfrac{1}{2} C_2H_6 + CO_2$$

The non-illuminated TiO_2 surface serves as the
cathode where Cu^{2+} ions are reduced with the deposition
of copper metal. The overall reaction in these two cases
can be written as follows:

$$Cu^{2+} + H_2O \longrightarrow Cu + \tfrac{1}{2} O_2 + 2H^+ , \quad \Delta G > 0 \quad (9.9a)$$

$$Cu^{2+} + 2CH_3COO^- \longrightarrow Cu + C_2H_6 + 2CO_2 , \quad \Delta G < 0 \quad (9.9b)$$

The first reaction (9.9a) is an example of photoelectrolysis, while the second reaction (9.9b) is an example of
photocatalysis.

It should be pointed out that the reverse reaction
to the photoelectrolysis step (Eq. 9.9a) occurs spontaneously. For this reason we may expect a steady state
to be established when the system is illuminated. In
the light, reactions on the semiconductor surface of the
type given in Eq. (9.9a) will be driven in the forward
direction (uphill). However simultaneously, the resultant products will spontaneously regenerate the initial
reactants outside the "reaction zone" (in the bulk of

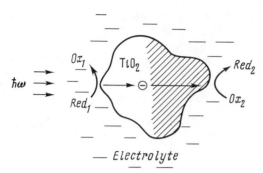

Fig. 9.13. The scheme of electrochemical reactions on the surface of an illuminated semiconductor particle in an electrolyte solution. The side of the particle in the shade is shown hatched.

the solution), always provided that the reverse thermal reaction is not kinetically inhibited. For photocatalytic reactions this cannot happen because the reaction proceeds in the same direction both in the light and in the dark.

The solvent photodecomposition reaction, which is an undesirable side process in this case, can naturally compete with the photoelectrochemical reaction of dissolved species. Thus, in aqueous solutions, in addition to the desired product, oxygen is also evolved on the photoanode as a result of the photooxidation of water. In the simplest case, the proportion of the overall photocurrent which goes towards the photoelectrosynthesis of the desired product is determined, as already stated, by the redox potentials of the two reactions: electrosynthesis and solvent decomposition (246, 501).

The study of photoelectrosynthesis at semiconductor electrodes began with the examination of the photooxidation reactions at the TiO_2 electrode. Photooxidation of acetate ion (see above) leads to the formation of ethane and CO_2 in aqueous and acetonitrile solutions (502-504). On TiO_2 photoanodes it is possible to oxidize methanol (505, 506), aromatic amines (507), and CN^- ions (508); on $MoSe_2$ photoanodes Cl^- and Br^- ions can be oxidized (509). It has been noted (510) that the radicals produced as the primary products of photooxidation may initiate polymerization, for example, of methylmethacrylate in concentrated acetic acid ("photopolymerization").

Among photoreduction reactions, let us mention the reduction of CO_2 on p-type GaP photocathodes with the formation of formic acid, formaldehyde and methanol (511-513). The efficiency of solar energy conversion in this reaction is close to 1 per cent; as could be expected from the values for ΔG, this is somewhat higher

than for water photoelectrolysis. p-Type GaP photocathodes may also be used for the photoelectrochemical
fixation of molecular nitrogen in a cell with an aluminum anode and a solution of $AlCl_3$ and titanium tetraiso-
propoxide in 1,2-dimethoxyethane. Under illumination
the dissolved nitrogen is reduced to NH_4^+ at the cathode,
and the aluminum anode dissolves. The reaction between
Al and N_2 is photocatalytic. It is also thermodynamically
possible in the dark (514).

   For all the photoelectrochemical reactions mentioned
above the oxidation potential turns out to be much less
positive (the reduction potential is less negative) than
the potential at which the same reaction proceeds in the
dark on a metal electrode (for example, platinum). This
gain (sometimes called the underpotential) brought about
by the consumption of the light energy reaches 1.5-2 V
in some systems.

Ξ 9.5. The Prospects for the
        Practical Application
        of Semiconductor
        Photoelectrochemical
        Cells

   Let us formulate the principal advantages of photoelectrochemical cells, both regenerative and photoelec-
trosynthetic, as compared with solid-state semiconductor
solar cells.

   (i) Simplicity and ease of fabrication: the photosensitive cell is produced by simply immersing the semiconductor electrode in the electrolyte solution. Therefore, there is no need to manufacture a p-n junction and
electrical contact on the illuminated surface of the
photocell (the auxiliary electrode plays this role).

   (ii) Semiconductor materials which only exist in
one conductivity type can be utilized effectively. For
these materials it is impossible to create a p-n junction
(these include TiO_2, $SrTiO_3$, and WO_3 amongst others).

   (iii) The possibility of using polycrystalline,
rather than single crystal, materials.

   These advantages suggest that, in the future,
photoelectrochemical cells may be far cheaper than
those solid-state solar cells in which single crystal
semiconductor materials are still used. It is precisely
high cost (rather than any exploitation, design or other

deficiency) that is at present the major obstacle to
the widespread use of solid-state cells. True, the maximum efficiency achieved for photoelectrochemical cells
with polycrystalline electrodes remains much lower than
the average efficiency of standard solid-state solar
cells. However, the relatively low efficiency of the photocells can be compensated for by the low cost of their
production and operation. On the whole, the prospects
for the use of photoelectrochemical cells should be
viewed in the light of progress in the sphere of solid-
state solar energy conversion, and in particular with
regard to the use of cheap polycrystalline or amorphous
semiconductor materials for this purpose (344, 345).

As for the probability of the practical use of
photoelectrochemical cells for power production in the
future, photocells of the regenerative type are apparently very promising, always provided that their stability can be ensured. There are also some technical
problems to overcome such as the hermetic sealing of the
cell, mass transfer, adequate transparency of the electrolyte, the use of readily available, non-toxic substances, the optimization of thermal conditions, and so
on.

As far as water photoelectrolysis is concerned,
researchers are increasingly coming to the conclusion
that the two-stage, rather than the direct, method for
hydrogen production is the most promising. This method
consists of a conventional water electrolysis cell fed
by the electrical current from solar cells (either
solid-state, or liquid-junction). This two-stage method
avoids the main shortcoming inherent in the one-stage
method; it is no longer necessary to combine good photosensitivity and high electrocatalytic activity in the same
electrode material. Furthermore, it is relatively easy
to obtain the required voltage for the electrolysis of
water by connecting several solar cells in series.
Finally, from a purely technological viewpoint, it is
much more convenient to use a compact electrolyzer
handling a large area of solar cells at one time than
to disperse photoelectrochemical cells over the same
area and use gas pipe-lines to collect the hydrogen. A
model installation of this kind is described in (515).
Figure 9.14 shows schematically a method for matching
the working characteristics of a panel of photocells at
a preset illumination intensity to the electrolyzer (the
area of the electrode surface is varied by changing the
number of separate electrode sections connected in parallel). To achieve the maximum efficiency the two

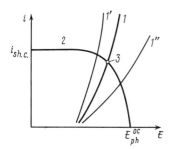

Fig. 9.14. Scheme for matching the
characteristics of a photocell and
an electrolyzer: 1, 1', 1") the
working characteristics of the electrolyzer as its electrode area is
varied; 2) the working characteristic of the photocell; 3) the point
on curve 2 corresponding to the
maximum power output.

curves should intersect at the working point on the
characteristic of the photocell at which the magnitude
of the power output, $i_{ph} \times E_{ph}$, is at its largest.   At
each value of illumination intensity, which changes
continuously throughout the day, this problem is solved
by selecting the appropriate number of solar cells and/
or sections of the electrolyzer.

Photoassisted electrolysis also holds out some hope
of application.

There is little doubt that electrochemical methods
will have an important role to play in solving the task
of utilizing solar energy.

Chapter 10

Light-Sensitive Etching of Semiconductors

The inherent participation of minority carriers in
the anodic dissolution and corrosion reactions of semiconductors, and the consequently strong dependence of
the rate of these processes on the conductivity and conductivity type of the material, underlies a number of
methods for the anodic and chemical etching of semiconductors used both in research and in the production of
semiconductor devices (see, for example, (104, 516)).
In this context it is often necessary to solve one of
the two opposing problems:

(i) to increase the selectivity of etching, for
example, to create a preset relief, to reveal heterogeneities in the material, and so on;

(ii) to secure a uniform etching regardless of the
local properties of the sample, for example, with the
aim of obtaining a highly polished surface.

Anodic and chemical etching of semiconductors exposed to light, so-called light-sensitive etching, is
applied in solving these problems. The change in the
concentration of minority carriers, brought about as
a result of illuminating the sample with light of a
quantum energy exceeding the width of the bandgap, is
used in this context.

In recent years, the study, on the one hand, of
the etching processes of semiconductors under conditions
of non-uniform illumination and, on the other hand, the

297

intensive development of the methods of quantum optics
has resulted in the emergence of new areas in the physical chemistry of semiconductor surfaces. Foremost among
these is the laser electrochemistry of semiconductors, an
area in which the work of two research groups has made an
essential contribution (see the reviews (517, 518)).
Laser electrochemistry embraces a range of problems related to laser stimulated electrochemical processes
occurring on the surfaces of semiconductors and metals.
In particular, light-sensitive etching using the coherence of laser emission may be utilized to record optical
information on the semiconductor surface, for example
to obtain holograms,[†] and also for the micro-profiling
of the surfaces of crystals and films. The photoelectrochemical profiling of semiconductor surfaces in interfering light beams was carried out for the first time in
(520, 521). The simplest holograms - surface phase diffraction gratings - were obtained by anodic etching of
silicon using an argon laser. The light-sensitive
etching of a number of semiconductor materials has been
applied with the same aim in subsequent work.

Below the processes of light-sensitive etching are
examined in detail chiefly with regard to their use in
the holographic recording of information; the concluding
section of this chapter contains a brief account of
other aspects of the light-sensitive etching effect.

10.1. Processes Occurring at the Semiconductor/Electrolyte Interface Under Non-Uniform Illumination

The influence of illumination on the anodic dissolution and corrosion of a semiconductor was analyzed
in Chapters 6 and 7. Let us now examine the specific
features which are introduced in the course of the
light-sensitive etching of homogeneous semiconductors
by non-uniform illumination of the surface. The uniform
illumination of non-homogeneous samples is discussed in
Section 10.7.

When the illumination is non-uniform, three major
cases can be singled out.

[†]Emission is called coherent if the monochromatic oscillations
have a constant phase displacement in time; a hologram is a light
interference pattern registered on a flat surface, which reproduces
a three-dimensional optical image (519).

(i) <u>Illumination under anodic polarization</u>. Under
the regime of a limiting anodic current the rate of
etching is limited by the rate of arrival of holes at
the electrode surface. The main source of holes in the
dark in n-type electrodes is generation in the neutral
bulk of the sample and/or in the space charge region.
Under steady-state conditions a certain, in general very
low, rate of dissolution is established. Illumination
serves as an additional source of holes (photogeneration),
and the illuminated areas of the surface dissolve faster
than the non-illuminated ones.

(ii) <u>Illumination in indifferent electrolyte under
open-circuit conditions</u>. The illumination of a semicon-
ductor/electrolyte interface is accompanied by the emergence of a photopotential (see Section 6.1). Thus when
an isolated sample, which does not react with the solution in the dark, is illuminated non-uniformly, gradients
in both the electrode potential and the concentration of
minority carriers are set up along the surface. As a
result, on the illuminated and non-illuminated areas of
the surface the conditions for the electrochemical reactions occurring between the sample and solution are
different.

When an n-type semiconductor is illuminated, its
stationary potential becomes more negative, and the
polarization curve shifts on the whole as shown in Fig.
10.1. On this basis, using the condition (see Eq. (7.3))
$i^a = -i^c$, it is easy to establish that the illuminated
areas of the n-type sample play the role of local anodes,
and that here the semiconductor is dissolved. On the
other hand, the non-illuminated areas play the role of
cathodes, and the evolution of hydrogen from water, the
reduction of dissolved oxygen, etc., occurs at these
areas. The electrolyte solution serves as a conductive
medium completing the circuit between the local "cathodes" and the local "anodes".

On p-type samples the non-illuminated areas serve
as anodes, while the illuminated areas serve as cathodes (see, for example, (522)). This distinction in
the behavior of n-type and p-type samples is caused by
the fact that the sign of the photopotential is determined by the initial (before illumination) sign of the
band bending. As a rule a depletion layer is established at open-circuit in wide bandgap semiconductors.
Consequently the signs of ϕ_{sc} and hence of the photopotential turn out to be opposite for n- and p-type
samples.

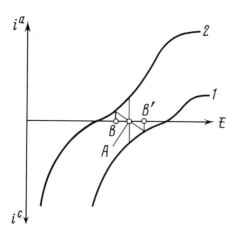

Fig. 10.1. Determination of
the photoetching characteristics of a homogeneous n-type
semiconductor sample when the
illumination is non-uniform.
Polarization curves for 1) the
dark and 2) the illuminated
semiconductor. Stationary
potentials are shown for (A)
a negligibly small and (B,B')
finite resistance in the
system (B is the potential of
illuminated, and B' the potential of dark areas of the
surface).

On the whole, the sample behaves as a corroding galvanic
couple (see, for example, (226)); the only difference is
that the anode and cathode of the couple emerge not as a
result of differences in the local chemical composition,
but as a result of differences in the local illumination
of the homogeneous sample. A method for the graphical
determination of the open-circuit potential was described
on p. 265 (cf. Figs. 10.1 and 9.5). If the total Ohmic
resistance R of the system is insignificant, then the
open-circuit potential over the entire surface of the
sample is the same (point A in Fig. 10.1). If the resistance R is finite, the illuminated and non-illuminated
areas will be at different potentials (points B and B'),
the difference being iR.

    (iii) <u>Illumination under conditions of corrosion</u>.
If the electrolyte contains an oxidant, whose equilibrium potential is more positive than the potential for
the anodic dissolution of the semiconductor, the solution is no longer indifferent. Corrosion will then
occur with simultaneous dissolution of the semiconductor and reduction of the oxidant. The simplest case is
that in which the cathodic process occurs exclusively
with the participation of electrons from the conduction
band and the anodic process involves only holes. Then,
as demonstrated in Section 7.2, the "dark" corrosion
rate for samples of any conductivity type is limited by
the supply of minority carriers, taking part in one of
the two conjugated reactions, to the semiconductor surface. Illumination accelerates the corresponding partial reaction and, hence, the overall corrosion process.
The case under consideration is similar to case (i) but
with the difference that the role of the externally

applied polarization is played by the "chemical" polarization achieved by adding an oxidant to the solution.

The simultaneous participation of carriers from both bands in at least one of the conjugated reactions introduces complications in this process. For example, the reduction of an oxidant can occur with the participation of both free and valence band electrons, i.e., via the valence band. In the dark this creates an additional source of holes for the anodic reaction, thereby diminishing the sensitivity to light of corrosion at large. As a consequence, the difference in the rate of etching on illuminated and non-illuminated areas is drastically reduced. In other words, the selectivity of etching decreases.

Comparing the various light-sensitive etching processes described above, it is possible to identify certain similarities and differences.

a) The rate of etching in the dark in cases (i) and (ii) is very small. This allows accurate control over the quantity of material removed from the surface in the course of light-sensitive etching. In contrast, in case (iii) there is always some "dark" etching rate which is not insignificant.

b) As a rule, the rate of etching in cases (i) and (iii) is proportional to the intensity of illumination. This makes it possible, by varying the local illumination, to obtain a predetermined profile during etching. In case (ii) this control is lost. This difference stems from the fact that in case (ii), due to the absence of any external polarization or oxidant in the solution, the supply of minority carriers is not rate limiting (as it usually is in cases (i) and (iii)).

c) In case (ii), unlike cases (i) and (iii), etching only occurs as a result of light energy conversion; no energy is provided by the external circuit or by the chemical energy of an oxidant in solution.

Etching under the polarization of an externally applied voltage (case (i)), referred to as photoanodic (photoelectrochemical) etching, and etching in an oxidizing solution (case (iii)), called photochemical etching, are the most commonly used of the three types of light-sensitive etching examined above.

Ξ 10.2. T h e   T h e o r y   o f   L a s e r   E t c h i n g

Let us consider a semiconductor/electrolyte system
for which the influence of light on the rate of the electrochemical process is primarily determined by the change
in the surface concentration of current carriers. The
following stages should be examined to describe the
kinetics of the formation of a microrelief during light-
sensitive (both photoanodic and photochemical) etching
of the semiconductor:

(i)   light absorption in the semiconductor,

(ii) diffusion of the photogenerated carriers
      from the semiconductor bulk to and along
      the surface,

(iii) subsequent heterogeneous reaction involving
      these carriers, the surface atoms of the
      semiconductor, and reagents from the solution.

The complicating side reactions occurring as a
result of the high intensity of the laser emission are
insignificant if the intensity of the radiation and the
corresponding fields are not too big. Some of these
complications will be considered in Section 10.6; for
the moment we shall assume that they are insignificant.

Suppose that two coherent monochromatic light beams
which lie in the same plane and have the same intensity
and wavelength λ (see Fig. 10.2) strike the initially
flat surface of an n-type semiconductor at the same
angle, Θ. We shall examine the case when the direction
of oscillation of the electric vector of the light wave
is perpendicular to the plane of incidence (s-wave)
because it is under these conditions that holograms are
most efficiently recorded.

In view of the fact that the period of oscillation
of the light wave is of many orders of magnitude less
than the time spent on the formation of the surface relief, we can examine the electromagnetic field of the
incident and scattered waves using a quasi-stationary
approximation. That is to say we shall assume that the
interface is immobile. The distribution of the non-zero
component of the electric field strength of the monochromatic wave in both media is then described by the
scalar Helmholtz equation (see, for example, (523)),
and depends, in the general case, on all three coordinates,

$$\left(\frac{\partial^2}{\partial x^2} + \frac{\partial^2}{\partial y^2} + \frac{\partial^2}{\partial z^2}\right)\xi_j + \frac{\omega^2}{c^2}[n_j]^2\,\xi_j = 0 \qquad (10.1)$$

where n is the complex refractive index of the medium (see Section 1.3 [†]), and the index j takes two values: j = 1 corresponds to the electrolyte; j = 2 corresponds to the semiconductor. The solution of Eq. (10.1) in both media, together with the matching conditions at the interface (524), fully describes the distribution of ξ_j. The motion of the interface X (y,z,t), occurring as a result of the etching of the semiconductor, changes the distribution, and determining the form of the relief X (y,z,t) (where y and z are coordinates along the surface and t is time) is the ultimate goal of these calculations. The rate of destruction of the crystal (to be more exact, the velocity of motion of the interface) v is related to the characteristic parameters of the electrochemical reaction by

$$v(y,z,t) = \frac{k_{el}\Omega}{\upsilon}\, p\,[X(y,z,t),y,z] \qquad (10.2)$$

where k_{el} is the rate constant for the electrode reaction (see Section 4.1), Ω is the volume occupied in the crystal by each dissolving particle (atom or molecule) and υ is the number of holes expended on the transfer of each of these particles to the solution. Specific examples of such reactions are given in Section 7. The quantity p (X,y,z) is the surface concentration of holes; it can be found from an equation of the type of Eq. (1.68). In this case, the rate of photogeneration, g_p (x,y,z), is proportional to the density of the local light flux J (see Eq. (1.69)). In its turn, J is equal to the square of the modulus of the electric field of the light wave in the semiconductor, ξ_2. As a result, taking into account the fact that this is a three dimensional problem, we obtain for the quasi-stationary case

$$D_p\left(\frac{\partial^2 p}{\partial x^2} + \frac{\partial^2 p}{\partial y^2} + \frac{\partial^2 p}{\partial z^2}\right) - \frac{p-p_o}{\tau_p} + \alpha\,\frac{c}{n_2}\,|\xi_2(x,y,z)|^2 = 0 \qquad (10.3)$$

Here the value of ξ_2 is found from the solution of Eq.

[†] It is evident that the expression for the field $\xi(x)$ given in Section 1.3 is a particular solution of Eq. (10.1).

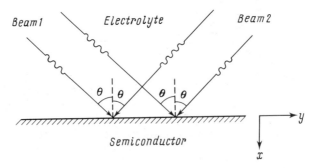

Fig. 10.2. The schematic arrangement of laser
beams with respect to the semiconductor surface
for holographic light-sensitive etching.

(10.1). At the interface the following relation serves
as the boundary condition for Eq. (10.3):

$$D_p(r\nabla p)\big|_{X = X(y,z)} = -(k_{el} + s_{eff})p(X,y,z) \qquad (10.4)$$

where r is a unit vector normal to the interface and s_{eff}
is the effective hole recombination velocity determined
by processes other than those associated with electrode
reaction, for example, surface recombination, and also
recombination within the space charge region (525). It
should be pointed out that the quasi-steady-state conditions for Eq. (10.3) are observed when the "etching
velocity" is much lower than the average velocity of
diffusional transfer, that is, when $v \ll L_p/\tau_p$. This

inequality is, in any case, observed at the low light
fluxes.

    Finally, the change in the relief $X(y,z,t)$ is re-
lated to the velocity $v(y,z,t)$ by an equation ensuing
from differential geometry

$$\frac{\partial X}{\partial t} = \frac{v(y,z,t)}{\sqrt{1 + (\partial X/\partial y)^2 + (\partial X/\partial z)^2}} \qquad (10.5)$$

In the one-dimensional case (for a flat interfacial sur-
face) $\partial X/\partial y = \partial X/\partial z = 0$ and therefore $\partial X/\partial t = v$.

    The system of equations (10.1) and (10.3), together
with the relations (10.2), (10.4) and (10.5) and the
matching conditions for the field, ξ_j, at the moving
interface, $X(y,z,t)$, contains the complete mathematical
formulation of the self-consistent problem.

Finding an accurate solution is very difficult, and it can only be achieved by resorting to numerical methods. It is possible, however, to get analytical expressions which describe the initial stage of the process. In this case the interface is still very similar to a plane. Thus the diffraction of the light by the emerging relief can be neglected, and Eq. (10.5) can also be simplified. Furthermore, if the incident light beams are flat, the initial coordinate system can be selected in such a way (see Fig. 10.2) that the problem contains no dependences on the z coordinate. The solution of the system of equations in this simplified case shows that a sine-shaped relief emerges under illumination and that its form is described by (517):

$$X(y,t) = 2J_o \frac{\Omega}{\upsilon} bt \left(1 + q \, cos \left\{ \frac{2\pi y}{d} \right\} \right) \tag{10.6}$$

where J_o, as before, is the density of the light flux entering the semiconductor; the quantity d is the spatial period of the emerging relief, where $d = \lambda/2ncos\theta$ and n is the refractive index of the semiconductor ($n = Ren_j$ when $j = 2$); the dimensionless parameters and q characterize the intensity of the light-sensitive etching (b) and the quality of the interference pattern recorded (q). Physically, the parameter q is the ratio of the amplitude of the component of the etching rate spatially modulated by light to the mean value of the etching rate. The parameters b and q included in Eq. (10.6) are given by

$$b = \frac{\alpha L_p}{1 + \alpha L_p} \frac{k_{el}}{k_{el} + s_{eff} + D_p/L_p}; \quad q = \frac{1 + \alpha L_p}{\gamma + \alpha L_p} \frac{k_{el} + s_{eff} + D_p/L_p}{k_{el} + s_{eff} + \gamma D_p/L_p} \tag{10.7}$$

where

$$\gamma = \sqrt{1 + (2\pi L_p/d)^2} \tag{10.8}$$

and characterizes the degree of "blurring" of the periodic distribution caused by the diffusion of holes. Let us note that in the initial stages of the process $X(y,t)$, according to Eq. (10.6), grows linearly with the time, t.

Under conditions of weak absorption of the light ($\alpha L_p \ll 1$), a small recombination velocity, s_{eff}, and a low electrochemical reaction rate, k_{el}, it follows from

Eq. (10.7) that $q = \gamma^{-2}$. This result was first obtained in (526). Irrespective of the value of αL_p, in the limiting case when $\gamma \to \infty$ ($L_D \gg d$), from Eq. (10.7), we get $q = 0$, in other words, the formation of a relief does not take place. In the opposite limiting case $\gamma \to 1$ ($L_p \ll d$) the parameter q attains its maximum value: $q \cong 1$. These results have a clear physical meaning. The quality of the recorded interference pattern is obviously reduced by any lateral spread in the charge carriers created by the light at a certain restricted area of the semiconductor. The effective distance over which this spreading occurs is determined by the rates of recombination in the bulk and on the surface. A re-duction in L_p as a result of an increase in the bulk recombination rate leads to an increase in the resolu-tion of the light-sensitive etching process. At the same time a reduction in L_p decreases the concentration of non-equilibrium minority carriers and, as a result, the rate of etching. The decrease in the value of b with decreasing L_p corresponds mathematically with this relationship.

Under certain conditions it is, at the same time, possible to substantially reduce the influence of relief erosion due to the diffusion of holes even when L_p is large. According to Eq. (10.7), when the inequalities $\alpha L_p \gg \gamma$ and $k_{el} \gg \gamma$ (D_p/L_p) are observed, the values of γ and L_p cancel in the expressions for q and b. In this particular limiting case we have $q = 1$ and $b = k_{el}/(k_{el} + s_{eff})$. The physical meaning of this result is also clear: if the holes are generated in the immediate region of the surface (for "surface-absorbed" light, α is large) and are consumed rapidly (for fast electrochemical reaction, k_{el} is large), then the dif-fusion of photogenerated holes beyond the boundaries of the illuminated surface areas can be neglected. In other words, the resolving power of light-sensitive etching is not limited by the diffusion length of minority carriers, as might have been assumed *a priori*.

In order to achieve greatest intensity of photo-etching and a high quality recording of the interfer-ence pattern it is necessary:

(i) to use semiconductor materials with reasonable values of L_p and, as far as possible, with small values of s_{eff} so that recombination does not reduce the inten-sity of the process;

(ii)   to use strongly absorbed light to decrease
the relative role of recombination of the carriers in
the bulk of the semiconductor;

(iii) to select the composition of the electrolyte
solution so as to ensure a sufficiently rapid rate for
the electrode reaction; in some cases other specific demands will be made on the composition of the electrolyte
(see Section 10.5).

In addition to these constraints, there are some
additional requirements for the light-sensitive etching
process when diffraction gratings of a high spatial frequency, corresponding to a small value of the spatial
period, are required. These are examined in the next
section.

Note that the solution obtained to describe the
dynamics of formation of the microrelief in its initial
stage is not self-consistent. In particular, it takes
no account of the effect of light diffraction on the
emerging relief and the ensuing redistribution of the
field ξ_j (j = 1,2). It also takes no account of the influence of the surface curvature on the transport of
the holes. The interplay of the emerging relief and
the conditions for its growth may, given sufficiently
long periods of time, lead to qualitatively new effects.
These include the saturation of the depth of the relief,
non-linear distortions and the generation of higher
spatial harmonics.

As has already been noted, the self-consistent solution can only be obtained in the general case by numerical integration. However, under conditions where the
relief being produced has a relatively large value of the
period d it may be possible to examine the properties of
the self-consistent solution analytically. This analysis
has been carried out (517) with the assumption that
$d > \lambda_n$ and $d \gg h$, where h is half the depth of the relief. These calculations are based on the expansion of
the fields ξ_1 and ξ_2 near the surface as infinite Fourier
series in the x and y directions (Rayleigh expansion)
followed by subsequent simplification of the terms in
the light of the inequalities above. Without going into
the details of the treatment let us summarize the main
conclusions. For small angles of incidence of the recording beams (corresponding to low spatial frequencies
d^{-1}) and strongly absorbed light (corresponding to large
values of α) the local etching rate remains a sine function of the coordinate even when the relief is well de-

veloped. Hence the diffraction of light under these conditions is not a source of significant non-linear distortions in the process of recording the information.

Other distortions can occur which are physically related to tangential etching within the pits of the relief
(see Eq. (10.5)). Mathematical solution by the method of
consecutive approximations shows that the effect of the
tangential etching is both to decrease the average rate
of surface etching and the amplitude of the main frequency. In addition, higher spatial harmonics of the form
$\cos (4\pi y/d)$, $\cos (6\pi y/d)$, and so on, arise whose amplitude
is proportional to the dimensionless parameter $\frac{1}{3}(\pi h/d)^2$.
Thus, substantial linear distortions, caused by tangential etching, are supposed to occur at $h \lesssim d$.

Ξ 10.3. E v a l u a t i o n   o f   t h e   R e s o l u t i o n
        o f   L a s e r   E t c h i n g

There is one additional effect, not examined above,
which limits the physically attainable spatial frequency
of holographic diffraction gratings. This is the effect
of the drift spreading of photogenerated minority carriers along the semiconductor (527, 528). However,
before we analyze this in detail, let us evaluate the
maximum attainable spatial frequency, d^{-1}, for these
gratings. To achieve the maximum value of d^{-1} it is
apparently necessary for all of the light to be absorbed
in the space charge region, which requires $\alpha L_{sc} \gg 1$, and
for the rate of etching to be large. Given these conditions photogenerated carriers do not become concentrated at the interface and the resolving power of the
etching process is determined by the spreading of the
charge carriers in the time (t) taken for them to reach
the surface. The diffusional deviation, δL, occurring
in the time t is of the order of $\sqrt{D_p t}$. The time taken
to reach the interface is $t \simeq \alpha^{-1}/v_d$, where v_d is the
drift velocity in the space charge field, and α^{-1} determines the depth of photogeneration of electron-hole
pairs. Thus:

$$\delta L \simeq \sqrt{D_p/\alpha v_d} \qquad\qquad (10.9)$$

The value of $(\delta L)^{-1}$ can be regarded as the maximum
spatial frequency characterizing the resolution of the
method. Assuming that $D_p = 10 \text{ cm}^2/\text{s}$, $\alpha = 10^5 \text{ cm}^{-1}$ and
$v_d = 10^7 \text{ cm/s}$ we obtain, from Eq. (10.9), a value of
$(\delta L)^{-1}$ of the order of $30,000 \text{ mm}^{-1}$. Such high spatial
frequencies are not achieved experimentally. This is
because of the limitations determined by the optical

characteristics of the interference pattern which occur
at the wavelength of the light used in the etching
process. In this case, the maximum value of d^{-1} is
determined by Eqs. (10.6)-(10.8).

When the rate constant for the electrode reaction
is less than the drift velocity of the minority carriers,
they accumulate in the illuminated areas. As a result,
the total surface charge density will change, and an
additional potential difference arises which tends to
repel the carriers to the dark areas and thus to level
out their concentration. This effect (of self-induced
drift) leads to a blurring of the diffraction pattern.
A mathematical analysis of this problem, using a number
of simplifying assumptions, has been carried out (527).

Let us assume that in the absence of illumination
a depletion layer is established in the semiconductor.
Then, in accordance with the Boltzmann distribution,
practically all of the holes are concentrated within a
thin layer near the surface. The thickness of this
layer, $L_{sc}^{(p)}$, inside the depletion layer is given by

$$L_{sc}^{(p)} \simeq \frac{kT}{e|\xi_{sc}|} \simeq \frac{kTL_{sc}}{2e|\phi_{sc}|} \qquad (10.10)$$

In Eq. (10.10), if $|\phi_{sc}| \gg kT/e$, then $L_{sc}^{(p)} \ll L_{sc}$, and
thus the holes account for only an insignificant part
of the space charge region. For this reason it is possible to use a one dimensional model, in which the x
coordinate is neglected, to describe the hole current
for spreading along the surface:

$$i_p(y) = eu_{p,s}p\xi(y) - eD_p \frac{dp}{dy} \qquad (10.11)$$

where $p(y)$ is the surface concentration of non-equilibrium holes measured in cm^{-2} (not be confused with p_s
measured in cm^{-3}) $u_{p,s}$ is the surface mobility of holes,
and $\xi(y)$ is the self-induced electric field periodically
changing in the y-direction: $\xi(y) = -d\phi(y)/dy$. The
light-induced surface potential is equal to $\phi(y)=Q(y)/C_{sc}$,
where $Q(y) = ep(y)$ is the additional surface charge,
induced by the light, and C_{sc} is the depletion layer
capacity and is independent of $p(y)$.

Let us assume, for simplicity, that the intensity
of the recording light beam incident normal to the
surface ($\theta = 0$) is given by

$$J(y) = J_o \left(1 + \zeta \cos \frac{2\pi}{d} y \right) \qquad (10.12)$$

where ζ is the coefficient of spatial modulation of the light. If the value of α^{-1} is small then all the holes generated by the light escape recombination and reach the surface. Then, from Eqs. (10.11) and (10.12), taking into account the continuity equation, we obtain the following equation determining the concentration, $p(y)$, of non-equilibrium holes

$$\frac{d}{dy} \left(\frac{eu_{p,s}p(y)}{C_{sc}} \frac{dp}{dy} + D_p \frac{dp}{dy} \right) + J_o \left(1 + \zeta \cos \frac{2\pi}{d} y \right) - \frac{s_{el}}{L_{sc}^{(p)}} p(y) = 0 \qquad (10.13)$$

The last term is proportional to the "electrochemical recombination velocity" s_{el} (see Eq. (7.9)) and describes the effect of the flux of holes, generated by light, from the semiconductor to the reagent in solution as a result of the electrode reaction.

To take into account the spreading effect due to the diffusion of holes during their motion towards the surface we must replace ζ in Eq. (10.13) by the effective value $\zeta' = \zeta \exp\{-2\pi^2 (\delta L)^2/d^2\}$, where $\overline{(\delta L)^2}$ is the root mean square value of δL (this is of the order of δL^2; δL is given by Eq. (10.9)).

It is possible to carry out the analysis of Eq. (10.13) with the assumption that the variable component, $\tilde{p}(y)$, is small when compared with the constant background component, $\bar{p}$. Then, assuming that in Eq. (10.13) $p(y)=\bar{p}+\tilde{p}$ and $|\tilde{p}| \ll p$, we get $\bar{p} = J_o L_{sc}^{(p)} s_{el}^{-1}$, and $\tilde{p}(y)$ is the solution of the linearized equation

$$D_{eff} \frac{d^2}{dy^2} \tilde{p} - \frac{s_{el}}{L_{sc}^{(p)}} \tilde{p} + J_o \zeta \cos \left(\frac{2\pi}{d} y \right) = 0 \qquad (10.14)$$

where

$$D_{eff} = D_p + \frac{eu_{p,s}}{C_{sc}} \bar{p}$$

is the effective diffusion coefficient for the diffusion of holes along the surface. This equation is readily solved if we look for $\tilde{p}$ of the form $\tilde{p} \sim \cos (2\pi y/d)$. Then as a result we obtain, for the total excess concentration

$$p(y) = \frac{J_o L_{sc}^{(p)}}{s_{el}} \left[1 + q \cos \left(\frac{2\pi}{d} y \right) \right] \qquad (10.15)$$

where q is the coefficient of modulation of the holographic relief and characterizes the quality of the recording process under these conditions; in its physical significance it is very similar to the coefficient q in Eq. (10.6):

$$q = \frac{\zeta}{1 + 4\pi^2 D_{eff} L_{sc}^{(p)}/d^2 s_{el}} \qquad (10.16)$$

The solution obtained for p(y) by this method is, strictly speaking, only valid for $q \ll 1$; this condition is known to apply when $\zeta \ll 1$, which corresponds to weak modulation of the illumination intensity. It is much more important, however, that for high spatial frequencies, d^{-1}, even if $\zeta \approx 1$, $q \ll 1$. Physically this stems from the fact that for large d^{-1} the spreading of the non-equilibrium holes is so strong that their spatial modulation, q, becomes very small. In other words, the solution obtained above is correct under those conditions which are the most interesting experimentally: when the resolution of light-sensitive etching approaches its maximum value.

Using Eqs. (10.15) and (10.16), let us formulate the parameters for the semiconductor and the electrolyte solution which determine the value of q, since this is essential to assess the potential of the methods under consideration. From the definitions of $L_{sc}^{(p)}$, given in Eqs. (10.10) and (3.22b), we have

$$L_{sc}^{(p)} \sim (|\phi_{sc}|N_D)^{-\frac{1}{2}}$$

Then from Eq. (10.16), it follows that the light-sensitive etching is possible at greater spatial frequencies when

(i)   the semiconductor is highly doped,

(ii) the band bending is large,

(iii) the rate of the electrochemical reaction is high.

At the same time, from Eq. (10.16), an increase in the intensity of the incident light, leading to an increase in p and thus D_{eff}, may impair the resolution of the method. Below it will be shown that these results have been confirmed experimentally.

10.4. Photoanodic Recording of Holograms

As already discussed, light-sensitive etching is applied in two modifications: either as photoanodic or as photochemical etching. These two methods, as applied to the recording of holographic information (the etching of diffraction gratings on to semiconductor surfaces), will be examined below.

The photoanodic (photoelectrochemical) method for recording information was elaborated in (529-534). As an example, let us examine the production of a periodic relief on the surface of a cadmium sulfide crystal by anodic etching using two interfering laser beams of equal intensity. For this purpose the light beam from a helium-cadmium laser (λ = 441.6 nm) was first expanded to a diameter of 15 mm and then, with the help of the reflection from two mirrors, divided into two beams which intersected on the surface of the electrode submerged in an electrolyte solution (Fig. 10.3, cf. Fig. 10.2).

The anodic reaction is

$$CdS + 2h^+ \rightarrow Cd^{2+} + S \qquad (10.17)$$

(cf. Fig. 7.9) and occurs with 100 per cent current efficiency. A layer of amorphous sulfur is formed on the surface of the CdS. This layer does not passivate the surface, and therefore, does not hinder the further dissolution of the crystal but it does cause the emergence of an undesirable, fortuitous microrelief. To prevent this, etching is carried out in electrolyte solution in which sulfur is either dissolved (NaOH) or oxidized (HNO_3).

The quality of the relief obtained was directly determined in the course of etching by using the fact that light reflected from the etched grating creates a diffraction pattern which continuously characterizes the depth and form of the grating profile. The intensity of the reflected light was also measured; both the light from the helium-cadmium laser, used to stimulate the

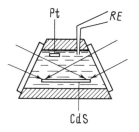

Fig. 10.3. Schematic view of the cell for holographic photoanodic etching. RE is the electrolytic contact to the reference electrode (517).

photoetching, and from an auxiliary helium-neon laser ($\lambda = 632.8$ nm), to which the surface layer of CdS is practically transparent so that it does not perturb the etching process, were used for this purpose. This method is highly sensitive: the emerging relief can be detected on the surface when its depth is about 0.01 µm.

A characteristic feature of the etching of a diffraction grating is the existence of an "optimum" amount of charge, Q_{optim}, which should be passed to develop a relief most closely resembling a sinusoidal shape. This is seen in the dependence of the intensity of the diffraction pattern (J_{diffr}) on the amount of charge, Q, passed. J_{diffr} passes through a maximum (curve 1 in Fig. 10.4) when $Q = Q_{optim} = i^a t_{optim}$, where t_{optim} is the optimum etching time for an average anodic photocurrent density of i^a. Experiments show that Q_{optim} is independent of both the spatial frequency of the etched grating d^{-1} (within the interval $10^2 - 10^3$ mm^{-1}) and the intensity of the laser beams used to record the grating (within the interval 0.1-10 mW/cm^2). Q_{optim} is about 5×10^{-2} C/cm^2. In the initial recording phase when $t < t_{optim}$ the depth of the etched grooves, in accordance with Eq. (10.6), is approximately proportional to the etching time t; when $t > t_{optim}$ the depth of the relief tends towards its limiting value (curve 3 in Fig. 10.4). The depth of the etched grooves of the relief (2h) on a CdS electrode under optimum conditions is about 110 nm; for a CdSe electrode the corresponding depth is about 170 nm. The profile of the relief closely approximates to a sinusoid. This is confirmed by studies of the diffraction and interference patterns (517), and for some other semiconductors (for example silicon), by electron microscope analysis of the cleavage planes of the crystal perpendicular to the direction of the grooves (521). At the same time, random (noncoherent) decomposition of the surface also takes place, and this is manifested both in the relatively low diffraction intensity and in the decrease in the reflectivity of the surface (curve 2 in Fig. 10.4). The etching process is observed to be highly sensitive to the homogeneity of the CdS crystal.

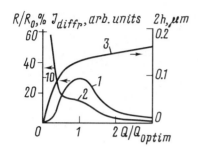

Fig. 10.4. Holographic photoanodic etching of CdS. The dependence of the diffraction pattern intensity J_{diffr} (curve 1), of the ratio of the specular reflection coefficients for the etched (R) and non−etched (R_o) surfaces (curve 2) and of the depth of the relief 2h (curve 3) on the ratio of the charge passed (Q) to the optimum charge (Q_{optim}) (517).

The scope of the methods is shown in Table 7 which gives the maximum values of the spatial frequency (d^{-1}) of the recorded gratings achieved, the wavelengths of the recording light (λ) used, and the diffusion length (L_p) of the minority carriers in the specimens.

On the basis of the data given we conclude that the resolving power of the photoelectrochemical process for recording information (defined as the maximum number of grooves in the diffraction gratings) is rather high (more than 6,000 mm^{-1}). For some systems that have been studied (for example, CdS) this has already reached the "optical limit" (cf. p. 308) set by the wavelength of the light used and the refractive index of the solu−tion. This limit is achieved when q = 1 which, from Eq. (10.7), occurs when the light is absorbed at the surface ($\alpha L_p > 1$) and the rate of discharge of the photogenera−ted holes at the illuminated electrode surface is fast ($k_{el} \gg \gamma (L_p/\tau_p)^{\frac{1}{2}}$). It should be stressed that for semi−conductor materials with a large minority carrier dif−fusion length L_p (germanium, silicon) the resolving power, as predicted in our analysis of Eq. (10.7) above, was found to be much better than could be expected on the basis of the value of L_p.

From a practical viewpoint, it is important that the gratings recorded by the photoelectrochemical method are found to possess the property of linearity. This means that when several gratings of different spatial frequency or orientation are recorded simultaneously on one and the same crystal surface, the information carried by each grating can be read independently. In other words, the separate gratings do not interfere with one another. In addition, the relation Q = i^a t = const is observed, so that the intensity of the image depends upon the product of the exposure time, t, and the select−ed anodic current density, i^a. This is called the principle of mutual replaceability.

TABLE 7. The Characteristics of Diffraction Gratings Obtained by Light-Sensitive Etching

Semiconductor	d^{-1}, mm^{-1}	λ, nm^a	L_p, cm	Reference
	Photoanodic etching			
CdSe (single crystal)	2,600	632.8	10^{-4}	(529)
CdSe (film)	2,000	632.8	10^{-5}	(534)
CdS	6,150	441.6	10^{-4}	(530)
Si	2,500	514.5	10^{-2}	(520, 521)
As_2Se_3	2,600	632.8	–	(535)
	Photochemical etching			
GaAs	5,000-6,000	441.6	10^{-4}	(527)
GaAlAs	2,500	632.8	–	(540)
$GaAsP^b$	2,700	514.5	–	(541)
As_2S_3	3,500	441.6; 514.5	–	(541)
$As_2Te_3{}^c$	2,600	632.8	–	(535)
Ge	1,000	632.8	10^{-1}	(240)
Cu_2O	2,600	–	–	(535, 542)
GeO	3,000	441.6	–	(543)

a) At these wavelengths the coefficient of light absorption is $(1-6) \times 10^4$ cm^{-1}.

b) Composition of the etchant: $HNO_3+HCl+H_2O$ (4:1:1).

c) Composition of the etchant: 3N $K_2Cr_2O_7+10$ N H_2SO_4.

The properties of etched gratings mentioned above make it possible, among other things, to carry out more complicated two-dimensional profiling of the surface by turning the sample at a certain angle while recording. Thus, recording two gratings with equal frequency whose grooves are turned through 90° with respect to each other results in a surface structure of symmetry C_4 (a square grating). Similarly, by turning through 60° surface structures of symmetry C_3 (a hexagonal grating) are obtained. Even more complicated structures can be generated as in, for example, a Fresnel plane lens.

Ξ 10.5. P h o t o c h e m i c a l R e c o r d i n g o f H o l o g r a m s

The photochemical method has been used (518, 528, 535-542) to etch diffraction gratings on a number of single

crystal semiconductors (Ge, Si, GaP, CdS, CdSe) and semiconductor films (including polycrystalline CdSe, CdTe,
As_2Te_3, InSe, GaSe, Cu_2O, and glassy Ag_2Se_3).

When selecting the composition of the electrolyte,
and in particular the oxidant required for the photochemical etching, a number of particular features of the
process should be taken into consideration. First and
foremost the solution should be transparent to the light
used for etching. Usually the blue/green region of the
spectrum is used; this means that a large number of
oxidants such as Br_2, I_2, Fe^{3+}, $Fe(CN)_6^{3-}$ are scarcely
fit for the purpose. Furthermore, the cathodic evolution
of hydrogen should not occur on the non-illuminated areas
of the sample since gas evolution is incompatible with
precision etching. It is therefore advisable to use
oxidants which are more readily reduced than H^+ ions or
water.

In addition, the following specific demands are
made on the composition of the solution (518).

(i) The oxidant should be reduced with the participation of only conduction band electrons in order to
avoid "electrochemical generation" of holes. This keeps
the "dark" dissolution to a minimum.

(ii) A high rate of photoetching should be ensured
in order to achieve a high resolution against the background of the "dark" dissolution of the sample. For
this reason relatively strong oxidants should be used at
sufficiently high concentration to avoid diffusional
limitations.

(iii) Gas evolution and slime formation must not
occur in the course of etching. This means that oxidants
such as HNO_3, H_2O_2 and others, which are very suitable
in other respects, are of limited use. It should be
noted that in general the selection of oxidants has been
conducted empirically.

The compositions of solutions used for photochemical
etching are given in (518). The characteristics of holograms obtained on some semiconductors using the photochemical method are given in Table 7. The highest spatial
frequency for a relief (5,000-6,000 mm^{-1}) was obtained
on an epitaxial film of GaAs with a donor concentration
of $10^{17} - 10^{18}$ cm^{-3} using a helium-cadmium laser.

The resulting relief profile of the grating is usually almost sinusoidal. However, if the single crystal is anisotropic with respect to etching, triangular and trapezoidal forms result. For example, on the (100) plane of gallium arsenide the grooves running parallel to the (110) axis have a triangular profile with a right angle at the apex. The depth of the etched grooves is usually roughly equal to half the distance between the neighboring grooves of the grating (for low spatial frequencies it becomes somewhat bigger). Apart from incidental violations of the etching regime (vibration of the apparatus, defects and dislocations in the semiconductor) the depth of etching is limited by the low rate of mass transfer in the solution in the deep grooves.

As in the case of photoanodic etching, the form of the relief is controlled in the course of etching by measuring the intensity of the light reflected from the surface of the sample. Once again there is an optimum time for photochemical etching, during which the etched relief reaches its most correct form. This time is characterized by a maximum effectiveness of the diffraction pattern. The etching process can also be characterized by the quantity of light energy received by the sample during the optimum etching time. For the materials quoted above this is approximately 1 J/cm^2. A minimum illumination intensity at which the rate of photochemical etching considerably exceeds the rate of the dark etching is about 10^{-4} W/cm^2.

In conclusion, let us deal with the etching of diffraction gratings on Cu_2O films (542). At first sight the mechanism for this process does not fall within any of the categories examined in Section 10.1. The photochemical decomposition of Cu_2O occurs under illumination in non-oxidizing solutions (dilute hydrochloric or oxalic acid) without external polarization. The relief is produced by a disproportionation type reaction

$$Cu_2O + 2H^+ \rightarrow Cu + Cu^{2+} + H_2O \qquad (10.18)$$

Figure 7.10 helps us to understand the mechanism of the processes occurring at the illuminated Cu_2O/solution interface. The electrochemical potentials for the anodic (with the participation of holes) and cathodic (with the participation of the conduction band electrons) decomposition of Cu_2O turn out to be very close to one another. This predetermines the extreme instability of Cu_2O with respect to photocorrosion. Under illumination, a situation is easily achieved in which the quasi-level

for holes reaches the anodic decomposition level, while
the quasi-level for electrons reaches the cathodic decomposition level. The two decomposition reactions
occur simultaneously on the surface of Cu_2O, and together they result in a photocorrosion process or the
type in Eq. (10.18) which formally resembles disproportionation. This process differs from the photocorrosion
and light-sensitive etching of other semiconductors in
that no special oxidant is required in the solution.
(The role of the "oxidant" is played by Cu_2O itself.)
Photoetching of As_2Se_3 in alkaline solutions probably
follows the same pattern.

Ξ 10.6. T h e   L a s e r   E l e c t r o c h e m i s t r y
         o f   S e m i c o n d u c t o r s :
         P r o b l e m s   a n d   P r o s p e c t s

Comparing the two methods for recording holograms,
photoanodic and photochemical etching, one can see
(Table 7) that as far as their resolution is concerned
they produce roughly comparable results. Photochemical
etching is a more universal method since both n- and
p-type semiconductors can be processed. It is also simpler technologically because it requires no external
power supply and no Ohmic contact to the semiconductor.
In its turn, photoanodic etching makes less rigid demands on the choice of electrolyte composition.

As for the comparison of these methods with the most
widely used non-electrochemical methods, such as the
recording of holograms on photographic plates and also
photolithography (etching through a mask obtained with the
use of a photoresist) (544, 545), all of the methods
possess a similar resolving power. The highest spatial
frequencies achieved are about 6,000 mm^{-1}, and these are
now limited by the possibilities of high spatial frequencies of the light beam, rather than the recording process.
It is obvious that as this parameter increases, the photoelectrochemical methods of etching should become preferred;
inasmuch as the relief is formed right on the semiconductor surface rather than, in the case of photolithography,
in the auxiliary photoresist layer. For this reason it
is also easier to secure a correct sinusoidal profile
for the diffraction grating.

The high quality of the recorded information produced
by the light-sensitive etching of semiconductor materials
warrants the hope that this method will find application.
It should be borne in mind that the gratings obtained
(unlike those recorded on photographic emulsion, electro-

chromic or magnetic films and other materials) have semiconductor properties in addition to their preset optical
characteristics. This provides quite new spheres of
application for such gratings. In particular, a distributed feedback laser (537) was constructed based on epitaxial GaAs films where the surface of the thin-film
waveguide, etched to a sinusoidal relief, performed the
function of the distributed mirrors of the optical quantum generator. In its characteristics, this laser excels
traditional optical quantum generators based on Fabry–
Perot resonators. Diffraction gratings produced by
etching are used as input and output couplers in integrated optics (see (518) and references therein).

Holographic recording of information is only one
example of the use of lasers in electrochemistry. Although most impressive results have been obtained in
light-sensitive etching, these in no way exhaust the
scope of the laser electrochemistry of semiconductors.
The study of processes stimulated by laser emission
presents a number of problems of interest to the theore-
tician (examination of the kinetics of photoelectrochemical reactions in strong fields, multi-quanta electron
transfer in the space charge region with the participation of surface states, local heating phenomena, and so
on) as well as being of broad prospective practical importance.

It should be stressed that semiconductor surfaces
are extremely convenient subjects for the study of laser-
stimulated electrochemical processes. This is primarily
due to the diversity of effects which can be controlled
by light (photoetching, photoanodization, photocatalysis,
and so on) and which are characterized by their high
sensitivity. The problems of laser electrochemistry,
unlike those of traditional semiconductor photoelectrochemistry, include the following processes (517).

(i) Threshold electrochemical reactions stimulated
by powerful laser emission. These reactions may proceed in different ways and give qualitatively different
products as a result of participation of both highly excited particles from the solution and the non-equilibrium electron-hole plasma. Such reactions open up
opportunities to develop new chemical technological
processes.

(ii) Photoelectrolysis at relatively weak light
intensities using the coherence of the laser emission.
Among processes of this type, light-sensitive etching,
described above, is particularly important.

(iii)   Emission of coherent light stimulated by electrochemical reaction. This type of emission could form the basis of a new type of laser using electrochemical pumping.

Progress in laser electrochemistry is largely dependent upon progress in a number of adjacent areas, primarily in the photoelectronics and physics of semiconductor surfaces. A rigorous description of laser-stimulated chemical processes at the semiconductor surfaces should take into account, among other things, the complicated structure of the wave field near the interface produced by interference, diffraction, the excitation of surface electromagnetic waves, and the specific behavior of light in the thin layer near the surface. The particular interactions of powerful laser emission with solids should also be taken into account, for example the creation of a largely degenerate electron-hole plasma, the overheating of the surface layer, the changeover of the system to a non-equilibrium thermodynamic state (generation of non-equilibrium phonons, shock waves). In addition, multi-photon absorption can lead to direct excitations which are impossible under ordinary conditions. These include excitation of the solvent molecules, the solute ions, and adsorbed particles. This considerably changes the energetic and kinetic characteristics of the electrochemical reactions. The list of problems discussed above, although in no way exhaustive, demonstrates the scope of the challenge of the laser electrochemistry of semiconductors.

Ξ 10.7. P h o t o a n o d i c   E t c h i n g   u n d e r   U n i f o r m   I l l u m i n a t i o n

Anodic etching is widely used in processing the surface of semiconductor samples. In particular, the method is used to remove the outer layer of the material in which the crystalline structure has been distorted in the course of mechanical processing (the damaged layer). Since n-type semiconductor anodes dissolve only very slowly in the dark, they should be illuminated, the rate of etching of the sample being set by the intensity of the illumination. The latter fact is used in a number of methods for the precision etching of semiconductor materials (see, for example, (516, 546, 547)).

Under sufficiently strong illumination, anodic etching is not limited by the transport of minority carriers but proceeds under kinetic control. For this reason local photoelectrochemical inhomogeneities are

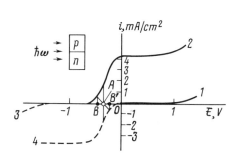

Fig. 10.5. Determination of the
photoetching characteristics of
a sample with a p-n junction. Polarization curves for a homogeneous
n-type GaAs electrode (1) in the
dark and (2) in the light, and for
a p-type GaAs electrode (3) in the
dark and (4) in the light; the
dots denote the stationary potentials on photoetching a sample
with a p-n junction when the ohmic
resistance in the system is (A)
negligible and (B,B') finite (B
is the potential of an n-type area,
and B' the potential of a p-type
area) (549). (Reprinted by permission of the Publisher, The
Electrochemical Society, Inc.)

largely levelled out and etching is practically uniform
even when the sample is not perfectly homogeneous with
respect to its conductivity type and magnitude (see,
for example, (548)). This promotes the production of
samples with highly polished surfaces.

Even so, photoanodic etching can be applied precisely
to reveal local inhomogeneities and this is often possible without an external voltage source. For example,
consider a sample with a p-n junction, which is immersed
in an indifferent electrolyte and illuminated (Fig. 10.5).
Such a system can be viewed as two short-circuited electrodes, one of them being n-type and the other p-type
(549). Due to the difference in photocurrent-voltage
characteristics of the two electrodes (these are shown
in Fig. 10.5), an electric current flows between the n-
and p-type regions; the magnitude of the current and the
stationary potential of the sample can be readily obtained by the familiar method of analyzing the behavior
of galvanic couples (cf. Fig. 10.1). As shown by Fig.
10.5 the n-type region turns out to be the anode and
this region is dissolved, whereas the p-type region is
the cathode where hydrogen is evolved. The magnitude
of the etching current is determined from Fig. 10.5
starting from the condition that the anodic and cathodic
currents must balance: $i^a = -i^c$. A step is formed as a
result of the photoetching at the boundary between the
n- and p-type regions; these steps thus show the places
where p-n junctions emerge on the crystal surface. This
very principle is used to reveal inhomogeneities of any

other sort which influence the current-voltage characteristic of the semiconductor, for example, inhomogeneities
in the electrical conductivity, the diffusion length of
holes, and so on (548, 550).

The sensitivity of this method for revealing inhomogeneities can be enhanced by complementing the local
dissolution at the local anodes by the electrodeposition of metals at the local cathodes (replacing the
evolution of hydrogen). With this aim in view, metal
salts, for example, of copper and silver, are added to
the electrolyte solution. The boundary of the metallic
deposit accurately marks the position of the p-n junction
(see (5)).

Chapter 11

SELECTED TOPICS IN THE PHOTOELECTROCHEMISTRY

OF SEMICONDUCTORS

This concluding chapter deals with some trends in the photoelectrochemistry of semiconductors which, due to different reasons of a sometimes transient character, have yet to be developed on a large scale. These areas of research are, however, not only of considerable scientific interest but also, in some cases, of considerable practical importance. In this connection it is expedient to elucidate the problems and the most significant results.

Ξ 11.1. P h o t o e l e c t r o c h e m i c a l   N o i s e

Noise in electrochemical systems (551, 552) is caused by spatial and temporal fluctuations: accidental (chaotic) changes in the parameters which determine the behavior of these systems. Both internal, for example, the Brownian movement of microparticles, and external, for example, fortuitous changes of temperature and illumination, factors inevitably lead to electrochemical fluctuations. These are registered as noise in measurements of the basic macroscopic characteristics of electrochemical systems such as, for example, current and potential.

There are a number of important reasons for studying electrochemical noise. Firstly, electrochemical fluctuations are an immediate manifestation of elementary microprocesses and, as such, their quantitative analysis lays the groundwork for the noise method for investigating

a number of kinetic microcharacteristics of electrode
reactions. Secondly, in accordance with the general
principles of statistical mechanics, the study of fluctuations holds the key to the quantitative description of
the degree of reversibility of macroprocesses occurring
in the system and to an evaluation of the stability of the
steady-state as a whole. Finally, the study of electrochemical noise is of considerable applied importance,
making possible an evaluation of the threshold of sensitivity of diverse electrochemical devices.

The main difficulties in the study of electrochem-
cial fluctuations are not of a fundamental nature. Their
origin is in the smallness of the quantities to be measured.
In recent years, however, due to advances in instrumen-
tation, an opportunity has arisen not only to reliably
observe electrochemical noise at a level of 10^{-8} - 10^{-9} V,
but also to thoroughly investigate it. This has made
possible the study of fluctuation electrochemistry on a
regular basis, and the research was started by Tyagai
and coworkers (553, 554). The wide range of problems related to the diversity of fluctuation effects in electrochemical systems has been reviewed (552). We shall confine ourselves to the description of some aspects of
fluctuation electrochemistry which are directly connected
with photoprocesses at semiconductor/electrolyte interfaces.

In order to illustrate the principles and possibilities of the fluctuation method let us examine the results of studying the noise of maximum photocurrent,
i_{ph}^{max}, at semiconductor electrodes (555). Let us assume
that the generation of minority carriers, using holes as
our example, taking part in the electrode reaction is the
rate determining stage of the photoprocess in question.
Then the maximum steady state photocurrent which can be
registered in the system is

$$i_{ph}^{max} = eJ_o YM \qquad (11.1)$$

where J_o is the light flux entering the semiconductor, Y
is the quantum yield for the photogeneration of electron-
hole pairs, and M is the anodic current multiplication
coefficient (see Sections 4.5 and 6.2). In accordance
with Eq. (11.1), temporal fluctuations of the maximum
current consist of the fluctuations in the photon density
of the incident light, fluctuations in the quantum yield,
and fluctuations of the multiplication coefficient. Each
of these fluctuations is the result of physically indepen-

dent causes and can be independently calculated as a root-mean-square deviation from the mean.

It should be borne in mind that the root-mean-square deviation, $\delta(u^2)$, of the quantity u from its mean value $\bar{u}$ is, by definition,

$$\delta(\overline{u^2}) = \overline{(u - \bar{u})^2} \tag{11.2}$$

where the bar denotes the average value in accordance with the preset statistical distribution. Multiplying out the expression in parentheses and bearing in mind that $\bar{\bar{u}} = \bar{u}$, we can write Eq. (11.2) as follows

$$\delta(\overline{u^2}) = \overline{u^2} - 2\bar{u}\,\bar{\bar{u}} + \overline{\bar{u}^2} = \overline{u^2} - \bar{u}^2 \tag{11.3}$$

In order to calculate the fluctuations determined by irradiation it is more convenient to use the power, $P(\omega)$, of the light flux entering the semiconductor. Assuming that the density of photons in the light flux is in compliance with Bose-Einstein statistics and performing the averaging in accordance with Eq. (11.3), we obtain an expression (556) for the root-mean-square fluctuation, $\delta(\overline{P^2})$, within the range of frequencies Δf

$$\delta(\overline{P^2}) = 2\hbar\omega P[1 - e^{-\hbar\omega/kT}]^{-1} \Delta f \tag{11.4}$$

For the root-mean-square fluctuation of the current, $\delta(\overline{i_p^2})$, caused by the photon fluctuations, we get, in conformity with Eq. (11.3),

$$\delta(\overline{i_p^2}) = (eYM)^2 \delta(\overline{J_o^2}) = \frac{1}{(\hbar\omega)^2}(eYM)^2 \delta(\overline{P^2}) \tag{11.5}$$

Combining Eqs. (11.1), (11.4) and (11.5), we obtain

$$\delta(\overline{i_p^2}) = 2ei_{ph}^{max} YM[1 - e^{-\hbar\omega/kT}]^{-1} \Delta f \tag{11.6}$$

Equations (11.4) and (11.6) only hold for "equilibrium" radiation. In particular, they can be used as approximations if, for example, a filament lamp serves as the source of light. However, they must be modified when laser emission is used.

The influence of fluctuations in the quantum yield, Y, can be easily taken into consideration using the assumption that each incident quantum generates one hole

with the probability Y and that it is not photoactive
with the probability (1-Y). Such a probability law is
described by the binomial distribution (Bernoulli distribution) (557). If the average number of photogenerated holes per second is proportional to J_0Y, then the
fluctuation from this value, according to Eq. (11.3), $\overline{}$
is proportional to $J_0^2\delta(\overline{Y^2})$. The calculation of $J_0\delta(\overline{Y^2})$
by averaging on the basis of the binomial distribution
shows that it is proportional to $J_0Y(1-Y)$. As a result,
taking Eq. (11.1) into account and within the range of
frequencies Δf, we obtain a relation for the fluctuation
in the current caused by the fluctuation in the number
of holes generated:

$$\delta(\overline{i_Y^2}) = 2ei_{ph}^{max}(1-Y)M\Delta f \qquad (11.7)$$

In order to take account of the fluctuations in the
multiplication coefficient M, let us assume (551) that
each photogenerated hole reacting at the interface
causes n elementary charges to flow in the external circuit with the probability β, and the flow of m elementary
charges with the probability $(1-\beta)$. Then the mean value
of M is given by

$$M = \beta n + (1-\beta)m \qquad (11.8)$$

and the value of $\delta(\overline{M^2})$, as for $\delta(\overline{Y^2})$ above, can be found
with the help of the Bernoulli distribution. Calculating
the fluctuations in the value of i_{ph}^{max} resulting from
chaotic changes in the multiplication coefficient, we get

$$\delta(\overline{i_M^2}) = 2ei_{ph}^{max}[\beta(1-\beta)(n-m)/M]\Delta f \qquad (11.9)$$

Then, with due account of the statistical independence
of the fluctuations P, Y and M, we find, for the resul-
tant noise $\delta(i_{ph}^{max})^2$, with the help of Eqs. (11.6), (11.7)
and (11.9)

$$\delta(i_{ph}^{max})^2 = 2e\gamma i_{ph}^{max}\Delta f \qquad (11.10)$$

where
$$\gamma = M[1-Y+Y(1-e^{-\hbar\omega/kT})^{-1}] + \frac{n-m}{M}\beta(1-\beta) \qquad (11.11)$$

For the frequencies, ω, of the exciting light used in
photoelectrochemistry $\hbar\omega/kT \gg 1$, and so Eq. (11.11) can
be written in the following approximate form

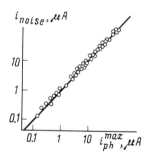

Fig. 11.1. The dependence of the equivalent noise current, in the 1-20 kHz frequency band, on the magnitude of the anodic current for CdS photooxidation, i_{ph}^{max}, in 1 M KCl at E = 1.5 V (against Normal Calomel Electrode) (555).

$$\gamma = M + (n-m)\beta(1-\beta)/M \qquad (11.12)$$

Thus, from Eq. (11.12), the generator of noise current, i_{noise}, in this particular case does not depend on the frequency; this type of noise is called "white noise."

Experimental studies of photoelectrochemical noise have been conducted (555) for the anodic oxidation of cadmium sulfide single crystals illuminated by light in the region of intrinsic absorption. The theoretical interpretation is unambiguous in the potential region where the overall noise measured considerably exceeds the thermal noise which is not directly related to the process under consideration. This is precisely the situation that is realized for the photoanodic oxidation of CdS.

Figure 11.1 shows the experimentally determined dependence of the equivalent noise current, i_{noise}, on current, i_{ph}^{max}. The resultant straight line through the origin supports the main concepts of the theory. The value of γ determined by the slope of this line is $\gamma = 0.97 \pm 0.1$. A comparison of this value with Eq. (11.12), using Eq. (11.8), shows that $\gamma = 1$ may only be realized if $n = m = 1$. Hence $M = 1$, and in this particular case there is no multiplication of the photoanodic hole current.

Electrochemical noise for reactions with the participation of minority carriers under conditions where the reaction rate is controlled by optical excitation has also been investigated (558-561). Of great interest here is the study of those systems where there is photocurrent multiplication and where noise measurements may be used directly to calculate the multiplication coefficient. In particular, the current multiplication effect can be expected in the anodic oxidation of substances capable of forming unstable intermediate products. For example, when oxidizing formaldehyde, CH_2O, photocurrent multiplication may result from the following consecutive stages in the electrode process (cf. Eq. (6.34)):

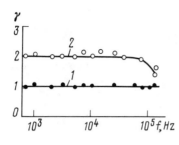

Fig. 11.2. The frequency
dependence of the noise
coefficient for CdSe photooxidation in the region of
the maximum anodic photo-
current. Solution: 1)
2 M KCl; 2) 2 M KCl +
1 M KOH + 1 M CH_2O (558).

$$CH_2O + h^+ + OH^- \rightarrow HCOOH + H^\bullet$$

(11.13)

$$H^\bullet \rightarrow H^+ + e^-$$

where $H^\bullet$ is an adsorbed hydrogen atom which is an intermediate product of the reaction.

These concepts were verified when studying the noise
from photoanodic reactions on cadmium selenide electrodes
(558). The frequency characteristics of the noise generator is shown in Fig. 11.2. In an indifferent electrolyte (KCl) the value of γ, in accordance with Eq. (11.12),
is independent of frequency and, as in the case of CdS
photooxidation, is practically unity. The introduction
of formaldehyde into the solution results in an increase in
γ, and over the range $1 < f < 10^5$ Hz, the magnitude of γ
is still independent of frequency but is now equal to
two. Using Eq. (11.12), these results may be interpreted
in such a way that two single pulses of photocurrent are
strictly correlated with each other. Therefore, the current multiplication coefficient M is exactly equal to two,
as would be expected from the scheme given above. The
decrease in γ at higher frequencies points to a finite
value for the correlation time of the two consecutive
electron transfers, hence, at high frequencies the correlation weakens and M diminishes (558, 559). This
effect may be used, among other things, to study the time
characteristics of two-stage (or multistage) electrochemical processes.

It should be pointed out that photocurrent doubling
has also been observed in photoelectrochemical noise
studies of germanium (559) (cf. Section 4.5) and zinc
oxide (560, 561) photoanodes.

Ξ 11.2. <u>E l e c t r o n   P h o t o e m i s s i o n   f r o m</u>
<u>S e m i c o n d u c t o r   E l e c t r o d e s</u>
<u>i n t o   S o l u t i o n</u>

The photoelectrochemical phenomena in the preceding sections were based on the transfer of the photoexcited electrons directly from a semiconductor to localized states in solution. These localized electronic levels are connected with the individual ions and molecules of the solution. However, the transfer of electrons to delocalized states in solution is also possible and fundamentally different.

This type of photoreaction is called electron emission (562, 563). The energy levels corresponding to delocalized states in solution can be approximately represented as a band of allowed states. Let us denote the lower boundary of this band as E_{deloc}. This band is formed by interactions with all of the solvent molecules and is similar to the conduction band in solids (23). Photoemission takes place to this band. After emission and thermalization, the emitted delocalized electron is further localized, thus forming a solvated (or in water, hydrated) electron. The energy level of the solvated electron, E_{solv}, lies below the bottom of the band of allowed delocalized states in the solution. Finally, the electron may pass from the solvated state to an even lower local energy level, E_A, associated with an acceptor (also called a scavenger, see below) in solution (Fig. 11.3).

The study of electron photoemission (the external photoeffect) at the electrode/solution interface is not only an inalienable part of contemporary photoelectrochemistry but is also one of its most interesting and promising trends. Recent decades have seen the publication of quite a few papers dealing with different aspects of emission phenomena at the electrode/solution interface, with most of the work being concentrated on metal electrodes (563). The considerable achievements in the photoelectrochemistry of metals have encouraged the development of theoretical and experimental studies to uncover the basic mechanisms of photoemission from semiconductor electrodes to solutions (564-568). These studies are indispensable to an understanding of not only the quantitative, but also the qualitative relationships of a broad range of processes occurring at the semiconductor/electrolyte interface. In particular, they make it possible to determine a number of parameters relating to the energy structure of the semiconductor/

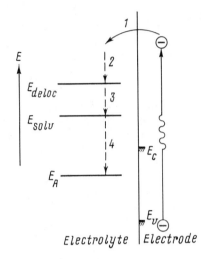

Fig. 11.3. The scheme of electron transitions stemming from photo-emission into solution: 1) electron photoexcitation in the emitter electrode and the act of photoemission; 2) thermalization of photo-emitted electron in solution; 3) solvation (hydration) of thermalized electron; 4) capture of the solvated electron by an acceptor (A) in solution.

electrolyte interface: the value of the Volta potential (see Section 2.1), the magnitude of band bending, the energy of the delocalized electron in solution, etc.

The studies of the external photoeffect on metals have also demonstrated that photoemission to solution is a rather effective and flexible method for the study of the structure of the interface proper and of the different physical and chemical processes proceeding at, and in close proximity to, it.

The photoemission current, I_p, for the photoemission of electrons from simple metals to solutions with a sufficiently high electrolyte concentration is given by

$$I_p = A(\hbar\omega - \hbar\omega_o - eE)^{5/2} \qquad (11.14)$$

This is known as the "five-halves power law" (563, 569). Here ω is the frequency of the incident light, ω_o is the so-called threshold frequency (the red boundary) of the external photoeffect, and E is the electrode potential. For photoemission from metals the value of ω_o is determined by $\hbar\omega_o = w_T{}^{ms}$, where $w_T{}^{ms}$ is the thermodynamic work function characterizing the metal/solution system at a potential, E, regarded as zero. If at $E = 0$ the irradiation frequency, ω, is such that $\omega < \omega_o$, then $I_p = 0$. Finally, in Eq. (11.14) A is a quantity which is independent of E and is proportional to the intensity of the incident light. Equation (11.14) describes the current for a single photon photoemission, that is, of the process proceeding with the absorption, in its

elementary act, of only one quantum of light. Usually
the currents relating to multiphoton processes are several
orders of magnitude smaller and may become prominent only
when $\omega < \omega_0$ (for details see (38)).

It should be emphasized that, as shown by Eq. (11.14),
the potential, E, applied to the system (in the case of a
metal electrode, this is completely localized in the Helmholtz layer) directly changes the magnitude of the work
function of the metal in solution (or, which is the same
thing, it shifts the red boundary of the external photoeffect).

The specific features of electron photoemission from
semiconductors arise from the fact that the electrochemical potential F, whose position determines the value of
the thermodynamic work function (see Section 2.1), is
located within the bandgap (heavily degenerate semiconductors, which are not considered here, may be an exception). Correspondingly, these materials are completely
devoid of electrons with an energy coincident with the
value of F. Thus, unlike metals, the minimum quantum
energy $\hbar\omega_0$ corresponding (at T = 0) to the possibility
of single photon emission does not coincide with the
thermodynamic work function of the semiconductor. The
characteristics of the semiconductor/electrolyte interface, which are of prime importance for photoemission,
are shown in Fig. 11.4. The concentration of electrons
in the conduction band (and also on the donor and acceptor
levels in the bandgap) is insignificant when compared with
their concentration in the valence band, and the photoemission current due to the photoexcitation of electrons
from the conduction band may also be regarded as negligible. (This is a cardinal difference between semiconductor emitters and metals; photoemission in the latter case
is determined precisely by the conduction band electrons.)

Thus, the threshold frequency $\hbar\omega_0$ is simply determined by the possibility of photoemission from the valence
band and, as shown in Fig. 11.4, is given by $\hbar\omega_0 = E_g + \chi_s$
where χ_s is the difference between the potential energy
level for the delocalized electron outside the semiconductor and the energy of the bottom of the conduction
band at the surface $E_{c,s}$. Note that χ now (cf. Section 2.1)
depends on the properties of both the semiconductor and the
surrounding medium into which electrons are emitted. The
threshold quantum energy $\hbar\omega_0$ is called the photoemission (non-
equilibrium) work function w_p. It is obvious that
w_p is larger than the thermodynamic equilibrium work
function w_T. It should be also pointed out that due to

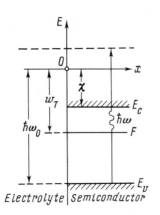

Fig. 11.4. Energy diagram of
the semiconductor/electrolyte
junction at photoemission.
The potential energy of the
delocalized electron in solution is regarded as the
reference point.

polarizing interactions of the emitted electrons with the
medium and also to dipole potential drops at the semicon-
ductor/solution interface the values of χ and w_p for
photoemission into the solution may differ considerably
from the corresponding "vacuum" values.

Bulk photoexcitation of electrons from the valence
band, which plays the major role in the generation of the
electrons to be emitted, is certainly accompanied by
interband transitions.   This explains another essential
feature of the external photoeffect at semiconductors.
The quantum energy $\hbar\omega_0 = E_g + \chi$ is only sufficient for
photoemission to occur in the case of electron photogeneration by the mechanism of indirect transitions (see
Fig. 11.5 and also Fig. 1.9).   In this case the threshold
frequency is equal to ω_0.   For direct transitions, which
are much more probable, it is necessary for the corresponding frequency of the absorbed quantum to be such that
$\hbar\omega_0' > \hbar\omega_0$ for the electron to attain the final energy
necessary for its photoemission (in Fig. 11.5 the length
of arrow 1 corresponds to the threshold energy $\hbar\omega_0'$).   The
difference $\hbar(\omega_0' - \omega_0)$ depends on the dispersion law $E(p)$
in the bands and on the orientation of the emitter sur-
face with respect to the crystallographic axes.   Thus,
the notion of a threshold frequency turns out to be connected with the mechanism of photogeneration of the
electron.

The regularities of photoemission from semiconductors into vacuum for various initial electronic states
and types of transition, taking scattering processes into
account, were investigated by Kane (570).

The problem of the motion of photoelectrons from the
crystal bulk toward the emitting surface is central to
the theoretical analysis of the photoemission of electrons.

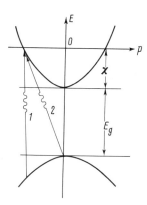

Fig. 11.5. Scheme for interband electron transitions at photoemission: 1) direct transition, 2) indirect transition.

In some papers it is assumed that this motion can be described, with some modifications, as a random walk of the point center in the medium (see, for example, (571, 572)). Another approach (564, 570) is based on the concepts of the propagation of a probability wave described by quantum mechanical equations. These two approaches conform to two different physical situations. The first approach is appropriate when the majority of photoelectrons undergo a considerable number of interactions after their generation. The second approach is correct when the emitted electrons leave the crystal without practically changing the quantum characteristics they obtained in the course of photoexcitation. For example, this is precisely the situation if the penetration depth of the exciting light, α^{-1}, is less than the free pathlength of the photoelectron.

The calculation of the photoemission current, I_p, from a semiconductor emitter to a concentrated electrolyte solution was performed in (564) (see also (563, 565, 568) within the framework of the quantum mechanical approach, neglecting dissipation processes. Furthermore, it was assumed that:

(i)   All the electronic states in the conduction band were empty while all those in the valence band were full.

(ii)   The radiation frequency was close to the threshold value so that $\hbar(\omega-\omega_o)/E_g < 1$.

(iii) There was a negigible contribution from processes occurring with the violation of the translational symmetry (resulting, for example, from heterogeneities).

The specific nature of the semiconductor/electrolyte interface makes itself felt, according to (564), in the shape of the surface potential barrier which is to be overcome by the emitted electrons. Of much importance in this context is the practical absence, given a sufficiently concentrated electrolyte solution, of long-range image forces.

As a result of calculating the electron photoemission current, we get the following expressions

$$
I_P = \begin{cases}
0 & \text{if} & \omega < \omega_o \\
B(\hbar\omega - \hbar\omega_o)^2 & \text{if} & \omega_o < \omega < \omega_o' \\
C(\hbar\omega - \hbar\omega_o')^{3/2} & \text{if} & \omega_o' < \omega
\end{cases}
\qquad (11.15)
$$

The two threshold frequencies ω_o and ω_o' included in (11.15) are given by

$$
\hbar\omega_o = E_g + \chi \quad ; \quad \hbar\omega_o' = E_g + \left(1 + \frac{m_c}{m_v}\right)\chi
\qquad (11.16)
$$

where m_c and m_v are the effective masses of the electron in the conduction band and of the hole in the valence band. B and C in Eq. (11.15) are independent of $\omega - \omega_o$ and $\omega - \omega_o'$ and are proportional to the probabilities of direct and indirect interband transitions respectively. From above, B << C so that the photocurrent determined by direct transitions, given comparable values of the differences $\omega - \omega_o$ and $\omega - \omega_o'$, is much larger than the photocurrent determined by indirect transitions (if B $\lesssim$ C then both terms should be taken into consideration in the frequency range $\omega \gtrsim \omega_o'$). It should also be pointed out that, if the simple approximation of effective masses m_c and m_v is not made, the expression $1 + m_c/m_v$ is replaced by some other quantity which also exceeds unity; the structure of Eqs. (11.15) and (11.16) is, however, preserved.

Let us now examine the influence of the potential difference E applied to the interface on the photocurrent. If the concentration of electrolyte is high and the effect of surface states negligible, practically all of the applied potential is dropped in the semiconductor (see Section 3.4). This leads to additional band bending, yet if the light is absorbed mainly within the space charge region ($\alpha^{-1} << L_{sc}$), the threshold frequencies ω_o and ω_o' do not change. Thus, unlike the metal/

electrolyte interface (see Eq. (11.4)), the application
of a potential difference to the semiconductor/electrolyte
junction under conditions of "surface" light absorption
does not directly influence the photoemission current. (For
weakly absorbed light ($\alpha^{-1} \gg L_{sc}$) the photoemission
threshold formally shifts by the value of the band bending
but here the necessary conditions to apply the above results may be violated.)

When there are many surface states at the interface
some part of the applied potential, δE, will be concentrated in the Helmholtz layer, $\delta E = \Delta \phi_H$. The electron
affinity, χ, changes by $e(\delta E)$ so that

$$\chi(E) = \chi(0) + e(\delta E) \tag{11.17}$$

where $\chi(0)$ is the value of χ at the semiconductor potential regarded as zero. In this case, from Eqs. (11.15)
and (11.16), we have, for $\alpha^{-1} \ll L_{sc}$

$$I_p = \begin{cases} B[\hbar\omega - \hbar\omega_o(0) - e(\delta E)]^2 & \text{if} \quad \omega_o < \omega < \omega_o' \\[2mm] C\left[\hbar\omega - \hbar\omega_o'(0) - e\left(1 + \dfrac{m_c}{m_v}\right)\delta E\right]^{3/2} & \text{if} \quad \omega > \omega_o' \end{cases} \tag{11.18}$$

where $\omega_o(0)$ and $\omega_o'(0)$ are the values of ω_o and ω_o' when
$E = 0$ and $\chi = \chi(0)$. Let us consider the different
coefficients for δE in the two frequency ranges described
by Eq. (11.18). According to Eq. (11.18), the dependence
of I_p on δE is relatively simple and is due to the shift
in the photoemission threshold of the type which occurs
for metal electrodes. In general, however, it should be
borne in mind that δE depends on E for semiconductor
electrodes and so the resultant dependence of I_p on E
may be rather complicated. The photoemission of electrons from semiconductor surface states to solution is
examined in (564).

When photoemission processes are studied experimentally (566, 567) it is necessary to separate them from
"ordinary" photoelectrochemical reactions - those in
which the photogenerated electrons are transferred from
the semiconductor on to local energy levels in the solution. Both types of process start with the interband
excitation of an electron, and both are threshold processes with respect to the radiation frequency, but the
threshold quantum energies corresponding to these two
processes are different. As shown above, the threshold

for emission exceeds the threshold for the photogeneration of electron-hole pairs, and consequently the threshold for photoelectrochemical reaction, by the electron
affinity of the semiconductor χ (see Fig. 11.4).   This
quantity is in some cases, at least in aqueous solution,
large enough to make it possible to differentiate quantitatively between these two types of photoprocesses.

The initial state of the emitted electron in solution (the delocalized electron) and its subsequent state
(the solvated electron) have relatively high energy
levels (Fig. 11.3). For this reason such electrons are
effectively captured by the semiconductor surface and
the resultant steady state current through the interphase
boundary is zero. To observe photoemission currents it
is necessary to trap the solvated electron quickly on a
lower energy level in the solution so that the corresponding state is stable with respect to capture by the
electrode surface. For this purpose special substances,
called scavengers, are added to the solution. These substances effectively capture the solvated electrons but
they do not capture electrons from the electrode, in
other words, they do not enter into electrochemical reactions (due, for example, to kinetic restrictions) over
some potential range. In aqueous solution hydrogen ions
or nitrous oxide are often used as scavengers. They react with the hydrated electron, e_{aq}^-, according to the
following equations

$$H^+ + e_{aq}^- \longrightarrow H^{\bullet}$$

$$N_2O + e_{aq}^- \longrightarrow [N_2O^-] \longrightarrow N_2 + OH^- + OH^{\bullet}$$

The resultant radicals $H^{\bullet}$ and $OH^{\bullet}$ are reduced on the
electrode when the potential is sufficiently negative
and as a result the measured photocurrent is equal to
double the photoemission current I_p. In this way the
photoemission current (or double it) is measured as the
difference in photocurrents in the presence and absence
of the scavenger.

The dependences of the photocurrent, I_p, for a
p-type gallium arsenide electrode in aqueous solution on
the quantum energy of the light (at fixed potential) and
on the potential (at fixed quantum energy) are shown in
Fig. 11.6. Both dependences (the second at more negative
potentials) are described by the "three-halves power law"
(see Eqs. (11.15) and (11.18), lower lines), showing
that they correspond to direct transitions. The photo-

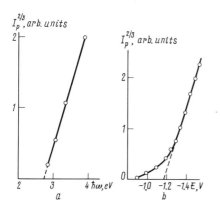

Fig. 11.6. The dependence of the photoemission current to the 2/3 power a) on the quantum energy of the light at fixed potential (-1.4 V against Saturated Calomel Electrode), and b) on the potential at a fixed quantum energy (3.4 eV). p-Type gallium arsenide ($N_A=6\times10^{18}$ cm^{-3}) (567).

current determined by indirect transitions (quadratic dependence of I_p on $\hbar\omega$ and E) has not been observed. As can be seen from Fig. 11.6, the threshold energy is 2.7 eV, a figure which is much larger than the bandgap for GaAs (1.43 eV). This is further confirmation that the photocurrent is due to photoemission and is not caused, for example, by direct photoelectrochemical reduction of the scavenger.

The very existence of a potential dependence for the photoemission current shows that a considerable portion of the applied voltage drops in the Helmholtz layer rather than in the semiconductor. This is probably explained by a high concentration of surface states. At more negative electrode potentials ($E < -1.2$ V) the potential drop in the space charge region can be viewed as approximately constant (in this case $\Delta E = \delta E = \Delta\phi_H$), while at $E > -1.2$ V it changes as a result of electrode polarization. This change in $\Delta\phi_{sc}$ can be evaluated on the basis of the deviation of the experimental data for $I_p^{2/3}$ as a function of E from the extrapolated linear plot (the dotted line in Fig. 11.6).

Apart from the photoemission of electrons from semiconductors into aqueous solutions, the photoemission into non-aqueous media (in particular from GaP and Si into liquid ammonia) is also being studied (573, 574).

An analysis (567) of the energy band diagram (Fig. 11.7) for the p-type germanium electrode has been performed as an example of the use of photoemission measurements to determine energy characteristics of the semiconductor/electrolyte junction. Figure 11.7 includes, for the sake of comparison, the energy diagram for a metal electrode in the same solution. Note that, as

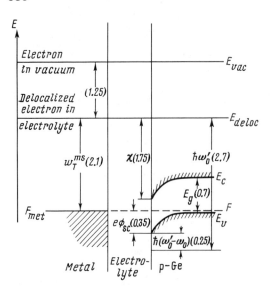

Fig. 11.7. Energy diagram for the p-type germanium/aqueous solution/metal cell. The absolute values of the energies are given in brackets (in electron-volts) (567).

already mentioned, the energy level of the delocalized electron in solution is much lower (by 1.25 eV (575)) than the vacuum level, E_{vac}, due to collective (and in particular, polarization) interactions of the electron and the solvent. At fixed potential (so that $F_{met} = F$) the work function of the metal in solution amounted to $w_T^{ms} = 2.1$ eV; thus the Fermi level in the metal (and, consequently, the Fermi level in the germanium electrode at the same potential) is 2.1 eV below the level for the delocalized electron in solution. The threshold energy for photoemission from germanium at the same chosen potential turned out to be $\hbar\omega_0' = 2.7$ eV. Comparing this value with w_T^{ms}, taking into account the band bending $e\phi_{sc} = 0.35$ eV (defined in accordance with the method above (Fig. 11.6)) and also assuming (to simplify the calculation) that $F \simeq E_v$ for a strongly doped semiconductor, it is possible to find the difference in the threshold energies for the direct and indirect transitions, $\hbar(\omega_0' - \omega_0)$. This is the "depth" (with respect to the top of the valence band) of the level from which the electrons are excited. The value of $\hbar(\omega_0' - \omega_0)$ was found to be 0.25 eV. The electron affinity for the junction in question can also be easily found (Fig. 11.7): $\chi = 1.75$ eV. Hence, the germanium electrode is indeed a potential well for delocalized electrons in solution and also for hydrated electrons.

It may be hoped that, with the help of photoemission measurements, it will be possible to study other properties of semiconductor electrodes, in particular those

determined by surface states. There is every ground to
believe that the potential of photoelectron emission as
an instrument for the study of semiconductor/electrolyte
junctions is far from exhausted.

Ξ 11.3. The Production of Photographic Images : Semiconductor Electrochemical Displays

Existing systems for recording photographic images
may be subdivided into two classes depending on whether
the initial image obtained under illumination is subjected to further intensification or not. Systems without
intensification of the primary image are characterized by
low photosensitivity, and the quantum yield of the process
does not exceed unity. For systems with intensification
of the primary image, the resultant quantum yield can be
effectively larger than unity by several orders of magnitude and this on the whole ensures that the photoprocess
has a much higher photosensitivity. Traditional materials
based on silver halides are characteristic representatives
of the latter class of photosensitive materials. Three functions are performed with the help of silver halides: 1)
photosensitivity is ensured; 2) a latent image is formed;
and 3) a visible image is formed.

The third stage in the photoprocess (the intensification
of the latent primary image) is performed by means of the
so-called "physical development." This usually consists
of the catalytic deposition of metal on a plate, the
centers of the latent image (the so-called "development
centers") serving as nuclei for crystal formation. In
silver halide emulsions the silver necessary to obtain a
visible image is present in the emulsions in great excess,
and after development the superfluous silver is removed.
As a result of the widespread use of photography today
this consequently wasteful process is accompanied by a
great loss of silver. As a result, since the late 1960s
the search has been underway for so-called "silverless"
methods for recording photographic images. The aim is
to preserve the principle of physical development since
this ensures a high photosensitivity of the photographic
process.

The underlying basis of the quest is as follows: to
distribute the three functions above, at present carried
out in ordinary photography using silver halides, among
two or three different substances.

The general scheme for the photoprocess can be
written as follows

$$A \xrightarrow{\hbar\omega} B \qquad\qquad (11.19a)$$

$$B + C \longrightarrow D \qquad\qquad (11.19b)$$

$$M^{z+} + Red \xrightarrow{(D)} M + Ox \qquad\qquad (11.19c)$$

where A is a photosensitive compound, B is the product of
its photochemical transformation, C is a compound creating
the centers of the latent image (D) by interaction with B, Red
is the reducing agent in the developer, Ox is its oxidized
form, M^{z+} is the metal ion to be reduced, and M is the
metal producing the visible image. In the reaction forming the visible image (Eq. (11.19b)) the latent image
centers (D) serve as a catalyst. In silver halide photomaterials, A, C, and M^{z+} are silver compounds.

In the methods of "silverless" photography being investigated at present silver compounds are only used for
the second stage. The third stage, which entails the
largest quantity of metal, is carried out with less ex-
pensive (or, at least, more abundant) metals. Among
those systems potentially capable of replacing silver
based compounds, great hopes are pinned on semiconductor
materials (see, for example, the review (576)). It
should be stressed that in photomaterials based on semiconductors the photosensitive stage in the overall process may be of electrochemical or some other nature. The
semiconductor/electrolyte interface is directly used in a
number of methods for producing latent and visible images.
A brief account of precisely how this works is given
below.

The photoelectrochemical deposition of metals is
often used to form images. The kinetics of electrodeposition are dealt with in a number of works (see
(578-580)). Without going into details, we shall only
emphasize that in the initial stages of deposition, when
the amount of metal on the semiconductor surface does not
exceed a monolayer, chemisorption should be taken into
consideration (581, 582) and also, probably, the phenomenon of the incorporation of deposited metal atoms into
the electrode's crystal lattice (583). These two phenomena are sometimes referred to by the single term "underpotential deposition." The atoms and clusters of the
deposited metal either create by themselves an image of
the necessary optical density (584-588) or serve as

nuclei in the consequent development (589). Photoelectrochemical reactions of organic dyes can also be used for
the same purpose (590-592).

Two schemes for forming an image on a semiconductor
electrode surface have been described. According to one
the semiconductor/electrolyte interface operates at
limiting (saturation) current. This means that at the
selected electrode potential a reaction with the participation of minority carriers occurs. Hence, in the dark
the current is very small, and is determined by the bulk
generation of the minority carriers (holes for an n-type
semiconductor and conduction band electrons for a p-type
semiconductor). When the optical image is projected on
to the semiconductor surface the concentration of minority
carriers in the illuminated areas increases, as does the
rate of the electrode reaction (cf. paragraph (i) in
Section 10.1). For example, the cathodic electrodeposition of a metal from a solution of its salt accelerates
on a p-type sample. Anodic reactions accelerate on n-
type materials, for example the oxidation of Pb^{2+} followed by the formation of a colored deposit of PbO_2 on
the electrode surface. Layers of PbSe, PbS, Si, or GaP
(587, 593) may be used as "photographic plates." Operating the semiconductor "photographic plate" in the regime
of a saturation current has the following advantage: the
photosensitivity is brought about by the imposed external
polarization and may be "switched off," to a considerable
extent, by switching off this polarization. Consequently,
this can be applied to reduce the veiling effect of the
ever present (background) thermal irradiation (594).

Another mode of operation of the photosensitive semi-
conductor/electrolyte interface is that of the photogalvanic couple (584-586, 589, 595). This method is based
on the potential difference which arises between the
illuminated and non-illuminated areas of the semiconductor
electrode surface (cf. paragraph (ii), Section 10.1). As
a result of the non-uniform illumination of the semiconductor, virtual local cathodes and anodes are formed on
its surface. It is known, for example (500, 596), that
semiconductor suspensions, for instance TiO_2, effectively
cleanse the solution of traces of metal ions when illuminated. On each TiO_2 particle the illuminated area of
the surface serves as a photoanode, where oxygen is
evolved, while the non-illuminated area serves as a
cathode, where metal is deposited (cf. Fig. 9.13). A
continuous polycrystalline film of TiO_2 or ZnO immersed
in a solution of metal salt and non-uniformly illuminated
acts in a similar way. This effect is used to form the

image: the difference in potential emerging between
illuminated and non-illuminated areas controls the non-
uniform metal deposition which is observed; see for
example (597-601).

Here there is a self-evident analogy with the photocorrosion behavior of semiconductors. The specific
feature in this case is that the cathodic partial reaction
is the reduction of metal ions from solution. It should
be noted that the ZnO/solution and TiO_2/solution interfaces have a memory: the latent image emerges when the
semiconductor is illuminated and there are no metal ions
in solution; it can be subsequently developed in the dark
when the exposed sample is put into a solution of a metal
salt. The mechanism for this effect is not clear at
present. It is probably connected with the charging of
slow surface states under illumination and with photoadsorption processes (599, 600, 602).

Finally, it should be noted that the semiconductor/
electrolyte interface may be used to form visible images,
even in the dark, using only the external polarization.
The method is based on the changes in the color of the
semiconductor electrode which result from electrochemical
reaction. For example, polycrystalline WO_3 films are
colored brightly by the passage of a certain amount of
charge (see, for example, (602)). This phenomenon is
called electrochromism. If the electrochemical reaction
which underlies it is easily reversible, then this coloring is a reversible effect. In particular, for electro-
chromic films of WO_3 it is possible to form and erase the
coloring many times over by simply cycling the potential
between two preset values. The operation of electrochemical semiconductor displays is based on this principle
(604).

The work examined in this section has still to leave
the laboratory but there are very good grounds to consider it promising.

Ξ 11.4. T h e   R a d i a t i o n   E l e c t r o c h e m i s t r y
         o f   S e m i c o n d u c t o r s

Let us now briefly examine the effect of electromagnetic radiation of much higher energies than the
visible and ultraviolet regions of the spectrum on the
semiconductor/electrolyte system. The energies of X-ray
and γ-ray quanta, and also those of high-energy electrons,
are of the order of hundreds of thousands to millions of
electron volts. These energies exceed the bandgap widths

of semiconductor materials by many orders of magnitude. Consequently irreversible changes occur in the system upon irradiation: the formation of defects in the crystalline structure of the semiconductor (605) and the radiolytic decomposition of the solvent and solute (606). However, in a number of cases, these changes are insignificant or, at least, exert no direct influence on the electrochemical behavior of the system. The effect of generation of non-equilibrium electron-hole pairs in the semiconductor then comes to the fore. This, in its turn, is manifest in the changes in the rates of electrochemical and corrosion reactions on the irradiated semiconductors in solution; see (607, 608). In particular, the titanium dioxide electrode has been thoroughly examined (362).

The extensive similarity in the behavior of electrodes when illuminated by ultraviolet light or by a beam of high-energy electrons or γ-quanta is significant. In both cases, the anodic evolution of oxygen from water occurs (in the absence of irradiation the rate of this reaction is negligible) and the shapes of the polarization curves are practically identical (see Fig. 11.8). The reaction rate is proportional to the radiation intensity $P = J_o \hbar \omega$, at least up to reasonable intensities (for example, up to $10^{18} - 10^{19}$ eV/(cm^{2}s)); see Fig. 11.9. All these observations indicate that the increase in the hole concentration in the valence band plays the decisive role in the radiation-electrochemical, as well as photoelectrochemical, behavior of the semiconductor. At the same time, long-term changes in the photoactivity of the electrode are also observed (382). These are apparently caused by the creation of donor levels under the influence of radiation (these effects are the same as those brought about by deliberate changes of TiO$_2$ stoichiometry; see Section 9.2.1).

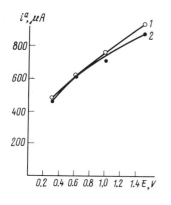

Fig. 11.8. Polarization curves for anodic oxygen evolution on a TiO$_2$ film electrode in 0.2 M KCl: 1) when illuminated by ultraviolet light; 2) when irradiated by electrons with an energy of 4.2 MeV (362).

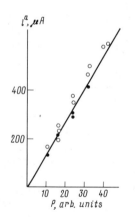

Fig. 11.9. The dependence of
the current for the anodic
evolution of oxygen on a TiO_2
electrode when irradiated by
electrons with an energy of
4.2 MeV on the irradiation
intensity. 0.2 M KCl,
$E = 1.5$ V (against Saturated
Calomel Electrode); maximum
irradiation intensity is
$P = 5 \times 10^{19} eV/(cm^2 s)$ (362).

    The particular features of the radiation-electro-
chemistry of semiconductors, as compared with their
photoelectrochemistry, stem from the greater penetration
of the hard radiation. The absorption depth for X-rays
or high-energy electrons is of the order of 1 - 10 cm.
Thus, there is generally no difference between the "illuminated" (towards the source of radiation) and the "non-
illuminated" electrode surfaces. Both are almost equally
active. This allows any particular processes to be inten-
sified by using multilayer or disperse electrode systems
absorbing more radiant energy than the usual (ordinary)
flat electrode.

    The acceleration of electrode processes on irradiated
semiconductors opens the door, at least theoretically, to
the direct conversion of the energy of ionizing radiation
into the chemical energy of electrolysis products (362,
609), a process similar to light energy conversion (see
Chapter 9). The efficiency of this process is, however,
still negligible for the following two reasons. Firstly,
for hard radiation quanta $\hbar\omega \gg E_g$, so that practically
all of the energy absorbed in the semiconductor is converted into heat, with only a small share being used to excite non-equilibrium carriers (though the yield of carriers
generated can be higher as a result of impact ionization;
see Section 1.5). Secondly, due to the high penetration
of ionizing radiation only a small amount of the energy
incident on the electrode surface is absorbed in the
"electrochemically active zone" - the depletion layer
(cf. Fig. 6.1). Most of the energy is wasted since it is
absorbed in the neutral semiconductor bulk, in the substrate material and in the solution where it is converted
into heat and/or the energy of defects and radiolysis.
The conversion efficiency will probably increase with the
help of disperse electrodes in which the bulk of the solid
"operates" more effectively.

Ξ 11.5. E l e c t r o g e n e r a t e d   L u m i n e s c e n c e

Photoelectrochemical processes are based on the conversion of light energy into electrical and/or chemical energy. The reverse process - the conversion of electrical energy into light - is seen in electrogenerated luminescence. In this process light emission occurs when a current passes across the semiconductor/electrolyte interface. In the course of the electrode reaction, excited products are formed, which are later deactivated, emitting light quanta. Thus the electrogenerated luminescence may be regarded as the converse of the processes of electrode and solute photoexcitation, examined in Chapters 5 and 6. This is a demonstration of the common nature of all photoelectrochemical phenomena; the interconversion of light and electrical (chemical) energy.

The excited intermediates produced in the course of electrochemical reactions are subdivided into two types: (i) excited ions and molecules in solution near (or at) the electrode surface and (ii) excited carriers (electrons and holes) in the semiconductor.

The first type of luminescence usually occurs in the course of the electroreduction of organic substances at semiconductor electrodes (610-612). The principle is illustrated by the scheme in Fig. 11.10a (cf. Fig. 5.4c). An electron from the conduction band of the semiconductor passes onto an excited state level (for example, a triplet [†]) of the organic substance, $^3R^*$, which is subsequently deactivated passing to the ground state, R, and emitting a light quantum. If an electron could only be transferred directly to the ground state of the organic substance there would be no luminescence (this is exactly what happens at metal electrodes; see Fig. 11.10b). However, in the case of the semiconductor electrode the ground state level is often in the bandgap, and the corresponding electron transfer is forbidden. As an example, let us cite the reduction of the bipyridyl complex of ruthenium (III) on CdS, ZnO, TiO_2, SiC or GaP electrodes in aqueous and acetonitrile solutions:

$$Ru(bipy)_3^{3+} + e^- \rightarrow [Ru(bipy)_3^{2+}]^* \rightarrow \hbar\omega + Ru(bipy)_3^{2+}$$

This has been found to be a source of luminescence (613, 614).

[†] For details of triplet states see the footnote on page p. 25.

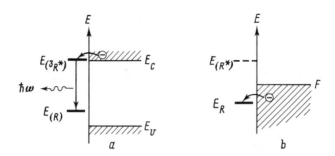

Fig. 11.10. Scheme for electrogenerated luminescence with the participation of photoexcited reagents in solution: a) semiconductor electrode; b) metal electrode. $E_{(3_R*)}$ is the energy of the excited triplet state.

The scheme for the second type of generation and deactivation of excited states is shown in Fig. 11.11 (cf. 1.10). Let us assume that, in the course of the electrode process (reduction of the substance A), minority carriers are injected into the semiconductor (for our purposes, holes into an n-type semiconductor). There they recombine, in particular according to the radiative recombination mechanism, with the majority carriers being supplied from the external circuit via the ohmic contact. Energy equal to the width of the bandgap is evolved, and the following relation holds for the emitted light quanta: $\hbar\omega = E_g$ (case 1 in Fig. 11.11). Yet more probable is recombination via surface (and sometimes bulk) levels within the bandgap, which capture an electron and a hole in succession. In this case the excess energy of the recombining carriers is lost in smaller packets and $\hbar\omega' < E_g$ (case II).

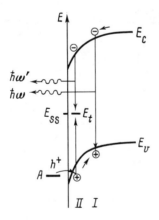

Fig. 11.11. Scheme for electrogenerated luminescence with the participation of photoexcited carriers in the semiconductor: (I) without and (II) with the participation of energy levels in the bandgap. E_{ss} and E_t are the energy of the surface and the bulk levels respectively.

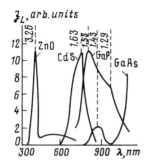

Fig. 11.12. Spectra of electrogenerated luminescence at photoexcited semiconductor electrodes. J_L is the intensity of luminescence. Energies corresponding to the maxima of the curves are given in electron volts (615).

Both recombination paths are seen in the luminescence spectra (Fig. 11.12) of n-type semiconductor electrodes under conditions of electrochemical hole injection (615). In the cases of ZnO and GaAs the energy of the emitted quanta is close to the width of the bandgap. On the other hand, for CdS and GaP the energy of emitted quanta is much smaller than the bandgap width. The luminescence associated with minority carrier injection from the electrolyte has been observed for other semiconductor electrodes (615-619).

Since the recombination of non-equilibrium carriers in the semiconductor is the source of the electrogenerated luminescence of this particular type, the use of luminescence measurements to evaluate the rate of recombination, often an undesirable process particularly in liquid-junction solar cells, has been proposed (620, 621).

Routine studies of electrogenerated luminescence on semiconductor electrodes have just begun. We can expect these processes to lay the groundwork for new methods for the study of the structure of the semiconductor/electrolyte interface and the kinetics of the reactions occurring on it.

Ξ 11.6. The Study of the Surface Properties of Semiconductor Electrodes by Electroreflectance

Until now we have confined ourselves to a discussion of the influence of light on those processes at semiconductor electrodes which are accompanied by the transfer of electrons across the interface. Nevertheless the illuminated semiconductor/electrolyte interface is also an object of investigation when no charge transfer occurs. These studies are connected with the examination of

electrooptic effects and of the electroreflection of
light in particular (6, 622-626).

The electroreflectance technique is based on the
fact that changes in the electrode potential are accompanied by changes in the coefficient of light reflection. Due to the relative simplicity of surface
potential modulation in semiconductor/electrolyte systems,
techniques for studying semiconductors by means of electroreflectance are among the most popular modulation
methods (frequency, temperature, and other modulations
(622)).

Electrooptic effects can serve as valuable sources
of information about the band structure of solids and
also about the structure of the surface region and the
strength of the internal electric field. From this
viewpoint, the study of electrooptic effects at semicon-
ductor/electrolyte interfaces is a new method which substantially augments the traditional methods for investigating the physics and chemistry of semiconductor
surfaces. It should be noted that electrooptic effects
on metal electrodes are already used to study adsorption-
desorption phenomena and the kinetics of electrode reactions (627, 628). Research on semiconductor systems
is not as advanced as yet, and electrooptics has not been
so closely related with photoelectrochemistry. It is to
be hoped that electrooptic effects will prove useful
tools for settling a number of purely electrochemical
problems in this area as well.

The change in the complex dielectric permeability of
the interfacial region of the semiconductor, under the influence of the electric field, is the physical cause of the
modulation of the coefficient of light reflection R. The
field may influence the magnitude of the dielectric permeability through the redistribution of free carriers near the
interface. A strong field tangibly influences interband
electronic transitions; it is as if the width of the
bandgap, E_g, were decreased (Franz-Keldysh effect (15,
16, 622)). In the presence of a field the probability
of indirect transitions increases; here the constant
field plays the role of a "third body" (see Section 1.3).
The field polarizes the electrons, thereby affecting the
electronic absorption spectra. The field can create
additional energy levels which contribute to light
absorption. In non-ideal crystals the field influences
the photoabsorption caused by impurities in the space
charge region. Furthermore, a change in adsorption
properties of the surface, which in turn, leads to a

change in the optical properties, is possible at electrode/electrolyte junctions under the influence of a field.

The heterogeneity of the change induced by the field in the characteristics of the medium material and, in particular, in the complex refractive index and complex dielectric permeability, $\hat{\epsilon} = \epsilon_1 + i\epsilon_2$ (see Eq. (1.28)), are a specific feature of electrooptic effects in the space charge region. The scale of these heterogeneities (10^{-4} - 10^{-5} cm) is comparable with the wavelength of the incident light. Therefore, the usual equations of geometric optics should be replaced with the more universal relationships of the wave optics of spatially heterogeneous media. On the other hand, the influence of the electric field on the coefficient of reflection R, which is experimentally observed, is usually very small: $\Delta R/R << 10^{-2}$. This makes it possible to apply the methods of perturbation theory. As a result, we obtain the following expression for the relative change in the coefficient of reflection, $\Delta R/R$, under the influence of the field (622, 625):

$$\frac{\Delta R}{R} = \alpha(\epsilon_1, \epsilon_2)\Delta\tilde{\epsilon}_1 + \beta(\epsilon_1, \epsilon_2)\Delta\tilde{\epsilon}_2 \qquad (11.20)$$

The partial coefficients α and β in Eq. (11.20) are given by $\alpha = d\ln R/d\epsilon_1$; $\beta = d\ln R/d\epsilon_2$ and, by definition, are independent of the potential but vary as a function of the radiation frequency. The real magnitudes of $\Delta\tilde{\epsilon}_1$ and $\Delta\tilde{\epsilon}_2$ are found from

$$\Delta\tilde{\epsilon}_1 + i\Delta\tilde{\epsilon}_2 = -\frac{2in\omega}{c}\int_0^\infty [\Delta\epsilon_1(x) + i\Delta\epsilon_2(x)]e^{2in\omega x/c}dx \qquad (11.21)$$

where $n = n + ik$ is the complex refractive index of the semiconductor, and $\Delta\epsilon_1$ and $\Delta\epsilon_2$ are the changes brought about by the field in the components ϵ_1 and ϵ_2 of the complex dielectric permeability. According to Eqs. (11.20) and (11.21), $\Delta R/R$, through α, β and $\Delta\tilde{\epsilon}_{1,2}$, is a rather complicated function of the frequency ω; in many cases the dependence has an oscillating character.

The calculation of the integral in Eq. (11.21) requires a knowledge of the behavior of the functions $\Delta\epsilon_{1,2}(x)$. In view of the fact that the dependence of the electric field, $\xi(x)$, on the coordinate in the space charge region can be taken as known, it is actually

necessary to know $\Delta\varepsilon_{1,2}(\xi)$. This dependence has been
examined for a number of particular mechanisms cited
above (622, 624).

Some important conclusions can already be drawn
simply on the basis of an analysis of the general structure of Eqs. (11.20) and (11.21). The situation in
which the wavelength of the light in the semiconductor,
$\lambda_n = 2\pi c/\omega n$, clearly exceeds the thickness of the space
charge region L_{sc} is of considerable interest. This
situation is realized in comparatively heavily doped
semiconductors where $L_D \lesssim 10^{-6}$ cm, and also in lightly
doped materials at high band bending (see Eq. (3.22)).
If the inequality $L_{sc} \ll \lambda_n$ holds, then in calculating
the integral we may assume that the oscillating factor
is unity. Taking into consideration the correlations
$dx = (d\phi/dx)^{-1}d\phi$ and $d\phi/dx|_{x=0} = -\xi_{sc}$, we find for the
magnitude of the differential electroreflectance, with
due account of Eq. (11.20),

$$\frac{1}{R}\frac{dR}{d\phi_{sc}} = \frac{4\pi}{\lambda}\xi_{sc}^{-1}\{(k\alpha-n\beta)\Delta\varepsilon_1(\phi_{sc})+(n\alpha-k\beta)\Delta\varepsilon_2(\phi_{sc})\} \quad (11.22)$$

where λ is the wavelength for light of the given frequency
in vacuum ($\lambda = \lambda_n n$).

In the opposite limiting case, $L_{sc} \gg \lambda_n$, the quantities $\Delta\varepsilon_{1,2}$ in the integral can be regarded as constants,
given by $\Delta\varepsilon_{1,2}|_{x=0} = \Delta\varepsilon_{1,2}(\phi_{sc})$; in accordance with Eq. (11.21),
$\Delta\tilde{\varepsilon}_{1,2} = \Delta\varepsilon_{1,2}(\phi_{sc})$. Differentiating Eq. (11.20) with
respect to ϕ_{sc} we find for the differential electrore-
flectance

$$\frac{1}{R}\frac{dR}{d\phi_{sc}} = \alpha\frac{d\Delta\varepsilon_1}{d\phi_{sc}} + \beta\frac{d\Delta\varepsilon_2}{d\phi_{sc}} \quad (11.23)$$

Thus, in the two limiting cases ($\lambda_n \ll L_{sc}$ and
$\lambda_n \gg L_{sc}$), to find the differential electroreflectance
it is sufficient to know the values of $\Delta\varepsilon_{1,2}$ when
$\phi = \phi_{sc}$; these can be calculated from the model of a
spatially constant electric field equal to ξ_{sc}. At the
same time, it should be stressed that the types of electroreflectance spectra described by Eqs. (11.22) and
(11.23) may considerably differ from one another.

As an example let us analyze the electroreflectance
determined by the heterogeneity of distribution of free
carriers in the space charge region of a semiconductor.

The reflection of light by free carriers is called plasma
reflection. This phenomenon is therefore called plasma
electroreflectance (626). Plasma electroreflectance can
be described with the help of Eq. (1.38). As a result
of the imposed electric field there is a change in the
spatial distribution of carrier concentration on which
the value of ω_p depends. This, in its turn, leads to a
spatial variation in the values of ε_1 and ε_2. Taking
for our example an n-type semiconductor and neglecting
the contribution made by the holes but taking into account
that the concentration of electrons in the space charge
region changes with the coordinate, from Eq. (1.38) we
get

$$\Delta\varepsilon_1 + i\Delta\varepsilon_2 = -\left(\frac{\omega_p}{\omega}\right)^2 \frac{\omega\tau}{i+\omega\tau}\left(e^{e\phi(x)/kT}-1\right) \qquad (11.24)$$

where ω_p is the plasma frequency, independent of potential and corresponding to a concentration $n = n_0$; the
potential $\phi(x)$ is determined independently from the self-
consistent Poisson-Boltzmann equation (see Section 3.2).
Substitution of Eq. (11.24) in Eq. (11.21) allows us to
calculate $\Delta R/R$.

    Various specific cases have been examined (626). In
particular, bearing in mind that we can usually assume
that $\omega\tau \gg 1$ given sufficiently long wavelength light
($\lambda_n \gg L_{sc}$), from Eqs. (11.20), (11.21) and (11.24) we
get the following expression for $\Delta R/R$ at an arbitrary
value of the band bending

$$\frac{\Delta R}{R} = 4\pi\left(\frac{\omega_p}{\omega}\right)^2 \frac{\Gamma_n}{\lambda n_0}(n\beta - k\alpha) \qquad (11.25)$$

where Γ_n is the surface excess of electrons (see p. 73)

$$\Gamma_n = n_0 \int_0^\infty (e^{e\phi(x)/kT} -1)\,dx$$

When measuring the differential plasma electroreflectance,
$R^{-1}dR/d\phi_{sc}$, over the range of positive potentials ϕ_{sc}, the
value of $d\Gamma_n/d\phi_{sc}$ may be expressed through the differential capacity C_{sc} of the space charge region of the semiconductor: $ed\Gamma_n/d\phi_{sc} \approx C_{sc}$.

    It is significant that, according to Eqs. (11.24)
and (11.25), the sign of the plasma electroreflectance

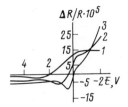

Fig. 11.13. The dependence
of the amplitude of electroreflectance from a silicon
electrode on potential (6).
$\hbar\omega$ = 3.42 eV. The thickness
of the oxide film (nm): 1)
120; 2) 35; 3) 17.

is determined by the sign of ϕ_{SC}. The quantity $\Delta R/R$ is
positive if $\phi_{SC} > 0$ (when the bands are bent downwards
with the formation of an accumulation layer) and negative
if $\phi_{SC} < 0$ (when the bands are bent upwards with the
formation of a depletion or inversion layer). In other
words, when $\Delta R/R$ is zero, ϕ_{SC} is also zero and thus it is
possible to determine the flat band potential of the
semiconductor being studied.

As an example, Fig. 11.13 shows the dependence of
the electroreflectance of a silicon electrode on its
potential. The amplitude of the electroreflectance
signal, $\Delta R/R$, passes through zero at the flat band potential (the latter was measured independently by the differential capacity method) in accord with the theoretical
predictions. Of importance is the dependence of the
silicon flat band potential on the thickness of the
oxide film.

Let us now give a brief account of some results of
studies, using the electroreflectance method, of the
structure of the surface layer of semiconductor electrodes. The dielectric properties of the surface semiconductor layer, which alter under the influence of an
electric field, are also dependent on the physical and
chemical condition of the sample and on its history. In
particular, interaction between the semiconductor and
its environment results in the enrichment (depletion) of
certain impurities near the surface. In addition crystallographic factors also play an important role. Due to
the influence of surface forces and to the mismatch in
the crystal lattices of the semiconductor and the oxide
covering it, internal stress occurs in the semiconductor,
and this also has a bearing on the optical properties.
This effect is concentrated in the so-called transition
layer which spreads from the surface over a distance of
several dozen Ångstroms. These effects may be "inten-
sified" by special processing of the surface, for example,
by mechanical polishing, bombardment with high-energy
particles, etc. As a result, such characteristics as
the width of the bandgap, the effective masses of electrons and holes, and their scattering conditions in the
semiconductor change.

It should be borne in mind that electroreflectance spectra provide information on the optical character- istics of the electrode averaged over the thickness of the region in which the electric field is modulated when the electroreflectance is measured. This distance is the space charge region. In order to precisely examine the properties of the transition layer using the electro- reflectance method, the thickness of this layer and that of the space charge region should be of the same order of magnitude. Usually, as already mentioned, the thick- ness of the transition layer is no more than 10^{-6} cm. For this reason heavily doped semiconductor samples have to be used in the research so that the magnitude of L_{sc} (see formula on p. 71) is sufficiently small. When, for some reason, it is necessary to use materials of comparatively low doping impurity content, the thickness of the layer is deliberately increased by using the special methods of surface treatment outlined above.

Let us now examine (629, 630) the results obtained for the analysis of the transition layer at the inter- face between semiconductor materials and aqueous solu- tions. Mechanical polishing, argon ion bombardment, and thermal oxidation under extreme conditions were used to modify the properties of the surface layer of the semiconductor. The electroreflectance spectra obtained when there is a transition layer differs greatly from the spectra of the same materials with an "unperturbed" surface (Fig. 11.14). Etching away the damaged layer of the semiconductor and electropolishing restores the electroreflectance spectra to that characteristic of the unperturbed semiconductor bulk.

In the literature quoted above the spectra were analyzed using a particular, simple model of the complex dielectric permeability. In this way the optical char- acteristics and thickness of the transition layer on Ge, Si, CdS and other semiconductor electrodes were determined (629, 630; see also (6)).

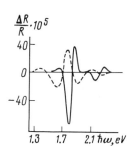

Fig. 11.14. The influence of the state of the CdSe surface on the electrore- flectance spectrum. The solid curve is for a per- fect surface (cleaved crystal), the dotted line is for a mechanically damaged surface (polished) (630).

From the changes in the electroreflectance spectra
in the course of the electrochemical treatment of electrodes it is possible to qualitatively follow the reversible and irreversible changes in the surface layer of
the electrode. For example, the partial decompostion
of ZnSe under anodic polarization (631) and the micro-
deposition of metals on to the ZnO surface under cathodic
polarization (632) have been studied in this way.

Conclusions

The pattern of photoelectrochemical behavior of semiconductor materials is now fairly comprehensive. The importance of semiconductor photoelectrochemistry is not, however, confined to this pattern. Firstly, the photoelectrochemistry of semiconductors has stimulated the study of photoprocesses at insulating electrodes, an area which is not traditional for electrochemistry. In this area the basic notions and mathematical formalism of the electrochemistry and photoelectrochemistry of semiconductors have been successfully applied. Secondly, the photoelectrochemistry of semiconductors provides an opportunity, in some cases a unique opportunity, to study the thermodynamic and kinetic characteristics of excited states both in solution and in the electrode, and the electron transfer processes in which they participate. Above all, this relates to the investigation of the energetics of photoexcited particles, and interesting results have already been obtained. The scope of this method, however, is far from exhausted. The investigation of photoelectrochemical processes stimulated by the excitation of the electron-hole ensemble of the semiconductor makes possible the experimental verification of the use of the concept of quasi-Fermi levels for describing charge transfer at the interface, and allows the range of its applicability to be determined. This is a fundamentally important problem, not only in the photoelectrochemistry of semiconductors but also in the physics and chemistry of semiconductors at large. (It should be borne in mind that the diverse quenching processes for photoexcited states at metal electrodes often hamper such studies.)

These considerations are also relevant, to a very large extent, on the investigation of the elementary act of electron transfer since semiconductor materials offer much broader opportunities for such studies than metals.

Of special significance are the problems of laser electrochemistry of semiconductors, an area which is in no way solely restricted to light-sensitive etching. First and foremost, mention should be made of the threshold electrochemical reactions stimulated by powerful laser emission, which may proceed by new routes and yield new products. Among other things, it opens up an opportunity to develop specific chemical and technological processes. Mention also should be made of the possibility of the reverse process: coherent light emission stimulated by electrochemical reaction. This may provide the basis for a new type of laser.

Data on the photoemission of electrons from semiconductors to solutions, together with data on the photoemission from metals to solutions and from solutions to the vapor phase, supplemented by equilibrium potential measurements, make possible the determination of the relative and absolute positions of the characteristic electronic levels in solution and in the semiconductor material.

While dealing with the directions of future studies, we should mention the electroreflectance technique which can be used to study the structure of the surface region of the semiconductor, as well as adsorption and electrode reactions.

In this connection, the important role of semiconductor properties in electrochemistry and photoelectrochemistry at large should be emphasized. For instance the abnormalities in the electrochemistry of such systems as metal electrodes covered with surface oxide and other films, polylayers of adsorbed organic species, and so on, are often explained in terms of their semiconductor nature. There is, however, a gap between the electrochemistry and photoelectrochemistry of electronic single crystal semiconductors, on the one hand, and the electrochemistry of oxide films possessing a usually amorphous structure and considerable ionic conductivity, on the other, which remains to be bridged. Besides, in many important cases the films involved are so thin (somewhere between a phase and an adsorbed layer) that the applicability of the traditional concepts of the physics and chemistry of semiconductors, in particular

band theory, remains doubtful. Thus, future researchers
face the task of extending the theoretical description
and methodological approach of the photoelectrochemistry
of semiconductors to oxide and other films on the surface
of metals.

Finally, a promising direction in the study of
photoelectrochemical properties of systems which, in
their behavior, are close to semiconductors is connected
with photobiology and, above all, with the conversion of
light energy in natural photosynthesizing systems. There
is no evidence today that these systems contain a semiconductor phase as such. It seems more probable that the
process of energy conversion proceeds through molecular
aggregates which include a relatively small number of
molecules. Even so the elementary stages of photosynthesis, especially those of an electrochemical nature, can
apparently be described, or at least simulated, within
the framework of the photoelectrochemistry of semiconductors.

As we have seen there are still many blank spots
on the map of the branch of science that we have dealt
with, and much that is new and interesting awaits the
researcher.

OPTICAL PROPERTIES OF SEMICONDUCTORS (BASED ON (18))

Semiconductor		Width of bandgap E_g $(300\,K)/$ eV	Type of transition on photoex- citation	Refractive index, n	Static dielectric permeability
	Si	1.11	indirect	3.44	11.7
	Ge	0.67	indirect	4.00	16.3
	α–SiC	2.8–3.2	indirect	2.69	10.2
	Se	1.74	direct	5.56	8.5
$A^{III}_B V$	GaP	2.25	indirect	3.37	10
	GaAs	1.43	direct	3.4	12
	GaSb	0.69	direct	3.9	15
	InP	1.28	direct	3.37	12.1
	InAs	0.36	direct	3.42	12.5
	InSb	0.17	direct	3.75	18
$A^{II}_B VI$	ZnO	3.2	direct	2.2	7.9
	α–ZnS	3.8	direct	2.4	8.3
	ZnSe	2.58	direct	2.89	8.1
	ZnTe	2.28	direct	3.56	9.7
	CdS	2.43	direct	2.5	8.9
	CdSe	1.74	direct	–	10.6
	CdTe	1.50	direct	2.75	10.9

Appendix 2

Pourbaix Diagrams of Semiconductors
in Aqueous Solutions (55)

(CB - bottom of the conduction band)

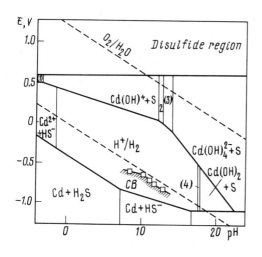

CdS

Region	Stable species
1	$Cd^{2+} + S$
2	$Cd(OH)_2 + S$
3	$Cd(OH)_3^- + S$
4	$Cd(OH)_2 + S^{2-}$

(CB – bottom of the conduction band)

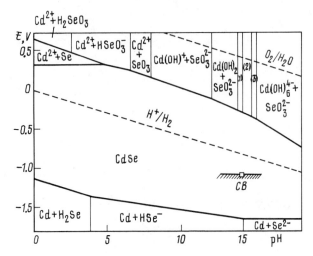

CdSe	
Region	Stable species
1	$Cd(OH)_3^- + SeO_3^{2-}$
2	$Cd(OH)_4^{2-} + SeO_3^{2-}$
3	$Cd(OH)_5^{3-} + SeO_3^{2-}$

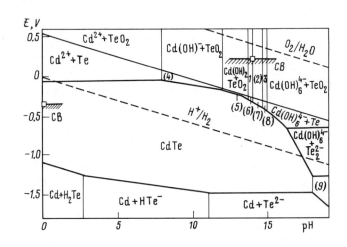

CdTe

Region	Stable species		Region	Stable species
1	$Cd(OH)_3^- + TeO_2$		6	$Cd(OH)_3^- + Te$
2	$Cd(OH)_4^{2-} + TeO_2$		7	$Cd(OH)_4^{2-} + Te$
3	$Cd(OH)_5^{3-} + TeO_2$		8	$Cd(OH)_5^{3-} + Te$
4	$Cd(OH)^+ + Te$		9	$Cd(OH)_6^{4-} + Te^{2-}$
5	$Cd(OH)_2 + Te$			

(CB - bottom of the conduction band)

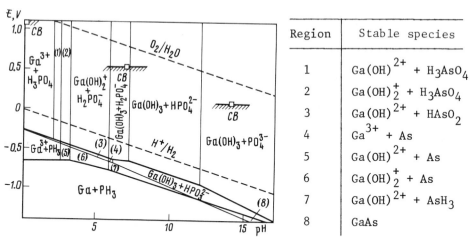

GaAs

Region	Stable species
1	$Ga(OH)^{2+} + H_3AsO_4$
2	$Ga(OH)_2^+ + H_3AsO_4$
3	$Ga(OH)^{2+} + HAsO_2$
4	$Ga^{3+} + As$
5	$Ga(OH)^{2+} + As$
6	$Ga(OH)_2^+ + As$
7	$Ga(OH)^{2+} + AsH_3$
8	$GaAs$

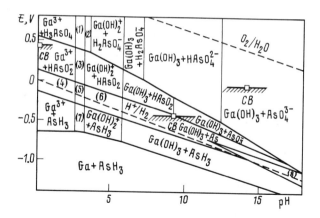

GaP

Region	Stable species	Region	Stable species
1	$Ga^{3+} + H_2PO_4^-$	5	$Ga(OH)^{2+} + PH_3$
2	$Ga(OH)^{2+} + H_2PO_4^-$	6	$Ga(OH)_2^+ + PH_3$
3	$Ga(OH)_2^+ + H_2PO_4^-$	7	$Ga(OH)_3 + PH_3$
4	$Ga(OH)_3 + H_2PO_3^-$	8	$Ga + HPO_3^{2-}$

TABLES OF FUNCTIONS $F(Y,\lambda)$ (BASED ON (64))

Function $F(Y,\lambda)$ is determined by the relation

$$F(Y,\lambda) = [\lambda(e^{-Y}-1) + \lambda^{-1}(e^{Y}-1) + (\lambda-\lambda^{-1})Y]^{\frac{1}{2}}$$

where

$$Y = e\phi_{sc}/kT, \quad \lambda = n_o/n_i, \quad \lambda^{-1} = p_o/n_i.$$

At $Y > 0$ when $e^{Y} \gg \lambda$, $F(Y,\lambda) = \lambda^{-\frac{1}{2}} e^{Y/2}$.

At $Y < 0$ when $e^{-Y} \gg \lambda^{-1}$, $F(Y,\lambda) = \lambda^{\frac{1}{2}} e^{-Y/2}$.

$F(Y,\lambda) = 0$ at all values of λ at $Y = 0$.

The tables have been compiled for values of Y from -20 to $+20$ and values of λ from 1 to 100. The values of λ correspond to an n-type semiconductor; the same tables can be used for a p-type semiconductor by replacing λ with λ^{-1} and taking due account of the fact that $F(Y,\lambda^{-1}) = F(-Y,\lambda)$.

The table contains the significant digits of the function $F(Y,\lambda)$; the actual value is achieved by multiplying by the factor given above the column. For example, $F(-13, 10) = 2.103 \times 10^3$.

Y \ λ	-20	-19	-18	-17	-16	-15	-14	-13	-12	-11
	$\times 10^4$						$\times 10^3$			
1	2.203	1.336	0.810	0.491	0.298	0.181	1.097	0.665	0.403	0.245
2	3.115	1.889	1.146	0.695	0.422	0.256	1.551	0.941	0.571	0.346
4	4.405	2.672	1.621	0.983	0.596	0.362	2.193	1.330	0.807	0.489
6	5.395	3.272	1.985	1.204	0.730	0.443	2.686	1.629	0.988	0.599
8	6.230	3.779	2.292	1.390	0.843	0.511	3.102	1.881	1.141	0.692
10	6.965	4.225	2.562	1.554	0.943	0.572	3.468	2.103	1.276	0.774
20	9.851	5.975	3.624	2.198	1.333	0.809	4.904	2.975	1.804	1.094
40	13.931	8.449	5.125	3.108	1.885	1.144	6.936	4.207	2.551	1.547
60	17.062	10.348	6.277	3.807	2.309	1.401	8.494	5.152	3.125	1.895
80	19.701	11.949	7.248	4.396	2.666	1.617	9.809	5.949	3.608	2.188
100	22.027	13.360	8.103	4.915	2.981	1.808	10.966	6.651	4.034	2.447

Y \ λ	-10	-9	-8	-7	-6	-5	-4	-3	-2	-1
	x 10²	x 10²	x 10²	x 10²	x 10²	x 10²	x 10	x 10	x 10	1
1	1.484	0.900	0.546	0.331	0.200	1.210	0.725	0.426	0.235	1.042
2	2.098	1.272	0.771	0.467	0.282	1.694	1.004	0.576	0.306	1.273
4	2.968	1.799	1.090	0.660	0.398	2.389	1.411	0.805	0.422	1.722
6	3.634	2.204	1.335	0.808	0.488	2.924	1.727	0.984	0.515	2.091
8	4.198	2.545	1.542	0.933	0.563	3.376	1.993	1.136	0.594	2.407
10	4.692	2.845	1.724	1.043	0.630	3.774	2.228	1.269	0.663	2.687
20	6.636	4.023	2.438	1.476	0.890	5.337	3.150	1.794	0.937	3.793
40	9.384	5.690	3.448	2.087	1.259	7.548	4.454	2.537	1.325	5.361
60	11.493	6.968	4.223	2.556	1.542	9.244	5.455	3.107	1.623	6.565
80	13.271	8.046	4.876	2.951	1.781	10.674	6.299	3.587	1.874	7.581
100	14.838	8.996	5.452	3.299	1.991	11.934	7.043	4.011	2.095	8.475

Y \ λ	1	2	3	4	5	6	7	8	9	10
	1	1			$\times\ 10$				$\times\ 10^2$	
1	1.042	2.350	0.426	0.725	1.210	2.004	3.309	0.546	0.900	1.484
2	1.046	2.113	0.348	0.555	0.890	1.443	2.359	0.387	0.637	1.050
4	1.285	2.375	0.350	0.495	0.719	1.091	1.721	0.278	0.453	0.744
6	1.525	2.747	0.387	0.514	0.691	0.980	1.475	0.232	0.374	0.610
8	1.742	3.103	0.429	0.551	0.706	0.946	1.357	0.207	0.328	0.531
10	1.937	3.434	0.470	0.593	0.737	0.947	1.300	0.192	0.298	0.479
20	2.719	4.788	0.647	0.793	0.934	1.095	1.321	0.170	0.238	0.358
40	3.838	6.747	0.908	1.104	1.280	1.449	1.635	0.188	0.229	0.302
60	4.699	8.258	1.110	1.349	1.558	1.751	1.945	0.217	0.248	0.301
80	5.426	9.533	1.281	1.556	1.795	2.013	2.222	0.244	0.272	0.315
100	6.066	10.657	1.432	1.739	2.005	2.245	2.472	0.270	0.297	0.335

λ \ Y	11	12	13	14	15	16	17	18	19	20
		$\times 10^2$					$\times 10^3$			
1	2.447	4.034	6.651	1.097	1.808	2.981	4.915	8.103	13.360	22.027
2	1.731	2.853	4.703	0.775	1.278	1.108	3.475	5.730	9.447	15.575
4	1.225	2.018	3.326	0.548	0.904	1.491	2.457	4.052	6.680	11.013
6	1.002	1.649	2.717	0.448	0.738	1.217	2.006	3.308	5.454	8.992
8	0.870	1.429	2.354	0.388	0.639	1.054	1.738	2.865	4.723	7.788
10	0.780	1.280	2.106	0.347	0.572	0.943	1.554	2.562	4.225	6.965
20	0.565	0.914	1.495	0.246	0.405	0.667	1.099	1.812	2.987	4.925
40	0.435	0.671	1.074	0.175	0.287	0.472	0.778	1.281	2.113	3.483
60	0.400	0.581	0.900	0.144	0.235	0.386	0.635	1.047	1.725	2.844
80	0.393	0.540	0.806	0.127	0.205	0.335	0.551	0.907	1.494	2.463
100	0.400	0.522	0.750	0.115	0.185	0.301	0.493	0.811	1.337	2.203

FLAT BAND POTENTIALS OF SEMICONDUCTOR ELECTRODES (633)

Semiconductor	Conductivity type[a]	Solution composition and pH	E_{fb} (against saturated calomel electrode)
Ge	intrinsic	pH7	-0.4
ZnO	n	1 N KCl + borate buffer (pH 8.8)	-0.42
CdS	n	1 N KCl	-0.9
CdSe	n	1 N KCl	-0.66
ZnSe	n	1 N KCl	-0.82
ZnTe	p	1 N KCl	-0.8
TiO_2	n	pH 7	-0.65
SnO_2	n	pH 6.8	0
GaAs	n	1 N KOH	-1.35
SiC	n	0.05 M H_2SO_4	-1.6
GaP	n	1 N H_2SO_4	-1.15
GaP	p	1 N H_2SO_4	+0.95
Si	n	1 N KCl	0 [b]

[a] Concentration of majority carriers in n- and p-type samples was usually 10^{15} - 10^{17} cm^{-3}.

[b] Obtained by extrapolation to zero thickness of the oxide layer.

REFERENCES

1. E. Becquerel, <u>Compt. Rend.</u>, 9:145 (1839).
2. V. I. Veselovsky, <u>Zh. Fiz. Khim.</u>, 20:1948 (1946).
3. W. H. Brattain and C. G. B. Garrett, <u>Bell System Techn. J.</u>, 34:129 (1955).
4. J. F. Dewald, <u>in</u>: "Surface Chemistry of Metals and Semiconductors", H. C. Gatos, ed., Wiley, New York (1960) p.205.
5. V. A. Myamlin and Yu. V. Pleskov, "Electrochemistry of Semiconductors", Plenum Press, New York (1967).
6. V. A. Tyagai and O. V. Snitko, "Elektrootrazhenie Sveta v Poluprovodnikakh", Naukova Dumka, Kiev (1980).
7. H. Gerischer, <u>J. Electrochem. Soc.</u>, 113:1174 (1966).
8. A. Fujishima and K. Honda, <u>Nature</u>, 238:37 (1972).
9. M. Archer, <u>J. Appl. Electrochem.</u>, 5:17 (1975).
10. J. C. Slater, "Insulators, Semiconductors and Metals", McGraw-Hill, New York, London (1967).
11. J. M. Ziman, "Principles of the Theory of Solids", Cambridge University Press, Cambridge (1972).
12. A. S. Davydov, "Teoriya Tvyordogo Tela", Nauka, Moscow (1976).
13. C. Kittel, "Introduction to Solid State Physics", Wiley, New York, London (1971).
14. R. A. Smith, "Semiconductors", Cambridge University Press, Cambridge (1978).
15. V. L. Bonch-Bruevich and S. G. Kalashnikov, "Fizika Poluprovodnikov", Nauka, Moscow (1977).
16. K. Seeger, "Semiconductor Physics", Springer-Verlag, Wien, New York (1973).
17. W. Shockley, "Electrons and Holes in Semiconductors", Van Nostrand, New York (1950).

18.   J. I. Pankove, "Optical Processes in Semiconductors",
      Prentice Hall, Englewood Cliffs (1971).
19.   "Light Scattering in Solids", M. Cardona, ed., Springer-
      Verlag, Berlin, Heidelberg, New York (1975).
20.   T. S. Moss, "Optical Properties of Semiconductors",
      Butterworths, London (1959).
21.   D. L. Greenaway and G. Harbeke, "Optical Properties and Band
      Structure of Semiconductors", Pergamon Press, Oxford (1968).
22.   F. Gutmann and L. Lyons, "Organic Semiconductors", Wiley,
      New York, London, Sydney (1967).
23.   N. F. Mott and E. A. Davis, "Electronic Processes in Non-
      Crystalline Materials", Clarendon Press, Oxford (1971).
24.   M. Cutler, "Liquid Semiconductors", Academic Press, New
      York, London, San Francisco (1977).
25.   A. A. Levin, "Solid State Quantum Chemistry", McGraw-Hill,
      New York, London, Paris (1977).
26.   L. D. Landau and E. M. Lifshits, "Statisticheskaya Fizika,
      Part I", Nauka, Moscow (1976).
27.   R. Kubo, "Statistical Mechanics", North-Holland, Amsterdam
      (1965).
28.   R. Kubo, "Thermodynamics", North-Holland, Amsterdam (1968).
29.   W. Ehrenberg, Proc. Phys. Soc., A63:75 (1950).
30.   R. Dingle and D. Arndt, Roy. S. Appl. Sci. Res., B6:114,
      (1956); B6:155 (1956).
31.   J. S. Blakemore, "Semiconductor Statistics", Pergamon Press,
      Oxford (1962) Appendix B.
32.   R. S. Knox and D. L. Dexter, "Excitons", New York, London,
      Sydney (1965).
33.   W. Shockley and W. Read, Phys. Rev., 87:835 (1952).
34.   A. A. Chernenko, Dokl. AN SSSR, 153:1129 (1963);
      Elektrokhimiya, 7:1737 (1971).
35.   R. Arndt and L. Roper, "Simple Membrane Electrodiffusion
      Theory", Publ. Division PBSM (1972).
36.   Yu. Ya. Gurevich and Yu. I. Kharkats, Elektrokhimiya, 16:870
      (1980).
37.   J. M. Houston and H. F. Webster, in: "Advances in
      Electronics and Electron Physics", L. Marton, ed., Academic
      Press, New York, London (1962) p.125.
38.   A. M. Brodsky and Yu. Ya. Gurevich, "Teoriya Elektronnoy
      Emissii iz Metallov", Nauka, Moscow (1973).
39.   S. Glasstone, "An Introduction to Electrochemistry," Van
      Nostrand, New York (1946).
40.   R. Parsons, in: "Modern Aspects of Electrochemistry", J.
      O'M. Bockris, ed., Butterworths, London (1954). p.1.
41.   R. Parsons, in: "Comprehensive Treatise of Electrochemistry",
      Vol. 1, J. O'M. Bockris, B. E. Conway and E. Yeager, eds.,
      Plenum Press, New York, London (1980) p.1.
42.   R. R. Dogonadze and Yu. A. Chizmadzhev, Dokl. AN SSSR, 145:849
      (1962).

43. H. Gerischer, in: "Physical Chemistry. An Advanced Treatise",
     Vol. 9, H. Eyring, ed., Academic Press, New York, London
     (1970) p.463.
44. H. Gerischer, Photochem. and Photobiol., 16:243 (1972).
45. R. Memming and F. Müllers, Ber. Bunsenges. Phys. Chem.,
     76:475 (1972).
46. L. I. Krishtalik, "Elektrodnye Reaktsii. Teoriya Elementarnogo
     Akta", Nauka, Moscow (1979).
47. H. Gerischer, in: "Solar Power and Fuels", J. Bolton, ed.,
     Academic Press, New York (1977) p. 77.
48. A. N. Frumkin, "Potentsialy Nulevogo Zaryada", Nauka, Moscow
     (1979).
49. A. N. Frumkin and B. B. Damaskin, Dokl. AN SSSR, 221:395
     (1975).
50. J. E. B. Randles, Trans. Faraday Soc., 52:1573 (1956).
51. E. A. Kanevsky, Zh. Fiz. Khim., 22:1397 (1948).
52. F. Lohmann, Z. Naturforsch., 22a:843 (1967).
53. W. M. Latimer, "The Oxidation States of the Elements and Their
     Potentials in Aqueous Solutions", Prentice Hall, New York
     (1952); M. Kh. Karapet'yants, "Thermodynamic Constants of
     Inorganic and Organic Compounds", Ann Arbor-Humphrey Science
     Publishers, Ann Arbor (1970).
54. M. Pourbaix, "Atlas of Electrochemical Equilibria", Pergamon
     Press, London (1966); "Lectures on Electrochemical Corrosion",
     Plenum Press, New York (1973).
55. S. Park and M. E. Barber, J. Electroanal. Chem., 99:67 (1979).
56. M. Green, in: "Modern Aspects of Electrochemistry", Vol. 2,
     J. O'M. Bockris, ed., Butterworths, London (1959).
57. J. O'M. Bockris and A. K. N. Reddy, "Modern Electrochemistry",
     Plenum Press, New York (1970).
58. H. Helmholtz, Ann. Phys. Chem. (Wiedemanns Ann). N. F., 7:337
     (1879)
59. G. J. Gouy, Anns. Chim. Phys., 8:291 (1906).
60. G. J. Gouy, Physique, 4:457 (1910).
61. D. L. Chapman, Phil. Mag., Ser. 6, 25:475 (1913).
62. A. Frumkin, O. Petrii and B. Damaskin, J. Electroanal. Chem.,
     27:81 (1970).
63. C. G. B. Garrett and W. H. Brattain, Phys. Rev., 99:376
     (1955).
64. "Fizika Poverkhnosti Poluprovodnikov", G. E. Pikus, ed.,
     Moscow (1959).
65. R. H. Kingston and S. F. Neustadter, J. Appl. Phys., 26:718
     (1955).
66. D. R. Frankl, J. Appl. Phys., 31:1752 (1960).
67. C. E. Young, J. Appl. Phys., 32:329 (1961).
68. R. Seiwatz and M. Green, J. Appl. Phys., 29:1034 (1958).
69. J. McDougall and E. C. Stoner, Phil. Trans. Roy. Soc., A237:67
     (1939).
70. R. F. Greene, in: "Solid State Surface Science", M. Green,
     ed., Marcel Dekker, New York (1969).
71. F. Stern, C. R. C. Crit. Rev. Solid State Sci., 4:499 (1974).

72. M. Jonson, J. Phys. Chem., 9:3055 (1976).
73. B. Vinter, Phys. Rev. B, 15:3947 (1977).
74. C. B. Duke, Phys. Rev., 159:632 (1967); 152:683 (1966).
75. D. Eger and Y. Goldstein, Phys. Rev. B, 19:1089 (1979).
76. Proc. of Intern. Conference on Electronic Properties of
    Quasi-2D System. Providence, USA, 1975; Surf. Sci., 58:1
    (1976).
77. Proc. of 2nd Intern. Conference on Electronic Properties of
    2D-system. Berchtesgaden, West Germany, 1977; Surf. Sci.,
    73:1 (1978).
78. V. A. Volkov, V. A. Petrov and V. B. Sandomirsky, Uspekhi Fiz.
    Nauk., 131:423 (1980).
79. B. B. Damaskin and O. A. Petrii, "Vvedenie v
    Elektrokhimicheskuyu Kinetiku", "Vysshaya Shkola" Publishing
    House, Moscow (1983).
80. A. N. Frumkin, V. S. Bagotsky, Z. A. Iofa and B. N. Kabanov,
    "Kinetika Elektrodnykh Protsessov", Moscow University Press,
    Moscow (1952).
81. P. Delahay, "Double Layer and Electrode Kinetics",
    Interscience, New York, London, Sydney (1965).
82. K. J. Vetter, "Elektrochemische Kinetik", Springer-Verlag,
    Berlin, Göttingen, Heidelberg (1961).
83. J. Newman, "Electrochemical Systems", Prentice-Hall, Englewood
    Cliffs (1973).
84. I. Tamm, Sow. Phys., 1:733 (1932); Z. Phys., 76:849 (1932).
85. S. G. Davison and J. D. Levine, Surface States, in: "Solid
    State Physics. Vol. 25", Academic Press, New York, London
    (1970).
86. W. Shockley, Phys. Rev., 56:317 (1939).
87. P. Y. Feibelman, J. Vac. Sci. Techn., 17:176 (1980).
88. J. A. Appelbaum and D. R. Hamann, Rev. Mod. Phys., 48:479
    (1976).
89. A. Ya. Belenky, Uspekhi Fiz. Nauk., 134:125 (1981).
90. M. Steslicka and Z. Perkal, Physica., BC96:327 (1979).
91. T. W. Nee and R. E. Prange, Phys. Letters., 25:582 (1967).
92. M. Khaikin, Uspekhi Fiz. Nauk., 96:409 (1968).
93. A. V. Rzhanov, "Electronnye Protsessy na Poverkhosti
    Poluprovodnikov", Nauka, Moscow (1971).
94. S. R. Morrison, "The Chemical Physics of Surfaces", Plenum
    Press, New York, London (1977).
95. P. J. Boddy and W. H. Brattain, J. Electrochem. Soc., 109:574
    (1962).
96. M. D. Krotova and Yu. V. Pleskov, Phys. Stat. Sol., 3:2119
    (1963).
97. O. V. Romanov, M. A. Sokolov and M. S. Grilikhes,
    Elektrokhimiya, 10:584 (1974).
98. R. De Gryse, W. P. Gomes, F. Cardon and J. Vennik,
    J. Electrochem. Soc., 122:711 (1975).
99. T. P. Birintseva and Yu. V. Pleskov, Izv. AN SSSR, Ser. Khim.,
    251 (1965).

100. V. A. Tyagai and G. Ya. Kolbasov, Elektrokhimiya, 11:1514 (1975).

101 Yu. V. Pleskov, Elektrokhimiya, 3:112 (1967).

102 A. Ya. Gokhshtein, "Poverkhnostnoe Natyazhenie Tveordykh Tel", Nauka, Moscow (1976).

103. L. J. Handley and A. J. Bard, J. Electrochem. Soc., 127:338 (1980).

104. D. R. Turner and J. I. Pankove, in: "Techniques of Electrochemistry", Vol. 3, E. Yeager and A. J. Salkind, eds., Wiley-Interscience, New York, Chichester, Brisbane, Toronto (1978) p.142.

105. R. Memming, Phil. Res. Rep., 19:323 (1964).

106. Yu. V. Pleskov and V. A. Tyagai, Dokl. AN SSSR, 141:1135 (1961).

107. J. F. Dewald, Bell System Tech. J., 34:615 (1960).

108. V. A. Tyagai, Izv. AN SSSR, Ser. Khim., 34 (1964).

109. P. J. Boddy, D. Kahng and Y. S. Chen, Electrochim. Acta, 13:1311 (1968).

110. M. Gleria and R. Memming, J. Electroanalyt. Chem., 65:163 (1975).

111. F. Möllers and R. Memming, Ber. Bunsenges. Phys. Chem., 76:469 (1972).

112. T. Watanabe, A. Fujishima and K. Honda, Chem. Lett., 897 (1974).

113. Z. A. Rotenberg, T. V. Dzhavrishvili, Yu. V. Pleskov and A. L. Asatiani, Elektrokhimiya, 13:1803 (1977).

114. R. Memming, J. Electrochem. Soc., 116:785 (1969).

115. R. Memming and G. Schwandt, Surface Sci., 5:97 (1966).

116. A. Frumkin and A. Gorodetzkaja, J. Phys. Chem., 136:451 (1928).

117. Yu. Y. Pleskov, Elektrokhimiya, 1:4 (1965).

118. V. A. Tyagai, G. Ya. Kolbasov, V. N. Bondarenko and O. V. Snitko, Fis. i Tekhn. Poluprovodn., 6:2325 (1972).

119. M. Hofmann-Perez and H. Gerischer, Z. Elektrochem., 65:771 (1961).

120. D. E. Yates and T. W. Healy, J. Chem. Soc., Faraday I, 76:9 (1980).

121. M. J. Sparnaay, Rec. Trav. Chim. Pays-Bas, 79:950 (1960).

122. M. A. Butler and D. S. Ginley, Chem. Phys. Lett., 47:319 (1977).

123. H. Minoura and M. Tsuiki, Electrochim. Acta, 23:1377 (1978).

124. T. Inoue, T. Watanabe, A. Fujishima and K. Honda, in: "Semiconductor Liquid-Junction Solar Cells", A. Heller, ed., The Electrochemical Society, Princeton (1977) p. 210.

125. R. Memming, in: "Topics in Surface Chemistry", E. Kay and P. S. Bagus, eds., Plenum Press, New York (1978) p. 1.

126. V. G. Baru and F. F. Volkenshtein, "Vliyanie Oblucheniya na Poverkhnostnye Svoistva Poluprovodnikov", Nauka, Moscow (1978).

127. R. I. Bickley, in: "Chemical Physics of Solids and Their Surfaces", Vol. 7, The Chemical Society, London (1978) p. 11.

128.   A. J. Nozik, in: "Semiconductor Liquid-Junction Solar Cells",
        A. Heller, ed., The Electrochemical Society, Princeton (1977)
        p. 272.
129.   D. M. Shub, A. A. Remnyov and V. I. Veselovsky,
        Elektrokhimiya, 9:676 (1973).
130.   Y. Nakato, A. Tsumura and H. Tsubomura, J. Electrochem. Soc.,
        127:1502 (1980).
131.   H. Tributsch, J. Electrochem. Soc., 125:1086 (1978).
132.   L. D. Landau and E. M. Lifshitz, "Quantum Mechanics", Pergamon
        Press, New York, London (1977).
133.   H. Reiss, J. Electrochem. Soc., 125:937 (1978).
134.   R. A. Marcus, in: "A Theory of Electron Transfer Processes at
        Electrodes. Trans., Symposium on Electrode Processes", E.
        Yeager, ed., Wiley, New York (1961); Electrochim. Acta.,
        13:995 (1968).
135.   R. R. Dogonadze and A. M. Kuznetsov, in: "Progress in Surface
        Science", Vol. 6, S. G. Davison, ed., Pergamon Press
        (1975) p. 1.
136.   H. Gerischer, Z. Phys. Chem., N. F., 26:223 (1960); 26:325
        (1960); 27:48 (1961).
137.   V. G. Levich, in: "Advances in Electrochemistry and
        Electrochemical Engineering", Vol. 4, P. Delahay, ed.,
        Interscience, New York, London (1966) p. 249.
138.   J. Ulstrup, "Charge Transfer Processes in Condensed Media",
        Springer-Verlag, Berlin, Heidelberg, New York (1980).
139.   J. O'M. Bockris and S. U. M. Khan, "Quantum Electrochemistry",
        Plenum Press, New York, London (1979).
140.   H. Gerischer, D. M. Kolb and J. K. Sass, Adv. Phys., 27:437
        (1978).
141.   B. Pettinger, H.-R. Schöppel and H. Gerischer, Ber. Bunsenges
        Phys. Chem., 77:960 (1973).
142.   W. Schmickler and J. Ulstrup, Chem Physics, 19:217 (1977).
143.   W. Schmickler, J. Electroanal. Chem., 82:65 (1977); 83:387
        (1977); 84:203 (1977).
144.   W. Schmickler, Ber. Bunsenges. Phys. Chem., 82:477 (1978).
145.   R. Memming, in: "Electroanalytical Chemistry", Vol. 11,
        A. J. Bard, ed., Marcel Dekker, New York, Basel (1979) p. 1.
146.   S. N. Frank and A. J. Bard, J. Amer. Chem. Soc., 99:4667
        (1977).
147.   P. A. Kohl and A. J. Bard, J. Amer. Chem. Soc., 99:7531
        (1977).
148.   D. Laser and A. J. Bard, J. Electrochem. Soc., 123:1828
        (1976); 123:1833 (1976); 123:1837 (1976).
149.   D. Laser, J. Electrochem. Soc., 126:1011 (1979).
150.   M. Nishida, J. Appl. Phys., 51:1669 (1980).
151.   B. Pettinger, H. R. Schöppel and H. Gerischer, Ber Bunsenges.
        Phys. Chem., 78:450 (1974).
152.   W. Lorenz and M. Handschuh, Electrochim. Acta., 28:293 (1980).
153.   W. Lorenz, Electrochim. Acta., 25:1111 (1980).
154.   V. A. Myamlin, Dokl. AN SSSR, 139:1153 (1961).
155.   D. R. Turner, J. Electrochem. Soc., 103:252 (1956).

156. R. D. Middlebroock, "An Introduction to Junction Transistor Theory", Wiley, New York (1957).

157. W. H. Brattain and C. G. B. Garrett, Physica, 20:885 (1954).

158. E. A. Efimov and I. G. Erusalimchik, Zh. Fis. Khim., 32:1103 (1958).

159. J. B. Flynn, J. Electrochem. Soc., 105:715 (1958).

160. A. E. Kuzmak, S. F. Timashov and S. A. Molchanova, Elektrokhimiya, 11:234 (1975).

161. Yu. Y. Pleskov, Dokl. AN SSSR, 132:1360 (1960).

162. P. Janietz, R. Weiche, J. Westfahl and R. Landsberg, J. Electroanal. Chem., 112:63 (1980).

163. Yu. Y. Pleskov, Dokl. AN SSSR, 126:111 (1959).

164. H. Gerischer and F. Beck, Z Phys. Chem., N.F. 13:389 (1957).

165. V. A. Tyagai, Elektrokhimiya, 1:387 (1965).

166. F. Beck and H. Gerischer, Z. Elektrochem., 63:500 (1959).

167. H. Gerischer, Faraday Disc. Chem. Soc., N58:219 (1974).

168. H. Gerischer, in: "Special Topics in Electrochemistry", P.A. Rock, ed., Elsevier (1977) p. 35.

169. H. Gerischer and F. Willig, in: "Topics in Current Chemistry", Fortschr. Chem. Forsch., 61:31 (1976).

170. Yu. Y. Gurevich and Yu. Y. Pleskov, "Elektrokhimiya", Vol. 18, (Itogi Nauki i Tekhniki), VINITI, Moscow (1982) p. 3.

171. I. D. Jackson, "Classical Electrodynamics", Wiley, New York, London (1962).

172. W. Mehl and J. M. Hale, in: "Advances in Electrochemistry and Electrochemical Engineering", Vol. 6, P. Delahay, ed., Interscience, New York (1967). p. 399.

173. R. Memming, Photochem. and Photobiol., 16:325 (1972).

174. R. Memming, Faraday Disc. Chem. Soc., N58:261 (1974).

175. M. Fujihira, T. Osa, D. Hursch and T. Kuwana, J. Electroanal. Chem., 88:285 (1978).

176. T. Osa and M. Fujihira, Nature, 264:349 (1976).

177. M. Fujihira, N. Ohnishi and T. Osa, Nature, 268:226 (1977).

178. M. Gleria and R. Memming, Z. Phys. Chem., N.F., 98:303 (1975).

179. R. Memming and F. Schröppel, Chem. Phys. Letters, 62:207 (1979).

180. A. Fujishima, T. Iwase, T. Watanabe and K. Honda, J. Amer. Chem. Soc., 97:4134 (1975).

181. A. Fujishima, T. Watanabe, O. Tatsuoki and K. Honda, Chem. Lett., 13 (1975).

182. T. Takizawa, T. Watanabe and K. Honda, J. Phys. Chem., 82:1391 (1978).

183. M. Matsumura, Y. Nomura and H. Tsubomura, Bull. Chem. Soc. Japan, 49:1409 (1976).

184. R. Schumacher, R. H. Wilson and L. A. Harris, J. Electrochem. Soc., 127:96 (1980).

185. H. J. Danzmann, K. Hauffe and Z. G. Szabo, Z. Phys. Chem., N.F. 104:95 (1977).

186. T. Watanabe, A. Fujishima, O. Tatsuoki and K. Honda, Bull. Chem. Soc. Japan, 49:8 (1976).

187. H. Gerischer and M. Lübke, Z Phys. Chem., N.F., 98:317 (1975).
188. U. Bode and K. Hauffe, J. Electrochem. Soc., 125:51 (1978).
189. T. Watanabe, A. Fujishima and K. Honda, Ber. Bunsenges. Phys. Chem., 79: 1213 (1975).
190. K. Hauffe and U. Bode, Faraday Disc. Chem. Soc., N58:281 (1974).
191. Yu. V. Pleskov and V. Yu. Filinovsky, "Rotating Disk Electrode", Consultants Bureau, New York (1976).
192. N. S. Lewis, A. B. Bocarsly and M. S. Wrighton, J. Phys. Chem., 84:2033 (1980).
193. T. Yamase, H. Gerischer, M. Lübke and B. Pettinger, Ber. Bunsengens Phys. Chem., 82:1041 (1978).
194. M. T. Spitler and M. Calvin, J. Chem. Phys., 66:4294 (1977).
195. M. Spitler, M. Lübke and H. Gerischer, Ber. Bunsenges. Phys. Chem., 83:663 (1979).
196. A. L. Lehninger, "Biochemistry", Worth Publishers, New York, (1972).
197. P. Fromherz and W. Arden, Ber. Bunsenges. Phys. Chem., 84:1045 (1980).
198. P. Fromherz and W. Arden, J. Amer. Chem. Soc., 102:6211(1980).
199. W. Arden and P. Fromherz, J. Electrochem. Soc., 127:370 (1980).
200. T. Miyasaka, T. Watanabe, A. Fujishima and K. Honda, Nature, 277:638 (1979).
201. T. Miyasaka, T. Watanabe, A. Fujishima and K. Honda, J. Amer. Chem. Soc., 100:6657 (1978).
202. W. W. Gärtner, Phys. Rev., 116:84 (1959).
203. P. P. Konorov and S. V. Shchegolikhina, Fiz. Tverd. Tela, 9:2117 (1977).
204. P. P. Konorov and Yu. A. Tarantov, Fiz. i Tekhn. Poluprovod., 7:1026 (1973).
205. R. H. Wilson, J. Appl. Phys., 48:4292 (1977).
206. J. Reichman, Appl. Phys. Lett., 36:574 (1980).
207. H. Dember, Phys. Z. (Leipzig), 32:554 (1931); 33:207 (1932).
208. M. A. Butler, J. Appl. Phys., 48:1914 (1977).
209. K. Colbow, D. J. Harrison and B. L. Funt, J. Electrochem. Soc., 128:547 (1981).
210. L. M. Peter, Electrochim. Acta., 23:1073 (1978).
211. Yu. V. Pleskov and V. V. Eletsky, Electrochim. Acta, 12:707 (1967).
212. V. V. Eletsky, Ya. Ya. Kulyavik and Yu. V. Pleskov, Elektrokhimiya, 3:753 (1967).
213. F. Cardon and W. P. Gomes, Surface Sci., 27:286 (1971).
214. T. Freund, Surface Sci., 33:295 (1972).
215. K. Micka and H. Gerischer, J. Electroanal. Chem., 38:397 (1972).
216. K. M. Sancier and S. R. Morrison, Surface Sci., 36:622 (1973).
217. R. A. L. Vanden Berghe, W. P. Gomes and F. Cardon, Z. Phys. Chem., N.F., 92:91 (1974).
218. E. C. Dutoit, F. Cardon and W. P. Gomes, Ber. Bunsenges. Phys. Chem., 80:1285 (1976).

219. M. Miyake, H. Yoneyama and H. Tamura, Denki Kagaku, 45:411 (1977).
220. S. R. Morrison, in: "Progress in Surface Science", Vol. 1, S. G. Davison, ed., Pergamon Press, Oxford (1972) p. 105.
221. R. R. Dogonadze and A. M. Kuznetsov, Elektrokhimiya, 1:1008 (1965).
222. F. Williams and A. J. Nozik, Nature, 271:137 (1978).
223. A. J. Nozik, Ann. Rev. Phys. Chem., 29:189 (1978).
224. D. S. Boudreaux, F. Williams and A. J. Nozik, J. Appl. Phys., 51:2158 (1980).
225. C. Wagner and W. Traud, Z. Elektrochem., 44:391 (1938).
226. V. V. Skorchelletti, "Teoreticheskie Osnovy Korrozii Metallov", Khimiya, Leningrad (1973).
227. H. Gerischer, in: "Corrosion Week", T. Farkas, ed., Akadémiae Kiadó, Budapest (1979) S. 68.
228. Yu. V. Pleskov, Zh. Fiz. Khim., 35:2576 (1961).
229. W. W. Harvey and M. C. Finn, Surface. Sci., 2:456 (1964).
230. A. P. Blokhina and S. O. Izidinov, Elektrokhimiya, 6:1409 (1970).
231. S. M. Repinsky and L. L. Sveshnikova, Elektrokhimiya, 14:1026 (1978).
232. E. M. Golubchik, T. A. Lozovik, L. G. Koshechko and Yu. M. Lugovykh, Elektrokhimiya, 16:768 (1980).
233. D. R. Turner, J. Electrochem. Soc., 107:810 (1960).
234. H. Gerischer and F. Beck, Z. Phys. Chem., N.F., 123:113 (1960).
235. H. Gerischer, J. Electroanal. Chem., 82:133 (1977).
236. A. J. Bard and M. S. Wrighton, J. Electrochem. Soc., 124:1706 (1977).
237. H. Gerischer, J. Vac. Sci. Technol., 15:1422 (1978).
238. R. Memming, Electrochim. Acta, 25:77 (1980).
239. R. Memming, J. Electrochem. Soc., 125:117 (1978).
240. L. V. Belyakov, D. N. Goryachev, L. G. Paritsky, S. M. Ryvkin and O. M. Sreseli, Fiz. I Tekhn. Poluprovod., 10:1142 (1976).
241. S. N. Frank and A. J. Bard, J. Amer. Chem. Soc., 97:7427 (1975).
242. M. E. Gerstner, J. Electrochem. Soc., 126:954 (1979).
243. A. N. Asanov, Dokl. AN SSSR, 225:838 (1975).
244. V. P. Pakhomov, D. M. Shub and M. M. Lukina, Elektrokhimiya, 7:1325 (1971).
245. A. Fujishima, T. Inoue, T. Watanabe and K. Honda, Chem. Lett., 357 (1978).
246. A. Fujishima, T. Inoue and K. Honda, J. Amer. Chem. Soc., 101:5582 (1979).
247. T. Inoue, T. Watanabe, A. Fujishima, K. Honda and K. Kohayakawa, J. Electrochem. Soc., 124:719 (1977).
248. M. J. Madou, K. W. Frese and S. R. Morrison, J. Electrochem. Soc., 127:987 (1980).
249. S. M. Repinsky, S. F. Devyatova, O. I. Semyonova and L. L. Sveshnikova, in: "Problemy Fizicheskoi Khimii Poverkhnosti Poluprovodnikov", A. V. Rzhanov, ed., Nauka, Novosibirsk (1978) p. 87.

250. H. Tributsch, Ber. Bunsenges. Phys. Chem., 81:361 (1977).

251. H. Tributsch and J. C. Bennett, J. Electroanal. Chem., 81:97 (1977).

252. R. Guittard, R. Heindl, R. Parsons, A. M. Redon and H. Tributsch, J. Electroanal. Chem., 111:401 (1980).

253. S. O. Izidinov, in: "Elektrokhimiya", Vol. 14, (Itogi Nauki i Tekhniki),VINITI, Moscow (1979) p. 208.

254. S. O. Izidinov, T. I. Borisova and V. I. Veselovsky, Dokl. AN SSSR, 145:598 (1962).

255. W. J. Albery and M. D. Archer, J. Electroanal. Chem., 86:1 (1978).

256. H. Gerischer, in: "Solar Energy Conversion", B. O. Seraphin, ed., Springer-Verlag (1979) p. 115.

257. A. J. Nozik, J. Crystal Growth, 39:200 (1977).

258. M. Tomkiewicz and H. Fay, Appl. Phys., 18:1 (1978).

259. L. A. Harris and R. H. Wilson, Ann. Rev. Mater. Sci.,8:99 (1978).

260. H. Maruska and A. K. Ghosh, Solar Energy, 20:443 (1978).

261. J. Rajeshwar, P. Singh and J. Dubow, Electrochim. Acta, 23:1117 (1978).

262. R. Memming, Philips Tech. Rev., 38:160 (1978/79).

263. A. T. Vasko, P. P. Pogorelsky and S. K. Kovach, in: "Elektronnye Protsessy V Vodnykh Rastvorakh", Naukova Dumka, Kiev (1979) p. 20.

264. Yu. V. Pleskov, Vestnik AN SSSR, No.6:69 (1979).

265. J. Manassen, D. Cahen, G. Hodes and A. Sofer, Nature, 263:97 (1976).

266. M. S. Wrighton, Chem. Eng. News, 57:29 (1979).

267. A. J. Bard, Science, 207:139 (1980).

268. M. A. Butler and D. S. Ginley, J. Mater. Sci., 15:1 (1980).

269. A. J. Nozik, Phil. Trans. Roy. Soc., London, A295:453 (1980).

270. H. Gerischer, Ber. Bunsenges. Phys. Chem., 80:1046 (1976).

271. M. S. Wrighton, Technol. Rev., 79:31 (1977).

272. K. Honda and T. Watanabe, Elektrokhimiya, 13:924 (1977).

273. J. O'M. Bockris and K. Uosaki, in: "Solid State Chemistry of Energy Conversion and Storage", J. B. Goodenough and M. S. Whittingham, eds., Amer. Chem. Soc., Washington (1977) p. 33.

274. A. J. Nozik, in: "Hydrogen Energy System", Vol. 3, T. N. Veziroglu and W. Seifritz, eds., Pergamon Press, Oxford (1978) p. 1217.

275. J. G. Mavroides, Mat. Res. Bull., 13:1379 (1978).

276. P. Clèchet, C. Martelet, J.-R. Martin and R. Olier, L'actualité Chimique (Soc. Chim. France, Soc. Chim. Ind.), No.9:17 (1978).

277. J. O'M. Bockris and L. Handley, Energy Conv., 18:1 (1978).

278. D. Cahen, J. Manassen and G. Hodes, Solar Energy Mater., 1:343 (1979).

279. S. Kar, K. Rajeshwar, P. Singh and J. DuBow, Solar Energy, 23:129 (1979).

280. V. Guruswamy and J. O'M. Bockris, Solar Energy Mater., 1:141 (1979).

281. R. E. Schwerzel, Radiat. Energy Convers. Space. Techn. Pap. 3rd NASA Conf., Moffett Field, Calif., 1978, New York (1978) p. 626.

282. M. D. Archer, Solar Energy, 20:167 (1978).

283. A. J. Bard, A. B. Bocarsly, F.-R. F. Fan, E. G. Walton and M. S. Wrighton, J. Amer. Chem. Soc., 102:3671 (1980).

284. F.-R. F. Fan and A. J. Bard, J. Amer. Chem. Soc., 102:3677 (1980).

285. A. B. Bocarsly, D. C. Bookbinder, R. N. Dominey, N. S. Lewis and M. S. Wrighton, J Amer Chem. Soc., 102:3683 (1980).

286. J. -N. Chazalviel and T. B. Truong, J. Electroanal. Chem., 114:299. (1980).

287. A. Heller and B. Miller, Electrochim. Acta., 25:29 (1980).

288. M. S. Wrighton, Accounts Chem. Res., 12:303 (1979).

289. G. Hodes, J. Manassen and D. Cahen, J. Electrochem. Soc., 127:544 (1980).

290. G. Hodes, J. Manassen and D. Cahen, J. Appl. Electrochem., 7:181 (1977).

291. A. B. Ellis, S. W. Kaiser and M. S. Wrighton, J. Amer. Chem. Soc., 98:6418 (1976).

292. A. B. Ellis, S. W. Kaiser and M. S. Wrighton, J. Amer. Chem. Soc., 98:6855 (1976).

293. A. B. Ellis, S. W. Kaiser, J. M. Bolts and M. S. Wrighton, J. Amer. Chem. Soc., 99:2839 (1977).

294. G. Hodes, Nature, 285:29 (1980).

295. G. Hodes, J. Manassen and D. Cahen, J. Amer. Chem. Soc., 102:5962 (1980).

296. G. Hodes, D. Cahen, J. Manassen and M. David, J. Electrochem. Soc., 127:2252 (1980).

297. B. Miller, A. Heller, M. Robbins, S. Menezes, K. C. Chang and J. Thomson, J. Electrochem. Soc., 124:1019 (1977).

298. R. N. Noufi, P. A. Kohl and A. J. Bard, J. Electrochem. Soc., 125:375 (1978).

299. M. Tsuiki, H. Minoura, T. Nakamura and Y. Ueno, J. Appl. Electrochem., 8:523 (1978).

300. M. Tsuiki and H. Minoura, Chem. Lett., 113 (1979).

301. C. -H. J. Lin and J. H. Wang, Appl. Phys. Lett., 36:852 (1980).

302. A. M. Redon, J. Vigneron and R. Heindl, J. Electrochem. Soc., 127:2033 (1980).

303. G. Hodes, J. Manassen and D. Cahen, Nature, 261:403 (1976).

304. J. Manassen, G. Hodes and D. Cahen, J. Electrochem Soc., 124:532 (1977).

305. B. Miller and A. Heller, Nature, 262:680 (1976).

306. L. M. Peter, J. Electroanal. Chem., 98:49 (1979).

307. B. Miller, S. Menezes and A. Heller, J. Electroanal. Chem., 94:85 (1978).

308. K. C. Chang, A. Heller, B. Schwartz, S. Menezes and B. Miller, Science, 196:1097 (1977).

309. A. B. Ellis, J. M. Bolts, S. W. Kaiser and M. S. Wrighton, J. Amer. Chem. Soc., 99:2848 (1977).

310. B. A. Parkinson, A. Heller and B. Miller, J. Electrochem. Soc., 126:954 (1979).
311. R. Noufi and D. Tench, J. Electrochem. Soc., 127:188 (1980).
312. A. Heller, B. Miller, S. S. Chu and Y. T. Lee, J. Amer. Chem. Soc., 101:7633 (1979).
313. A. Heller, H. J. Lewerenz, B. Miller, Ber. Bunsenges. Phys. Chem., 84:592 (1980).
314. J. M. Bolts, A. B. Ellis, K. D. Legg and M. S. Wrighton, J. Amer. Chem. Soc., 99:4826 (1977).
315. A. B. Ellis, S. W. Kaiser and M. S. Wrighton, J. Amer. Chem. Soc., 98:1635 (1976).
316. A. Heller, K. C. Chang and B. Miller, J. Electrochem. Soc., 124:697 (1977).
317. A. Heller, K. C. Chang and B. Miller, J. Amer. Chem. Soc., 100:684 (1978).
318. W. D. Johnson, H. J. Leamy, B. A. Parkinson, A. Heller and B. Miller, J. Electrochem. Soc., 127:90 (1980).
319. H. Gerischer and J. Gobrecht, Ber. Bunsenges. Phys. Chem., 82:520 (1978).
320. D. Cahen, G. Hodes and J. Manassen, J. Electrochem Soc., 125:1623 (1978).
321. A. Heller, G. P. Schwartz, R. G. Vadimsky, S. Menezes and B. Miller, J. Electrochem. Soc., 125:1156 (1978).
322. R. N. Noufi, P. A. Kohl, J. W. Rogers, J. M. White and A. J. Bard, J. Electrochem. Soc., 126:949 (1979).
323. K. T. L. De Silva and D. Haneman, J. Electrochem. Soc., 127:1554 (1980).
324. H. Tributsch, Ber. Bunsenges. Phys. Chem., 82:169 (1978).
325. J. Gobrecht, H. Tributsch and H. Gerischer, J. Electrochem. Soc., 125:2085 (1978).
326. J. Gobrecht, H. Gerischer and H. Tributsch, Ber. Bunsenges. Phys. Chem., 82:1331 (1978).
327. H. Tributsch, H. Gerischer, C. Clemen and E. Bucker, Ber. Bunsenges. Phys. Chem., 83:655 (1979).
328. H. Tributsch, Z. Naturforsch., 32a:972 (1977).
329. H. Tributsch, Solar Energy Mater., 1:257 (1979).
330. W. Kautek and H. Gerischer, Ber. Bunsenges. Phys Chem., 84:645(1980).
331. S. Menezes, F. J. Di Salvo and B. Miller, J. Electrochem., Soc., 127:1751 (1980).
332. L. F. Schneemeyer, M. S. Wrighton, A. Stacy and M. J. Sienko, Appl. Phys. Lett., 36:701 (1980).
333. L. F. Schneemeyer and M. S. Wrighton, J. Amer. Chem. Soc., 102:6964 (1980).
334. F. -R. F. Fan, H. S. White, B. Wheeler and A. J. Bard, J. Electrochem. Soc., 127:518 (1980).
335. F. -R. F. Fan, H. S. White, B. Wheeler and A. J. Bard, J. Amer. Chem. Soc., 102:5142 (1980).
336. K. D. Legg, A. B. Ellis, J. M. Bolts and M. S. Wrighton, Proc. Nat. Acad. Sci. USA, 74:4116 (1977).

337. M. S. Wrighton, R. G. Austin, A. B. Bocarsly, J. M. Bolts
     O. Haas, K. D. Legg, L. Nadio and M. C. Palazzotto, J. Amer.
     Chem. Soc., 100:1602 (1978).
338. J. M. Bolts and M. S. Wrighton, J. Amer. Chem. Soc., 101:6179
     (1979).
339. A. B. Bocarsly, E. G. Walton, M. G. Bradley and M. S. Wrighton,
     J. Electroanal. Chem., 100:283 (1979).
340. J. M. Bolts, A. B. Bocarsly, M. C. Palazzotto, E. G. Walton,
     N. S. Lewis and M. S. Wrighton, J. Amer. Chem. Soc., 101:1378
     (1979).
341. H. Yoneyama, Y. Murao and H. Tamura, J. Electroanal. Chem.,
     108:87 (1980).
342. J. M. Bolts and M. S. Wrighton, J. Amer. Chem. Soc., 100:5257
     (1978).
343. Y. Avigal, D. Cahen, G. Hodes, J. Manassen, B. Vainas and
     R. A. G. Gilson, J. Electrochem. Soc., 127:1209 (1980).
344. J. I. B. Wilson, J. McGill and D. Weaire, Adv. Phys., 27:365
     (1978).
345. J. I. B. Wilson and D. Weaire, Nature, 275:93 (1978).
346. J. O'M. Bockris, "Energy. The Solar-Hydrogen Alternative",
     Australia and New Zealand Book Company, Sydney (1975).
347. M. Wolf, Energy Conversion, 14:9 (1975).
348. S. N. Paleocrassas, in: "Hydrogen Energy", Part A, T. N.
     Verizoglu, ed., Plenum Press, New York, London (1975) p. 243.
349. P. R. Ryason, Energy Sources, 4:1 (1978).
350. G. Stein, Israel J. Chem., 14:213 (1975).
351. P. L. Kapitsa, Priroda, No.2:70 (1976).
352. N. N. Semenov, Vestnik AN SSSR, No.4:11 (1977).
353. K. I. Zamarayev and V. N. Parmon, Usp. Khimii, 49:1457 (1980).
354. T. S. Dzhabiev and A. E. Shilov, Zhurn. Vses. Khim. Ob-va Im.
     D.I. Mendeleeva, 25:503 (1980).
355. M. D. Archer, in: "Photochemistry", The Chemical Society,
     London, Vol. 6 (1975) p. 737; Vol. 7 (1976) p. 559.
356. D. Haneman, Proc. RACI, 44:37 (1977).
357. A. Yu. Borisov, Molek. Biol., 12:267 (1978).
358. J. R. Bolton, Solar Energy, 20:181 (1978).
359. R. T. Ross and T. -L. Hsiao, J. Appl. Phys., 48:283 (1977).
360. P. J. Boddy, J. Electrochem. Soc., 115:199 (1968).
361. A. V. Bulatov and M. L. Khidekel, Izv. AN SSSR, Ser. Khim.,
     1902 (1976).
362. M. D. Krotova, Yu. V. Pleskov and A. A. Revina, Elektrokhimiya,
     17:528 (1981).
363. A. B. Bocarsly, J. M. Bolts, P. G. Cummings and M. S. Wrighton,
     Appl. Phys. Lett., 31:568 (1977).
364. H. H. Kung, H. S. Jarrett, A. W. Sleight and A. Ferretti, J.
     Appl. Phys., 48:2463 (1977).
365. J. F. Juliao, F. Decker and M. Abramovich, J. Electrochem.
     Soc., 127:2264 (1980).
366. C. D. Jaeger and A. J. Bard, J. Phys. Chem., 83:3146 (1979).
367. D. Laser and S. Gottesfeld, J. Electrochem. Soc., 126:475
     (1979).

368.   R. H. Wilson J. Electrochem. Soc., 127:228 (1980).
369.   M. P. Dare-Edwards and A. Hamnett, J. Electroanal. Chem.,
       105:283 (1979).
370.   K. Yazawa, H. Kamagawa and H. Morisaki, Int. J. Hydrogen
       Energy, 4:205 (1979).
371.   E. C. Dutoit, F. Cardon, F. Vanden Kerchove and W. P. Gomes,
       J. Appl. Electrochem., 8:247 (1978).
372.   J. F. Houlihan, D. P. Madacsi, E. J. Walsh and L. N. Mulay,
       Mat. Res. Bull., 11:1191 (1977).
373.   D. Haneman and P. Holmes, Solar Energy Mater., 1:233 (1979).
374.   R. Wang and C. H. Henager, J. Electrochem. Soc., 126:83
       (1979).
375.   W. Gissler, P. L. Lensi and S. Pizzini, J. Appl. Electrochem.,
       6:9 (1976).
376.   K. L. Hardee and A. J. Bard, J. Electrochem. Soc., 122:739
       (1975).
377.   A. L. Asatiani, T. V. Dzhavrishvili and Z. A. Rotenberg,
       Elektrokhimiya, 13:309 (1977).
378.   D. S. Ginley and M. L. Knotek, J. Electrochem. Soc., 126:2163
       (1979).
379.   L. A. Harris, M. E. Gerstner and R. H. Wilson, J. Electrochem.
       Soc., 126:850 (1979).
380.   R. Schumacher, Ber. Bunsenges. Phys. Chem., 84:125 (1980).
381.   T. V. Dzhavrishvili, Z. A. Rotenberg and A. L. Asatiani,
       Elektrokhimiya, 16:868 (1980).
382.   M. D. Krotova, Yu. V. Pleskov and A. A. Revina, Elektrokhimiya,
       15:1396 (1979).
383.   L. A. Harris and R. Schumacher, J. Electrochem. Soc., 127:1186
       (1980).
384.   R. H. Wilson, L. A. Harris and M. E. Gerstner, J. Electrochem.
       Soc., 126:844 (1979).
385.   J. C. Pesant and P. Vennereau, J. Electroanal. Chem., 106:103
       (1980).
386.   J. Keene, D. H. Weinstein and G. M. Haas, Nature, 253:719
       (1975).
387.   G. S. Popkirov and K. D. Kochev, Compt. Rend. Acad. Bulg.
       Sci., 32:591 (1979).
388.   A. Aladjem, J. Mater. Sci., 8:688 (1973).
389.   H. Tamura, H. Yoneyama, C. Iwakura and T. Murai, Bull. Chem.
       Soc. Japan, 50:753 (1977).
390.   G. S. Popkirov and Yu. V. Pleskov, Elektrokhimiya, 16:238
       (1980).
391.   A. Fujishima, K. Kohayakawa and K. Honda, J. Electrochem.
       Soc., 122:1487 (1975).
392.   J. -L. Desplat, J. Appl. Phys., 47:5102 (1976).
393.   J. G. Mavroides, D. I. Tchernev, J. A. Kafalas and D. F.
       Kolesar, Mat. Res. Bull., 10:1023 (1975).
394.   T. Hirai, I. Tari and J. Yamaura, Bull. Chem. Soc. Japan,
       51:3057 (1978).
395.   A. I. Kulak, V. P. Pakhomov, V. V. Sviridov and G. L. Shchukin,
       Elektrokhimiya, 15:1380 (1979).

396. L. A. Harris and R. H. Wilson, J. Electrochem. Soc., 123:1010 (1976).

397. L. A. Harris, D. R. Cross and M. E. Gerstner, J. Electrochem. Soc., 124:839 (1977).

398. H. Yoneyama, T. Murai and H. Tamura, Ber. Bunsenges. Phys. Chem., 83:1294 (1979).

399. N. Getoff, S. Solar and M. Cohn, Naturwissenschaften, 67:7 (1980).

400. C. N. Sayers and N. R. Armstrong, Surface Sci., 77:301 (1978).

401. V. B. Kireyev, E. M. Trukhan and D. A. Filimonov, Zh. Fiz. Khim., 52:457 (1978).

402. M. A. Butler, J. Electrochem. Soc., 126:338 (1979).

403. A. Wold, Virginia J. Sci., 28:129 (1977).

404. H. Morisaki, T. Watanabe, M. Iwase and K. Yazawa, Appl. Phys. Lett., 29:338 (1976).

405. A. J. Nozik, Nature, 257:383 (1975).

406. A. J. Nozik, US Pat. No. 4011149 (1977).

407. K. Miyatani and I. Sato, US Pat. No. 4124464 (1978).

408. D. K. Cartmell and H. Witzke, US Pat. No. 4086398 (1978).

409. M. Nakamura, Japan Pat. No. 53-100988 (1978).

410. D. H. Bradhurst and G. Z. A. Stolarski, Austral. Pat. No. AU-B1 87902/75 (1979).

411. H. McKinzie and E. A. Trickett, US Pat. No. 4181754 (1980).

412. M. Okuda, K. Yoshida and N. Tanaka, Japan J. Appl. Phys., 15:1599 (1976).

413. R. D. Nasby and R. K. Quinn, Mater. Res. Bull., 11:985 (1976).

414. J. H. Kennedy and K. W. Frese, J. Electrochem. Soc., 123:1683 (1976).

415. M. S. Wrighton, A. B. Ellis, P. T. Wolczanski, D. L. Morse, H. B. Abrahamson and D. S. Ginley, J. Amer. Chem. Soc., 98:2774 (1976).

416. J. G. Mavroides, J. A. Kafalas and D. F. Kolesar, Appl. Phys. Lett., 28:241 (1976).

417. T. Watanabe, A. Fujishima and K. Honda, Bull. Chem. Soc. Japan, 49:355 (1976).

418. J. M. Bolts and M. S. Wrighton, J. Phys. Chem., 80:2641 (1976).

419. A. B. Ellis, S. W. Kaiser and M. S. Wrighton, J. Phys. Chem., 80:1325 (1976).

420. D. M. Schleich, C. Derrington, W. Godek, D. Weisberg and A. Wold, Mater. Res. Bull., 12:321 (1977).

421. G. Campet, M. P. Dare-Edwards, A. Hamnett and J. B. Goodenough, Nouv. J. Chim., 4:501 (1980).

422. M. A. Butler, R. D. Nasby and R. K. Quinn, Solid State Communs., 19:1011 (1976).

423. W. A. Gerrard, J. Electroanal. Chem., 86:421 (1978).

424. G. Hodes, D. Cahen and J. Manassen, Nature, 260:312 (1976).

425. W. Gissler and R. Memming, J. Electrochem. Soc., 124:1710 (1977).

426. L. -S. R. Yeh and N. Hackerman, J. Electrochem. Soc., 124:833 (1977).

427. A. F. Sammels and P. G. P. Ang, J. Electrochem. Soc., 126:1831 (1979).

428. J. S. Curran and W. Gissler, J. Electrochem. Soc., 126:56 (1979).

429. K. L. Hardee and A. J. Bard, J. Electrochem. Soc., 124:215 (1977).

430. K. L. Hardee and A. J. Bard, J. Electrochem. Soc., 123:1024 (1976).

431. R. K. Quinn, R. D. Nasby and R. J. Baughman, Mater. Res. Bull., 11:1011 (1976).

432. M. A. Butler, D. S. Ginley and M. Eibschutz, J. Appl. Phys., 48:3070 (1977).

433. J. H. Kennedy and K. W. Frese, J. Electrochem. Soc., 125:709 (1978).

434. J. H. Kennedy, R. Shinar and J. P. Ziegler, J. Electrochem. Soc., 127:2307 (1980).

435. J. H. Kennedy and K. W. Frese, J. Electrochem. Soc., 125:723 (1978).

436. S. M. Wilhelm, K. S. Yun, L. W. Ballenger and N. Hackerman, J. Electrochem. Soc., 126:419 (1979).

437. D. S. Ginley and M. A. Butler, J. Appl. Phys., 49:2019 (1977).

438. T. Pajkossy, I. Molnár, M. Pálfy and R. Schiller, Acta Chim. Acad. Sci. Hung., 101:93 (1979).

439. C. E. Derrington, W. S. Godek, C. A. Castro and A. Wold, Inorg. Chem., 17:977 (1978).

440. S. R. Morrison and T. Freund, Electrochim. Acta, 13:1343 (1968).

441. M. S. Wrighton, D. L. Morse, A. B. Ellis, D. S. Ginley and H. B. Abrahamson, J. Amer. Chem. Soc., 98:44 (1976).

442. M. Tomkiewicz and J. M. Woodall, Science, 196:990 (1977).

443. M. Tomkiewicz and J. M. Woodall, J. Electrochem. Soc., 124:1436 (1977).

444. J. O'M. Bockris and K. Uosaki, J. Electrochem. Soc., 124:98 (1977).

445. H. Yoneyama, A. Sakamoto and H. Tamura, Electrochim. Acta, 20:341 (1975).

446. M. A. Butler and D. S. Ginley, J. Electrochem. Soc., 127:1273 (1980).

447. R. M. Candea, M. Kastner, R. Goodman and N. Hickok, J. Appl. Phys., 47:2724 (1976).

448. S. Mayumi, C. Iwakura, H. Yoneyama and H. Tamura, Denki Kagaku, 44:339 (1976).

449. K. Ohashi, K. Uosaki and J. O'M. Bockris, Energy Res., 1:25 (1977).

450. H. S. Jarrett, H. H. C. Kung and A. W. Sleight, US Pat. No. 4144147 (1979).

451. T. Kawai, K. Tanimura and T. Sakata, Chem. Lett., 137 (1979).

452. F. -R. F. Fan, B. Reichman and A. J. Bard, J. Amer. Chem. Soc., 102:1488 (1980).

453. A. J. Nozik, Proc. 11th Intersoc. Energy Conv. Engng. Conf., Lake Tahoe, Ca., USA, Sept. 1976, p. 43.

454. J. F. Houlihan, D. B. Armitage, T. Hoovler, D. Bonaquist,
      D. P. Madacsi and L. N. Mulay, Mater. Res. Bull., 13:1205
      (1978); 14:915 (1979).

455. Ya. L. Kogan and A. M. Vakulenko, Solar Energy Mater., 3:357
      (1980).

456. J. Augustynski, J. Hinden and C. Stalder, J. Electrochem.
      Soc., 124:1063 (1977).

457. C. Stalder and J. Augustynski, J. Electrochem Soc., 126:2007
      (1979).

458. A. K. Ghosh and H. P. Maruska, J. Electrochem. Soc., 124:1516
      (1977).

459. G. Campet, J. Verniolle, J. -P. Doumera and J. Claverie,
      Mater. Res. Bull., 15:1135 (1980).

460. A. Monnier and J. Augustynski, J. Electrochem. Soc., 127:1576
      (1980).

461. H. P. Maruska and A. K. Ghosh. Solar Energy Mater., 1:237
      (1979).

462. Y. Matsumoto, J. Kurimoto, Y. Amagasaki and E. Sato, J.
      Electrochem. Soc., 127:2148 (1980).

463. P. D. Fleischauer and J. K. Allen, J. Phys. Chem., 82:432
      (1978).

464. W. D. K. Clark and N. Sutin, J. Amer. Chem. Soc., 99:4676
      (1977)

465. A. Hamnett, M. P. Dare-Edwards, R. D. Wright, K. R. Seddon and
      J. B. Goodenough, J. Phys. Chem., 83:3280 (1979).

466. H. Tsubomura, M. Matsumura, Y. Nomura and T. Amamiya, Nature,
      261:402 (1976).

467. H. Tsubomura, Y. Nakato and T. Sakata, Elektrokhimiya, 13:1689
      (1979).

468. M. Matsumura, S. Matsudaira, H. Tsubomura, M. Takata and H.
      Yanagida, Ind. and Eng. Chem. Prod. Res. and Develop., 19:415
      (1980).

469. A. Mackor and J. Schoonman, Rec. Trav. Chim. Pays-Bas, 99:71
      (1980)

470. S. Anderson, E. C. Constable, M. P. Dare-Edwards, J. B.
      Goodenough, A. Hamnett, K. R. Seddon and R. D. Wright, Nature,
      280:571 (1979).

471. D. C. Bookbinder, N. S. Lewis, M. G. Bradley, A. B. Bocarsly
      and M. S. Wrighton, J. Amer. Chem. Soc., 101:7321 (1979).

472. R. Memming, F. Schröppel and U. Bringman, J. Electroanal.
      Chem., 100:307 (1979).

473. Y. Nakato, T. Ohnishi and H. Tsubomura, Chem Lett., 883 (1975).

474. Y. Nakato, K. Abe and H. Tsubomura, Ber. Bunsenges. Phys.
      Chem., 80:1002 (1976).

475. R. H. Wilson, L. A. Harris and M. E. Gerstner, J. Electrochem.
      Soc., 124:1233 (1977).

476. S. Menezes, A. Heller and B. Miller, J. Electrochem. Soc.,
      127:1268 (1980).

477. L. A. Harris, M. E. Gerstner and R. H. Wilson, J. Electrochem.
      Soc., 124:1511 (1977).

478. Y. Nakato, S. Tonomura, H. Tsubomura, Ber Bunsenges. Phys. Chem., 80:1289 (1976).

479. H. Yoneyama, S. Mayumi and H. Tamura, J. Electrochem. Soc., 125:68 (1978).

480. S. Hara and S. Murakami, Japan. Pat. No. 53-31576 (1978).

481. H. Uchida, H. Yoneyama and H. Tamura, J. Electrochem. Soc., 127:99 (1980).

482. J. O'M. Bockris, in: "Trends in Electrochemistry", J.O'M. Bockris, D. A. J. Rand and B. J. Welch, eds., Plenum Press, New York, London (1977) p. 79.

483. S. Gourgaud and D. Elliot, J. Electrochem. Soc., 124:102 (1977).

484. M. Hara, M. Namida and S. Murakami, Japan. Pat. No. 53-23869 (1978).

485. C. D. Jaeger, F. -R. R. Fan and A. J. Bard, J. Amer. Chem. Soc., 102:2592 (1980).

486. H. Morisaki, H. Ono, H. Dohkoshi and K. Yazawa, Japan J. Appl. Phys., 19: L148 (1980).

487. H. P. Maruska and A. K. Ghosh, Solar Energy Mater., 1:411 (1979).

488. P. A. Kohl, S. N. Frank and A. J. Bard, J. Electrochem. Soc., 124:225 (1977).

489. S. Wagner and J. L. Shay, Appl. Phys. Lett., 31:446 (1977).

490. A. J. Nozik, Appl. Phys. Lett., 29:150 (1976).

491. K. Nobe, G. L. Bauerle and M. Braun, J. Appl. Electrochem., 7:379 (1977).

492. A. J. Nozik, US. Pat. No. 4094751 (1978).

493. A. Bourasse and G. Horowitz, J. Phys. Lett., 38:291 (1977).

494. K. Ohashi, J. McCann and J. O'M. Bockris, Nature, 266:610 (1977).

495. V. Guruswamy and J. O'M. Bockris, Solar Energy Mater., 1:441 (1979).

496. A. J. Nozik, Appl. Phys. Lett., 30:567 (1977).

497. G. W. Murphy, J. Electrochem. Soc., 128:1819 (1981).

498. G. W. Murphy, J. Electrochem. Soc., 127:2088 (1980).

499. B. A. Dreyfus, French Pat. No. 2410506 (1979).

500. H. Reiche, W. W. Dunn and A. J. Bard, J. Phys. Chem., 83:2248 (1979).

501. T. Inoue, T. Watanabe, A. Fujishima and K. Honda, Chem. Lett., 1073 (1977).

502. B. Kraeutler and A. J. Bard, J. Amer. Chem. Soc., 99:7729 (1977)

503. B. Kraeutler and A. J. Bard, Nouv. J. Chim., 3:31 (1979).

504. K. Hirano and A. J. Bard, J. Electrochem. Soc., 127:1056 (1980).

505. M. Miyake, H. Yoneyama and H. Tamura, Electrochim. Acta, 21:1065 (1976).

506. V. A. Bendersky, Ya. M. Zolotovitsky, Ya. L. Kogan, M. L. Khidekel and D. M. Shub, Dokl. AN SSSR, 222:606 (1975).

507. K. Nakatani and H. Tsubomura, Bull. Chem. Soc., Japan, 50:783 (1977).

508. K. Kogo, H. Yoneyama and H. Tamura, J. Phys. Chem., 84:1705 (1980).

509. C. P. Kublak, L. F. Schneemeyer and M. S. Wrighton, J. Amer. Chem. Soc., 102:6898 (1980).

510. B. Kraeutler, H. Reiche, A. J. Bard and R. G. Hocker, J. Polymer Sci., Polym. Lett. Ed., 17:535 (1979).

511. M. Halmann, Nature, 275:115 (1978).

512. T. Inoue, A. Fujishima, S. Konishi and K. Honda, Nature, 277:637 (1979).

513. V. Gurusvami and J. O'M. Bockris, Int. J. Energy Res., 3:397 (1979).

514. C. R. Dickson and A. J. Nozik, J. Amer. Chem. Soc., 100:8007 (1978).

515. E. N. Costogue and R. K. Yasui, Solar Energy, 19:205 (1977).

516. G. L. Schnable and P. F. Schmidt, J. Electrochem. Soc., 123:3110 (1976).

517. V. A. Tyagai, V. A. Sterligov and G. Ya. Kolbasov, in: "Problemy Fizicheskoi Khimii Poverkhnosti Poluprovodnikov", Novosibirsk, Nauka, (1978) p. 181.

518. L. V. Belyakov, D. N. Goryachev and O. M. Sreseli, in: "Problemy Physiki Poluprovodnikov", Leningrad, Phys. Techn. Inst., (1979) p. 5.

519. R. J. Collier, G. B. Burckhardt and L. H. Lin, "Optical Holography", Academic Press, New York, London (1971).

520. A. L. Dalisa, W. K. Zwicker, D. J. De Bitetto and P. Harnack, Appl. Phys. Lett., 17:208 (1970).

521. A. L. Dalisa and D. J. De Bitetto, Appl. Optics, 11:2007 (1972).

522. R. W. Haisty, J. Electrochem. Soc., 108:790 (1961).

523. P. Morse and H. Feshback, "Methods of Theoretical Physics", McGraw-Hill, New York, Toronto, London (1953) Chap. 9.

524. M. Born and E. Wolf, "Principles of Optics", Pergamon Press, Oxford (1964).

525. A. V. Sachenko, Ukrain. Fiz. Zh., 19:1585 (1974).

526. Zh. I. Alferov, D. N. Goryachev and S. A. Gurevich, Zh. Tekhn. Fiz., 46:1505 (1976).

527. L. V. Belyakov, D. N. Goryachev, S. M. Ryvkin, O. M. Sreseli and R. A. Suris, Fiz. I Tekhn. Poluprovod., 13:2173 (1979).

528. L. V. Belyakov, D. N. Goryachev, S. M. Ryvkin and O. M. Sreseli, Fiz. I Tekhn. Poluprovod., 12:2244 (1978).

529. V. A. Tyagai, V. A. Sterligov, G. Ya. Kolbasov and O. V. Snitko, Fiz. I Tekhn. Poluprovod., 7:632 (1973).

530. V. A. Sterligov, G. Ya. Kolbasov and V. A. Tyagai, Pisma v ZhTF., 2:437 (1976).

531. V. A. Tyagai, V. A. Sterligov and G. Ya. Kolbasov, Elektrokhimiya, 13:1556 (1977).

532. V. A. Tyagai, V. A. Sterligov and G. Ya. Kolbasov, Electrochim. Acta, 22:819 (1977).

533. V. A. Tyagai, V. A. Sterligov and G. Ya. Kolbasov, Elektrokhimiya, 15:188 (1979).

534. V. A. Sterligov and V. A. Tyagai, Pisma v ZhTF, 1:704 (1975).

535. L. V. Belyakov, D. N. Goryachev, Yu. I. Ostrovsky and L. G.
      Paritsky, Zh. Nauchn. i Prikl. Photogr. i Kinematogr., 19:54
      (1974).
536. L. V. Belyakov, D. N. Goryachev, M. N. Mizerov and E. D.
      Portnoy, Zh. Tekhn. Fiz., 44:1331 (1974).
537. Zh. I. Alferov, S. A. Gurevich and R. F. Kazarinov, Fiz. i
      Tekhn. Poluprovod., 8:2031 (1974).
538. D. N. Goryachev and L. G. Paritsky, Fiz. i Tekhn. Poluprovod.,
      7:1449 (1973).
539. L. V. Belyakov, D. N. Goryachev and O. M. Sreseli, Pisma v
      ZhTF, 3:922 (1977).
540. L. V. Belyakov, D. N. Goryachev and O. M. Sreseli, Zh. Tekhn.
      Fiz., 49:876 (1979).
541. Yu. A. Bykovsky, V. L. Smirnov and A. V. Shmalko, Zh. Nauchn.
      i Priklad. Photogr. i Kinematogr., 23:129 (1978).
542. L. V. Belyakov, D. N. Goryachev and O. M. Sreseli, Pisma v
      ZhTF, 2:976 (1979).
543. N. A. Vlasenko, F. A. Nazarenkov, V. A. Sterligov and V. A.
      Tyagai, Pisma v ZhTF, 4:1037 (1978).
544. W. T. Tsang and S. Wang, Appl. Phys. Lett., 28:44 (1976).
545. L. Comerford and P. Zory, Appl. Phys. Lett., 25:208 (1974).
546. M. Sullivan, L. D. Klein, R. M. Finne and L. A. Pompliano, J.
      Electrochem. Soc., 110:412 (1963).
547. C. R. Elliott and J. C. Regnault, J. Electrochem. Soc.,
      127:1557 (1980).
548. A. Yamamoto and S. Yano, J. Electrochem. Soc., 122:260 (1975).
549. L. Hollan, J. C. Tranchart and R. Memming, J. Electrochem.
      Soc., 126:855 (1979).
550. T. van Dongen, R. P. Tijburg and J. Bakker, J. Electrochem.
      Soc., 127:238 (1980).
551. V. A. Tyagai, Elektrokhimiya, 10:3 (1974).
552. V. A. Tyagai, in: "Elektrokhimiya", Vol. II (Itogi nauki i
      tekhniki), VINITI, Moscow (1976) p. 109.
553. V. A. Tyagai and N. B. Lukyanchikova, Elektrokhimiya, 3:316
      (1967).
554. V. A. Tyagai, Elektrokhimiya, 3:1331 (1967).
555. V. A. Tyagai and N. B. Lukyanchikova, Elektrokhimiya, 5:216
      (1969).
556. G. Spescha and M. Strutt, Sci. Electrica, 5:121 (1959).
557. S. M. Rytov, "Vvedenie v Statisticheskuyu Radiofiziku", Nauka,
      Moscow (1976) Chapter 1.
558. G. Ya. Kolbasov and V. A. Tyagai, Ukr. Khim. Zh., 37:1298
      (1971).
559. G. Ya. Kolbasov and V. A. Tyagai, Fiz. i Tekhn. Poluprovod.,
      6:964 (1972).
560. W. P. Gomes and F. Cardon, J. Solid-State Chem., 3:125 (1971).
561. F. Cardon, Physica, 57:390 (1972).
562. P. Görlich, Photoeffekte, Bd.1-3, Leipzig (1963-1966).
563. Yu. Ya. Gurevich, Yu. V. Pleskov and Z. A. Rotenberg,
      "Photoelectrochemistry", Consultants Bureau, New York (1980).
564. Yu. Ya. Gurevich, Elektrokhimiya, 8:1564 (1972).

565. Yu. Ya. Gurevich, in: "Elektronnye Protsessy Na Poverkhnosti Poluprovodnikov", Novosibirsk, Nauka (1974) p. 93.

566. M. D. Krotova and Yu. V Pleskov, Fiz. Tverd. Tela, 15:2806 (1973).

567. G. V. Boikova, M. D. Krotova and Yu. V. Pleskov, Elektrokhimiya, 12:922 (1976).

568. Yu. Ya. Gurevich, M. D. Krotova and Yu. V. Pleskov, J. Electroanal. Chem., 75:339 (1977).

569. Yu. Ya. Gurevich, A. M. Brodsky and V. G. Levich, Elektrokhimiya, 3:1302 (1967).

570. E. O. Kane, Phys. Rev., 127:131 (1962).

571. L. W. James and J. L. Moll, Phys. Rev., 183:740 (1969).

572. F. G. Baksht, V. G. Ivanov and V. Ya. Moizhes, Fiz. Tverd. Tela, 13:2896 (1971).

573. R. E. Malpas, K. Itaya and A. J. Bard, J. Amer. Chem. Soc., 101:2535 (1979).

574. C. E. Krohn and J. C. Thompson, Chem. Phys. Lett., 65:132 (1979).

575. R. R. Dogonadze, L. I. Krishtalik and Yu. V. Pleskov, Elektrokhimiya, 10:507 (1974).

576. V. V. Sviridov and V. A. Kondratyev, in: "Uspekhi Nauchnoy Fotografii", Vol. 19, Nauka, Moscow (1978), p. 43.

577. H. Kiess, in: "Progress in Surface Science, Vol. 9", S. G. Davison, ed., Pergamon Press, Oxford, New York (1979) p. 113.

578. D. M. Kolb, M. Przasnyski and H. Gerischer, J. Electroanal. Chem., 54:25 (1974).

579. P. Bindra, H. Gerischer and D. M. Kolb, J. Electrochem. Soc., 124:1012 (1977).

580. T. Yamaze, H. Gerischer, M. Lübke and B. Pettinger, Ber. Bunsenges. Phys. Chem., 83:658 (1979).

581. D. M. Kolb, in: "Advances in Electrochemistry and Electrochemical Engineering", Vol. 11, H. Gerischer and C. Tobias, eds., Wiley, New York (1978) p. 125.

582. O. V. Romanov, A. M. Yafyasov and M. I. Rudenko, Fiz. i Tekhn. Poluprovod., 9:464 (1975).

583. B. N. Kabanov, I. I. Astakhov and I. G. Kiseleva, in: "Kinetika Slozhnykh Elektrokhimicheskikh Reaktsii", Nauka, Moscow (1981) p. 200.

584. D. N. Goryachev, L. G. Paritsky and S. M. Ryvkin, Fiz. i Tekhn. Poluprovod., 4:1580 (1970).

585. D. N. Goryachev, L. G. Paritsky and S. M. Ryvkin, Fiz. i Tekhn. Poluprovod., 4:1582 (1970).

586. I. P. Akimchenko, V. S. Vavilov and A. F. Plotnikov, Fiz. i Tekhn. Poluprovod., 4:1841 (1970).

587. D. N. Goryachev, L. G. Paritsky and S. M. Ryvkin, Fiz. i Tekhn. Poluprovod., 6:1148 (1972).

588. N. I. Bochkaryova, L. G. Paritsky and S. M. Ryvkin, Fiz. i Tekhn. Poluprovod., 7:558 (1973).

589. J. J. Kelly and J. K. Vondeling, J. Electrochem. Soc., 122:1103 (1975).

590. M. Yamana, Appl. Phys. Lett., 29:570 (1976).

591. M. M. Nicholson, J. Electrochem. Soc., 119:461 (1972).

592. B. Reichman, F. -R. F. Fan and A. J. Bard, J. Electrochem. Soc., 127:333 (1980).

593. T. Inoue, A. Fujishima and K. Honda, Japan J. Appl. Phys., 18:2177 (1979).

594. L. G. Paritsky and S. M. Ryvkin, Fiz. i Tekhn. Poluprovod., 4:764 (1979).

595. Zh. G. Dokholyan and L. G. Paritsky, Fiz. i Tekhn. Poluprovod., 7:192 (1973).

596. M. S. Wrighton, P. T. Wolczanski and A. B. Ellis, J. Solid State Chem., 22:17 (1977).

597. F. Müllers, H. J. Tolle and R. Memming, J. Electrochem. Soc., 121:1160 (1974).

598. A. I. Kulak, V. P. Pakhomov and G. L. Shchukin, Elektrokhimiya, 15:1698 (1979).

599. V. N. Fateyev, E. S. Matveyeva, V. P. Pakhomov and V. A. Kondratyev, Elektrokhimiya, 14:1270 (1978).

600. V. N. Fateyev, E. S. Matveyeva, V. P. Pakhomov and V. A. Kondratyev, Elektrokhimiya, 15:206 (1979).

601. P. Clechet, C. Martelet, J. -R. Martin and R. Olier, Compt. Rend. Acad. Sci., C287:405 (1978).

602. R. S. Davidson, R. M. Slater and R. R. Meek, J. Chem. Soc., Faraday Trans. I, 75:2507 (1979).

603. B. Reichman and A. J. Bard, J. Electrochem. Soc., 126:583 (1979); 126:2133 (1979).

604. T. Inoue, A. Fujishima and K. Honda, J. Electrochem. Soc., 127:1582 (1980).

605. V. L. Vinetsky and G. A. Kholodar, "Radiatsionnaya Fizika Poluprovodnikov", Naukova Dumka, Kiev (1979).

606. A. K. Pikayev, "Impulsnyi Radioliz Vody i Vodnykh Rastvorov", Nauka, Moscow (1965).

607. A. V. Byalobzhesky, "Radiatsionnaya Korroziya", Nauka, Moscow (1967).

608. E. K. Oshe and I. L. Rozenfeld in: "Korroziya i Zashita ot Korrozii", Vol. 7 (Itogi Nauki i Tekhniki), VINITI, Moscow (1978) p. 111.

609. "Hydrogen Manufacture by Electrolysis. Thermal Decomposition and Unusual Techniques", M. S. Casper, ed., Nayes Data Corp., Park Ridge, New York (1978).

610. R. Dehmlow, P. Janietz and R. Landsberg, J. Electroanal. Chem., 65:115 (1975).

611. M. Rius de Riepen, J. Range and K. Hauffe, Z. Phys. Chem. N.F., 99:257 (1976).

612. J. D. Luttmer and A. J. Bard, J. Electrochem. Soc., 126:414 (1979).

613. M. Gleria and R. Memming, Z. Phys. Chem. N.F., 101:171 (1976).

614. J. D. Luttmer and A. J. Bard, J. Electrochem. Soc., 125:1423 (1978).

615. B. Pettinger, H. -R. Schüppel and H. Gerischer, Ber. Bunsenges. Phys. Chem., 80:849 (1976).

616. B. Pettinger, H. -R. Schüppel, T. Yokoyama and H. Gerischer, Ber. Bunsenges. Phys. Chem., 78:1024 (1974).

617. P. P. Konorov, Yu. A. Tarantov and E. V. Kasyanenko, in: "Problemy Fizicheskoi Khimii Poluprovodnikov", A. V. Rzhanov, ed., Nauka, Novosibirsk (1978) p. 247.

618. H. Morisaki and K. Yazawa, Appl. Phys. Lett., 33:1013 (1978).

619. V. A. Tyagai, M. K. Sheinkman, E. L. Shtrum, G. Ya. Kolbasov and M. K. Moiseeva, Fiz. i Tekhn. Poluprovod., 14:189 (1980).

620. B. R. Karas and A. B. Ellis, J. Amer. Chem. Soc., 102:968 (1980).

621. B. R. Karas, D. J. Morano, D. K. Bilich and A. B. Ellis, J. Electrochem. Soc., 127:1144 (1980).

622. M. Cardona, "Modulation Spectroscopy", Academic Press, New York (1969).

623. B. O. Seraphin, R. B. Hess and N. Bottka, J. Appl. Phys., 36:2242 (1965); Phys. Rev., 139:A560 (1965); 140:1716 (1965).

624. B. O. Seraphin, Surface Sci., 13:136 (1969).

625. V. A. Tyagai, Ukrain. Fiz. Zh., 15:1164 (1970).

626. N. Dmitruk and V. Tyagai, Phys. Stat. Sol. (b), 43:557 (1971).

627. R. M. Lazorenko-Manevich, in: "Elektrokhimiya", Vol. 8, (Itogi Nauki i Tekhniki), VINITI, Moscow (1972) p. 85.

628. A. G. Akimov and I. L. Rozenfeld, Usp. Khimii, 43:612 (1974).

629. V. I. Gavrilenko, A. M. Evstigneyev, V. A. Zuev, V. G. Litovchenko, O. V. Snitko and V. A. Tyagai, Fiz. i Tekhn. Poluprovod., 10:1076 (1976).

630. V. N. Bondarenko, A. M. Evstigneyev, V. A. Tyagai and O. V. Snitko, Izv. AN SSSR, Neorg. Materialy, 11:342 (1975).

631. P. Lemasson, J. Gautron and J. -P. Dalbera, Ber. Bunsenges. Phys. Chem., 84:796 (1980).

632. D. M. Kolb, Ber. Bunsenges. Phys. Chem., 77:891 (1973).

633. Yu. V. Pleskov, in: "Comprehensive Treatise of Electrochemistry", Vol. 1, J. O'M. Bockris, B. E. Conway and E. Yeager, eds., Plenum Press, New York, London (1980) p. 291.

ADDITIONAL REFERENCES

Chapter 3

F. Di Quarto and A. J. Bard, "Semiconductor electrodes. 38. Photoelectrochemical behaviour of n- and p-type GaAs electrodes in tetrahydrofuran solutions", J. Electroanal. Chem., 127:43 (1981).

J. Gobrecht and H. Gerischer, "Schottky barrier height, photovoltage and photocurrent in liquid-junction solar cells", Solar Energy Mater., 2:131 (1979).

H. Masuda, S. Morishita, A. Fujishima and K. Honda, "Study of photoelectrochemical reactions at an n-type polycrystalline ZnO electrode using photoacoustic detection", J. Electroanal. Chem., 121:363 (1981).

Y. Nakato, A. Tsumura and H. Tsubomura, "Surface intermediates of an n-type gallium phosphide electrode as related with the shifts of the surface band energy induced by oxidants in solution", J. Electrochem. Soc., 128:1300 (1981).

Y. Nakato, A. Tsumura and H. Tsubomura, "The concept of 'surface-trapped hole' as an intermediate of anodic reaction of a gallium phosphide semiconductor electrode", Chem. Lett., 127 (1981).

Y. Nakato, A. Tsumura and H. Tsubomura, "The concept of 'surface-trapped hole' in n-type semiconductors and the conditions for efficient and stable photoelectrochemical cells", Chem. Lett., 383 (1981).

P. Singh, R. Singh, R. Gale, K. Rajeshwar and J. DuBow, "Surface charge and specific ion adsorption effects in photoelectrochemical devices", J. Appl. Phys., 51:6286 (1980).

M. Tomkiewicz, "Surface states on chemically modified TiO_2 electrodes", Surface Sci., 101:286 (1980).

G. Van Amerongen, M. Herlem, R. Heindl and J. -L. Sculfort, "Electrochemical behavior of n- and p-type $Ga_xIn_{1-x}P$ electrodes in sodium iodide liquid ammoniate at room temperatures", J. Electroanal. Chem., 117:119 (1981).

C h a p t e r   4

H. Gerischer, N. Müller and O. Haas, "On the mechanism of hydrogen evolution at GaAs-electrodes", J. Electroanal. Chem., 119:41 (1981).
B. Vainas, G. Hodes, J. Manassen and D. Cahen, "Activation analysis of forward-biased CdS-electrolyte diode", Appl. Phys. Lett., 38:458 (1981).

C h a p t e r   5

P. A. Breddels and G. Blasse, "Spectral sensitization of $SrTiO_3$ electrodes with magnesium tetraphenylporphine", Chem. Phys. Lett., 79:209 (1981).
M. P. Dare-Edwards, J. B. Goodenough, A. Hamnett, K. R. Seddon and R. D. Wright, "Sensitization of semiconductor electrodes by Ru-based dyes", Faraday Discuss. Chem. Soc., N70:285 (1980).
M. Fujihira, T. Kubota and T. Osa, "Organo-modified metal oxide electrode. Part V. Efficiency of electron injection into conduction band from photo-excited dye molecule covalently attached to an SnO_2 surface", J. Electroanal. Chem., 119:379 (1981).
H. Hada, Y. Yonezawa and H. Inaba, "Spectral sensitization of ZnO film electrodes by the Y-aggregate of cyanine dyes", Ber. Bunsenges. Phys. Chem.,85:425 (1981).
M. Matsumura, K. Mitsuda, N. Yoshizawa and H. Tsubomura, "Photocurrents in the ZnO and TiO_2 photoelectrochemical cells sensitized by xanthine dyes and tetraphenylporphines. Effect of substitution on the electron injection processes", Bull. Chem. Soc. Japan, 54:692 (1981).
R. Memming, "Electron transfer reactions of excited ruthenium (II) complexes at semiconductor electrodes", Surface Sci., 101:551 (1980).
T. M. Mezza, C. L. Linkous, V. R. Shepard, N. R. Armstrong, R. Nohr and M. Kenney, "Improved photoelectrochemical efficiencies at phthalocyanine-modified SnO_2 electrodes", J. Electroanal. Chem., 124:311 (1981).
T. Miyasaka, T. Watanabe, A. Fujishima and K. Honda, "Photoelectrochemical study of chlorophyll-a multilayers on SnO_2 electrode", Photochem. Photobiol., 32:217 (1980).
T. Miyasaka, and K. Honda, "Photoelectrochemical studies on the monolayer assemblies of chlorophyll-a on the quantum efficiency of photocurrent generation", Surface Sci., 101:541 (1980).
A. H. A. Tinnemans and A. Mackor, "Spectral sensitization of $SrTiO_3$

photoanodes with binuclear 1,10-phenanthroline bis(2,2'-bipyridine) complexes of ruthenium (II) and tris-(2,2'-bipyridine) ruthenium (II)", Rec. Trav. Chim. Pays-Bas, 100:95 (1981).

T. Watanabe, T. Takizawa and K. Honda, "Reductive and non-reductive effects of co-sensitizers on the Rhodamine B-sensitized charge separation at CdS surface", Ber Bunsenges. Phys. Chem., 85:430 (1981).

C h a p t e r 6

W. J. Albery, P. N. Bartlett, A. Hamnett and M. P. Dare-Edwards, "The transport and kinetics of minority carriers in illuminated semiconductor electrodes", J. Electrochem. Soc., 128:1492 (1981).

P. Allongue and H. Cachet, "Photoelectrochemical behaviour of an n-type GaAs electrode studied by impedance measurements. Determination and simulation of the faradaic resistance", J. Electroanal. Chem., 119:371 (1981).

M. A. Butler and D. S. Ginley, "Surface treatment induced sub bandgap photoresponse of GaP electrodes", J. Electrochem. Soc., 128:712 (1981).

M. P. Dare-Edwards, J. B. Goodenough, A. Hamnett and N. D. Nicholson, "Photochemistry of Nickel (II) oxide", J. Chem. Soc. Faraday, 11:63 (1981).

F. DiQuarto, A. Di Paola and C. Sunseri, "Semiconducting properties of anodic WO_3 amorphous films", Electrochim. Acta, 26:1177 (1981).

H. Gerischer and M. Lübke, "The influence of crystallographic anisotropy on photocurrents at GaSe-electrodes" Ber. Bunsenges. Phys. Chem., 85:713 (1981).

S. Hinckley and D. Haneman, "Surface barrier heights and changes induced by ferrocene group addition on Ge and Si surfaces cleaved in electrolytes", Surface Sci., 101:180 (1980).

W. Kautek, H. Gerischer and H. Tributsch, "The role of carrier diffusion and indirect optical transitions in the photoelectrochemical behaviour of layer type d-band semiconductors", J. Electrochem. Soc., 127:2471 (1980).

V. B. Kireev, E. M. Trukhan and D. A. Filimonov, "On efficiency of light energy conversion in semiconductor photoelectrochemical cells", Elektrokhimiya, 17:344 (1981).

V. Maeda, A. Fujishima and K. Honda, "The investigation of current doubling reactions on semiconductor photoelectrodes by temperature change measurements", J. Electrochem. Soc., 127:1731 (1981).

Y. F. McCann and J. Pezy, "The measurement of the flatband potentials of n-type and p-type semiconductors by rectified alternating photocurrent voltammetry", J. Electrochem. Soc., 127:1735 (1981).

S. P. Perone, J. H. Richardson, S. B. Deutscher, J. Rosenthal and

J. N. Ziemer, "Laser-induced photoelectrochemistry: time-resolved coulostatic-flash studies of photooxidation at n-TiO$_2$ electrodes", J. Electrochem. Soc., 127:2580 (1980).

P. Salvador, "Determination of carrier concentration from photoelectrolysis spectra of semiconducting electrodes", Solid State Communs., 34:1 (1980).

P. Salvador, "Influence of pH on the potential dependence of the efficiency of water photo-oxidation at n-TiO$_2$ electrodes", J. Electrochem. Soc., 128:1895 (1981).

J. -L. Sculfort, "Reflexion sur la determination des potentiels de bandes plates par photoélectrochimie: application aux alliages semiconducteurs Ga$_x$In$_{1-x}$P", C.R. Acad. Sci., sér. 2, 292:295 (1981).

P. Singh and K. Rajeshwar, "Mechanism of charge transfer at the n-GaAs/room temperature molten salt electrolyte interface", J. Electrochem. Soc., 127:1724 (1981).

C h a p t e r   7

F. Cardon, W. P. Gomes, F. Vanden Kerchove, D. Vanmaekelbergh and F. Van Overmeire, "On the kinetics of semiconductor electrode stabilization", Faraday Discuss. Chem. Soc., N70: 153 (1980).

K. W. Frese, M. J. Madou and S. R. Morrison, "Investigation of photoelectrochemical corrosion of semiconductors", J. Phys. Chem.,84:3172 (1980).

K. W. Frese, M. J. Madou and S. R. Morrison, "Investigation of photoelectrochemical corrosion of semiconductors, II. Kinetic analysis of corrosion-competition reactions on n-GaAs", J. Electrochem. Soc., 128:1527 (1981).

K. W. Frese, M. J. Madou and S. R. Morrison, "Investigation of photoelectrochemical corrosion of semiconductors, III. Effects of metal layer on stability of GaAs", J. Electrochem. Soc., 128:1939 (1981).

A. Fujishima, T. Kato, E. Maekawa and K. Honda, "Mechanisms of the current doubling effect. 1. The ZnO photoanode in aqueous solution of sodium formate", Bull. Chem. Soc. Japan, 54:1671 (1981).

H. Gerischer, "Photodecomposition of semiconductors, thermodynamics, kinetics and application to solar cells", Faraday Discuss. Chem. Soc., N70:137 (1980).

B. H. Loo, K. W. Frese and S. R. Morrison, "The influence of surface oxide films on the stabilization of n-Si photoelectrode", Surface Sci., 109:75 (1981).

M. J. Madou, K. W. Frese and S. R. Morrison, "Photoelectrochemical corrosion as influenced by an oxide layer", J. Phys. Chem., 84:3423 (1980).

H. Tributsch, "Photoelectrochemistry of layer-type zirconium disulfide", J. Electrochem. Soc., 128:1261 (1981).

C h a p t e r   8

A. Aruchamy and M. S. Wrighton, "A comparison of the interface
    energetics for n-type cadmium sulfide/and cadmium telluride/
    nonaqueous electrolyte junctions", J. Phys. Chem., 84:2848
    (1980).
A. S. Baranski, W. R. Fawcett, A. C. McDonald, R. M. de Nobriga and
    J. R. MacDonald, "The structural characterization of cadmium
    sulfide films grown by cathodic electrodeposition", J.
    Electrochem. Soc., 128:963 (1981).
A. J. Bard, F. -R. F. Fan, A. S. Gioda, G. Nagasubramanian and H. S.
    White, "On the role of surface states on semiconductor
    electrode photoelectrochemical cells", Faraday Discuss. Chem.
    Soc., N70:19 (1980).
J. A. Bruce and M. S. Wrighton, "Study of textured n-type silicon
    photoanodes: electron microscopy, Auger and electrochemical
    characterization of chemically derivatized surface", J.
    Electroanal. Chem., 122:93 (1981).
Y. Mirovsky, D. Cahen, G. Hodes, R. Tenne and W. Giriat,
    "Photoelectrochemistry of the CuIn S_2/S_n^{2-} system", Solar Energy
    Mater., 4:169 (1981).
D. Cahen, B. Vainas and J. M. Vandenberg, "Changes in surface
    crystallinity and morphology of CdS and CdSe photoelectrodes
    under direct current polarization", J. Electrochem. Soc.,
    128:1484 (1981).
D. Canfield and B. A. Parkinson, "Improvement of energy conversion
    efficiency by specific chemical treatment of n-MoSe$_2$ and
    n-WSe$_2$ photoanodes", J. Amer. Chem. Soc., 103:1279 (1981).
F. Decker and B. A. Parkinson, "The suppression of GaAs
    photocorrosion in aqueous solutions by sulfonated
    anthraquinones", J. Electrochem. Soc., 127:2370 (1980).
K. T. L. de Silva, D. J. Miller and D. Haneman, "Structure of
    annealed and unannealed CdSe films for photoelectrochemical
    solar energy conversion", Solar Energy Mater., 4:233 (1981).
R. N. Dominey, N. S. Lewis and M. S. Wrighton, "Fermi level pinning
    of p-type semiconducting indium phosphide contacting liquid
    electrolyte solutions: rationale for efficient
    photoelectrochemical energy conversion", J. Amer. Chem. Soc.,
    103:1261 (1981).
M. Etman, H. Tributsch and E. Bucher, "Photovoltages exceeding the
    band gap observed with WSe$_2$/I$^-$-solar cells", J. Appl.
    Electrochem., 11:653 (1981).
F. -R. F. Fan and A. J. Bard, "Semiconductor electrodes. XXXVI.
    Characteristics of n-MoSe$_2$, n- and p-WSe$_2$ electrodes in
    aqueous solutions", J. Electrochem. Soc., 128:945 (1981).
F. -R. F. Fan, B. L. Wheeler, A. J. Bard and R. N. Noufi,
    "Semiconductor electrodes. 39. Techniques for stabilization
    of n-silicon electrodes in aqueous solution
    photoelectrochemical cells", J. Electrochem. Soc., 128:2042
    (1981).

R. J. Gale and J. Dubow, "Electrolyte properties for photoelectrochemical cells with emphasis on the molecular adaptation of metallocenes: a review", Solar Energy Mater., 4:135 (1981).

D. Haneman, D. J. Miller, K. T. L. de Silva and J. F. McCann, "Solar energy conversion by photoelectrochemical cells", J. Electroanal. Chem., 118:101 (1981).

A. Heller, H. J. Lewerenz and B. Miller, "Silicon photocathode behavior in acidic V(II)-V(III) solutions", J. Amer. Chem. Soc., 103:200 (1981).

A. Heller and B. Miller, "Photoelectrochemical solar cells. Chemistry of the semiconductor - liquid junction", in: Interfacial photoprocesses: Energy Conversion and Synthesis", M. S. Wrighton, ed., American Chemical Society, Washington (1980) p. 215.

M. S. Kazacos and B. Miller, "Method of making metal-chalcogenide photosensitive devices", US Pat. No. 4256544 (17.3.81).

G. Kline, K. Kam, D. Canfield and B. A. Parkinson, "Efficient and stable photoelectrochemical cells constructed with WSe_2 and $MoSe_2$ photoanodes", Solar Energy Mater., 4:301 (1981).

G. J. Liu, J. Olsen, D. R. Saunders and J. H. Wang, "Photoactivation of CdSe films for photoelectrochemical cells", J. Electrochem. Soc., 128:1224 (1981).

R. E. Malpas, K. Itaya and A. J. Bard, "Semiconductor electrodes. 32. n- and p-GaAs, n- and p-Si and $n-TiO_2$ in liquid ammonia", J. Amer. Chem. Soc., 103:1622 (1981).

J. F. McCann and M. Skyllas Kazacos, "The electrochemical deposition and formation of cadmium sulphide thin film electrodes in aqueous electrolytes", J. Electroanal. Chem., 119:409 (1981).

S. Menezes, L. F. Schneemeyer and H. J. Lewerenz, "Efficiency losses from carrier-type inhomogeneity in tungsten diselenide photoelectrodes", Appl. Phys. Lett., 38:949 (1981).

B. Miller, A. Heller, S. Menezes and H. J. Lewerenz, "Surface modification in semiconductor-liquid junction cells", Faraday Discuss. Chem. Soc., N70:223 (1980).

D. J. Miller and D. Haneman, "Preparation of stable efficient CdSe films for solar PEC cells", Solar Energy Mater., 4:223 (1981).

G. Nagasubramanian and A. J. Bard, "Semiconductor electrodes. XXXIV. Photoelectrochemistry of p-type WSe_2 in acetonitrile and the $p-WSe_2$/nitrobenzene cell", J. Electrochem. Soc., 128:1055 (1981).

Y. Nakato, M. Shioji and H. Tsubomura, "Photovoltage and stability of an n-type silicon semiconductor coated with metal or metal-free phthalocyanine thin films in aqueous redox solutions", J. Phys. Chem., 85:1670 (1981).

R. Noufi, A. J. Frank and A. J. Nozik, "Stabilization of n-type silicon photoelectrodes to surface oxidation in aqueous electrolyte solution and mediation of oxidation reaction by surface-attached organic conducting polymer", J. Amer. Chem. Soc., 103:1849 (1981).

R. Noufi, D. Tench and L. F. Warren, "Stabilization of n-CdSe photoanodes in nonaqueous $Fe(CN)_6^{3-/4-}$ electrolytes", *J. Electrochem. Soc.*, 127:2709 (1980).

B. A. Parkinson, T. E. Furtak, D. Canfield, K. Kam and G. Kline, "Evaluation and reduction of efficiency losses at tungsten diselenide photoanodes", *Faraday Discuss. Chem.*, N70:233 (1980).

D. R. Pratt, M. E. Langmuir, R. A. Boudreau and R. D. Rauh, "Chemically deposited CdSe thin films for photoelectrochemical cells", *J. Electrochem. Soc.*, 128:1627 (1981).

K. Rajeshwar, L. Thompson, P. Singh, R. C. Kainthla and K. L. Chopra, "Photoelectrochemical characterization of CdSe thin film anodes", *J. Electrochem. Soc.*, 128:1744 (1981).

K. Rajeshwar, P. Singh and R. Thapar, "Effect of temperature on the operation of a photoelectrochemical device: studies on the n-GaAs/room temperature molten salt electrolyte interface", *J. Electrochem. Soc.*, 128:1750 (1981).

Y. Ramprakash, S. Basu and D. N. Bose, "Studies on n-InP/redox electrolyte photoelectrochemical cells", *J. Indian Chem. Soc.*, 58:153 (1981).

G. Razzini, M. Lazzari, L. Peraldo Bicelli, F. Levy, L. De Angelis, F. Calluzzi, E. Scafè, L. Fornarini and B. Scrosati, "Electrochemical solar cells with layer-type semiconductor anodes. Performance of $n-MoSe_2$ cells", *J. Power Sources*, 6:371 (1981).

J. Reichman and M. A. Russak, "Properties of CdSe thin film electrodes for photoelectrochemical cells", *J. Electrochem. Soc.*, 128:2025 (1981).

M. A. Russak and J. Reichman, "Thin film CdSe electrodes for backwall photoelectrochemical cells", *J. Electrochem. Soc.*, 128:2029 (1981).

P. Singh, K. Rajeshwar, R. Singh and J. Dubow, "Estimation of series resistance losses and ideal fill factors for photoelectrochemical cells", *J. Electrochem. Soc.*, 128:1396 (1981).

P. Singh, R. Singh, K. Rajeshwar and J. Dubow, "Photoelectrical behaviour on n-GaAs electrodes in ambient temperature molten salt electrolytes: device characterization and loss mechanisms", *J. Electrochem. Soc.*, 128:1145 (1981).

T. Skotheim, I. Lundström and J. Prejza, "Stabilization of n-Si photoanodes to surface corrosion in aqueous electrolyte with a thin film of polypyrrole", *J. Electrochem. Soc.*, 128:1625 (1981).

M. Skyllas Kazacos, J. F. McCann and D. Haneman, "The effect of the temperature on the power outputs of two metal dichalcogenide liquid junction cells", *Solar Energy Mater.*, 4:215 (1981).

M. Skyllas Kazacos and B. Miller, "Electrodeposition of CdSe films from selenosulfite solution", *J. Electrochem. Soc.*, 127:2378 (1980).

R. Tenne, "The effect of some surface treatments on the

characteristics of the Cd-chalcogenide/polysulfide Schottky
barrier", Ber. Bunsenges. Phys. Chem., 85:43 (1981).

H. Tributsch, "Photoelectrochemical behaviour of layer-type
transition metal dichalcogenides", Faraday Discuss. Chem.
Soc., N70:255 (1980).

H. Tributsch, "Photo-intercalation: possible application in solar
energy devices", Appl. Phys., 23:61 (1980).

H. S. White, F. -R. F. Fan and A. J. Bard, "Semiconductor electrodes
XXXIII. Photoelectrochemistry of n-type WSe$_2$ in acetonitrile",
J. Electrochem. Soc., 128:1045 (1981).

M. S. Wrighton, A. B. Bocarsly, J. M. Bolts, M. G. Bradly, A. B.
Fischer, N. S. Lewis, M. C. Palazzotto and E. G. Walton,
"Chemically derivatized semiconductor photoelectrodes. A
technique for the stabilization of n-type semiconductors", in:
"Interfacial photoprocesses: Energy Conversion and Synthesis",
M. S. Wrighton, ed., American Chemical Society, Washington
(1980) p. 269.

Chapter 9

P. G. P. Ang and A. F. Sammells, "Photoelectrochemical systems with
energy storage", Faraday Discuss. Chem. Soc., N70:207 (1980).

M. D. Archer, G. C. Morris and G. K. Yim, "Electrochemical approaches
to solar energy conversion: a brief overview and preliminary
results obtained with n-type cobalt ferrite", J. Electroanal.
Chem., 118:89 (1981).

V. M. Arutyunyan, A. G. Sarkisyan, Zh. R. Panosyan, V. M. Arakelyan,
A. O. Arakelyan and G. E. Shakhnazaryan, "Water
photoelectrolysis at photoelectrodes based on titanium oxide",
Elektrokhimiya, 17:1051 (1981).

M. Ya. Bakirov and D. T. Efendiev, "Photoelectrochemical cells for
hydrogen production by splitting water", Geliotekhnika, N4:43
(1980).

D. C. Bookbinder, J. A. Bruce, R. N. Dominey, N. S. Lewis and M. S.
Wrighton, "Synthesis and characterization of a photosensitive
interface for hydrogen production: chemically modified p-type
semiconducting silicon photocathodes", Proc. Natl. Acad. Sci.
USA, 77:6280 (1980).

G. S. Calabrese and M. S. Wrighton, "Photoelectrochemical reduction
of 2-t-butyl-9,10-anthraquinone at illuminated p-type Si: an
approach to the photochemical synthesis of hydrogen peroxide",
J. Electrochem. Soc., 128:1014 (1981).

M. P. Dare-Edwards, A. Hamnett and J. B. Goodenough, "The efficiency
of photogeneration of hydrogen at p-type III/V semiconductors",
J. Electroanal. Chem., 119:109 (1981).

W. W. Dunn, Y. Aikawa and A. J. Bard, "Semiconductor electrodes.
XXXV. Slurry electrodes based on semiconductor powder
suspensions", J. Electrochem. Soc., 128:222 (1981).

W. W. Dunn, Y. Aikawa and A. J. Bard, "Characterization of
particulate titanium dioxide photocatalysts by

photoelectrophoretic and electrochemical measurements", J. Amer. Chem. Soc., 103:3456 (1981).

D. Duonghong, E. Borgarello and M. Grätzel, "Dynamics of light-induced water cleavage in colloidal systems", J. Amer. Chem. Soc., 103:4685 (1981).

T. V. Dzhavrishvili, A. L. Asatiani and Z. A. Rotenberg, "Photoelectrochemical MnO_2 deposition on polycrystalline TiO_2 electrode", Elektrokhimiya, 17:278 (1981).

J. Gautron, P. Lemasson and J. F. Marucco, "Correlation between the non-stoichiometry of titanium dioxide and its photoelectrochemical behaviour", Faraday Discuss. Chem. Soc., N70:81 (1980).

J. Gautron, J. F. Marucco and P. Lemasson, "Reduction and doping of semiconducting rutile (TiO_2)", Mater. Res. Bull., 16:575 (1981).

H. Gerischer, "Heterogeneous electrochemical systems for solar energy conversion", Pure and Appl. Chem., 52:2649 (1980).

J. B. Goodenough, A. Hamnett, M. P. Dare-Edwards, G. Campet and R. D. Wright, "Inorganic materials for photoelectrolysis", Surface Sci., 101:531 (1980).

D. H. Grantham, "Hydrogen gas generation utilizing a bromide electrolyte, an amorphous silicon semiconductor and radiant energy", USA Pat. No. 4236984 (2.12.80).

V. Guruswamy, P. Keillor, G. L. Campbell and J. O'M Bockris, "The photoelectrochemical response of the lanthanides of chromium, rhodium, vanadium and gold on a titanium base", Solar Energy Mater., 4:11 (1980).

H. Hada, K. Takaoka, M. Saikawa and Y. Yonezawa, "Energy conversion and storage in solid-state photogalvanic cells", Bull. Chem. Soc. Japan, 54:1640 (1981).

L. A. Harris and J. A. Hugo, "Thin platinum films on silicon", J. Electrochem. Soc., 128:1203 (1981).

A. Heller and R. G. Vadimsky, "Efficient solar to chemical conversion: 12% efficient photoassisted electrolysis in the p-type InP(Ru)/HCl-KCl/Pt(Rh) cell", Phys. Rev. Lett., 46:1153 (1981).

J. F. Houlihan, M. A. Petro and D. P. Madacsi, "Effects of processing variables on the optical-to-chemical conversion efficiencies of sintered TiO_2 photoanodes", Mater. Res. Bull., 16:31 (1981).

T. Inoue, C. Weber, A. Fujishima and K. Honda, "An investigation of the power characteristics in heterogeneous electrochemical photovoltaic cells for solar energy utilization", Bull. Chem. Soc. Japan, 53:334 (1980).

I. Izumi, F. -R. F. Fan and A. J. Bard, "Heterogeneous photocatalytic decomposition of benzoic acid and adipic acid on platinized TiO_2 powder. The Photo-Kolbe decarboxylative route to the breakdown of the benzene ring and to the production of butane", J. Phys. Chem., 85:218 (1981).

H. S. Jarrett, A. W. Sleight, H. H. Kung and J. L. Gillson,

"Photoelectrochemical properties of LuRhO$_3$", _Surface Sci._, 101:205 (1980).

H. S. Jarrett, A. W. Sleight, H. H. Kung and J. L. Gilson, "Photoelectrochemical and solid-state properties of LuRhO$_3$", _J. Appl. Phys._, 51:3916 (1980).

R. K. Kalia, M. F. Weber, L. Schumacher and M. J. Dignam, "Photoelectrolysis at the oxide-electrolyte interface as interpreted through the transition layer model", _Surface Sci._, 101:214 (1980).

K. Kalyanasundaram, E. Borgarello and M. Grätzel, "Visible light induced water cleavage in CdS dispersions loaded with Pt and RuO$_2$, hole scavenging by RuO$_2$", _Helvetica Chim. Acta_, 64:362 (1981).

M. L. Knotek, "Characterization of hydrogen species on metal-oxide surfaces by electron-stimulated desorption: TiO$_2$ and SrTiO$_3$", _Surface Sci._, 101:334 (1980).

T. Kobayashi, H. Yoneyama and H. Tamura, "Influence of the reactivity of reducing agents on anodic photocurrents at TiO$_2$ electrodes", _J. Electroanal. Chem._, 124:179 (1981).

K. D. Kochev and G. S. Popkirov, "Preparation and electrochemical sensitization of thin TiO$_2$ wafers", _C. R. Acad. Bulg. Sci._, 34:331 (1981).

M. Koizumi, H. Yoneyama and H. Tamura, "Electrochemical fixation of molecular nitrogen on p-type gallium phosphide photocathode", _Bull. Chem. Soc. Japan_, 54:1682 (1981).

A. Mackor and G. Blasse, "Visible-light induced photocurrents in SrTiO$_3$ - LaCrO$_3$ single-crystalline electrodes", _Chem. Phys. Lett._, 77:6 (1981).

Y. Matsumoto, J. Kurimoto, T. Shimizu and E. Sato, "Photoelectrochemical properties of polycrystalline TiO$_2$ doped with transition metals", _J. Electrochem. Soc._, 128:1040 (1981).

J. F. McCann and J. O'M. Bockris, "Photoelectrochemical properties of n-type In$_2$O$_3$", _J. Electrochem. Soc._, 127:1719 (1981).

H. McKinzie and E. A. Trickett, "Method of preparing modified titanium dioxide photoactive electrodes", US Pat. No. 4215155 (29.7.80).

R. G. Meyerand, "Hydrogen-bromine generation utilizing semiconducting platelets suspended in a vertically flowing electrolyte solution", US Pat. No. 4263110 (21.4.81).

G. W. Murphy, "Model systems in photoelectrochemical energy conversion", _Solar Energy_, 21:403 (1978).

V. S. Nguen, S. N. Subbarao, R. Kershaw, K. Dwight and A. Wold, "Preparation and photoelectronic properties of n-type Cd$_2$GeO$_4$", _Mat. Res. Bull._, 14:1535 (1979).

R. D. Rauh, J. M. Buzby, T. F. Reise and S. A. Alkaitis, "Design and evaluation of new oxide photoanodes for the photoelectrolysis of water with solar energy", _J. Phys. Chem._, 83:2221 (1979).

P. Salvador, "The behaviour of aluminium-doped n-TiO$_2$ electrodes in the photoassisted oxidation of water", Mat. Res. Bull., 15:1287 (1980).

D. E. Scaife, "Oxide Semiconductors in photoelectrochemical conversion of solar energy", Solar Energy, 25:41 (1980).

J. Schoonman, K. Vos and G. Blasse, "Donor densities in TiO_2 photoelectrodes", J. Electrochem. Soc., 128:1154 (1981).

E. Serwicka, R. N. Schindler and R. Schumacher, "An ESR study of TiO_2 treated with atomic hydrogen", Ber. Bunsenges. Phys Chem., 85:192 (1981).

M. Sharon and A. Sinha, "Reversible photovoltaic-electrochemical solar cell", in: "Hydrogen Energy Progress", Vol. 4, T. N. Veziroglu, K. Fueki and T. Ohta, eds., Pergamon Press, Oxford (1981) p. 2047.

A. A. Soliman and H.J.J. Seguin, "Reactively sputtered TiO_2 electrodes from metallic targets for water electrolysis using solar energy", Solar Energy Mater., 5:95 (1981).

Y. Takahashi, K. Tsuda, K. Sugiyama, H. Minoura, D. Makino and M. Tsuiki, "Chemical vapour deposition of TiO_2 film using an organometallic process and its photoelectrochemical behaviour", J. Chem. Soc. Faraday I, 77:1051 (1981).

F. T. Wagner, S. Ferrer and G. A. Somorjaj, "Photocatalytic hydrogen production from water over $SrTiO_3$ crystal surfaces, electron spectroscopy studies of adsorbed H_2O_2 and H_2O", Surface Sci., 101:462 (1980).

N. L. Weaver, R. Singh, K. Rajeshwar, P. Singh and J. Dubow, "An economic analysis of photoelectrochemical cells", Solar Cells, 3:221 (1981).

K. Yazawa, H. Kamagawa and H. Morisaki, "Semiconducting TiO_2 films for photoelectrolysis of water", in: "Hydrogen Energy Progress", Vol. 4, T. N. Veziroglu, K. Fueki and T. Ohta, eds., Pergamon Press, Oxford (1981) p. 2095.

Y. Yoneyama, T. Ohkubo and H. Tamura, "Photoelectrochemical properties of $CdSnO_3$ and $LaRhO_3$ electrodes in aqueous solutions" Bull. Chem. Soc. Japan, 54:404 (1981).

Chapter 10

M. A. Evseeva, T. S. Antropova and G. A. Kitaev, "Selected anodic etching of illuminated cadmium selenide films", Zh. Priklad. Khim., 54:1632 (1981).

J. P. Salerno, J. C. C. Fan and R. P. Gale, "Anodic dissolution technique for preparing large area GaAs samples for transmission electron microscopy", J. Electrochem. Soc., 128:1162 (1981).

A. Yamamoto, S. Tohno and C. Uemura, "Detection of structural defects on n-type InP crystals by electrochemical etching under illumination", J. Electrochem. Soc., 128:1095 (1981).

A. B. Ellis and B. R. Karas, "Luminescent properties of semiconductor photoelectrodes", in: "Interfacial Photoprocesses: Energy Conversion and Synthesis", M. S.

Wrighton, ed., American Chemical Society, Washington (1980) p. 185.

Chapter 11

A. B. Ellis, B. R. Karas and H. H. Streckert, "Luminescent tellurium-doped cadmium sulphide electrodes as probes of semiconductor excited-state deactivation processes in photoelectrochemical cells", Faraday Discuss. Chem. Soc., N70:165 (1980).

B. R. Karas, H. H. Streckert, R. Schreiner and A. B. Ellis, "Luminescent photoelectrochemical cells 5. Multiple emission from tellurium-doped cadmium sulfide photoelectrodes and implication regarding excited-state communication", J. Amer. Chem. Soc., 103:1648 (1981).

J. H. Richardson, S. P. Perone, L. L. Steinmetz and S. B. Deutscher, "Laser-induced photoelectrochemistry: time-resolved luminescence studies of CdS electrodes", Chem. Phys. Lett., 77:93 (1981).

H. H. Streckert, B. R. Karas, D. J. Morano and A. B. Ellis, "Luminescent photoelectrochemical cells 4. Electroluminescent properties of undoped and tellurium-doped cadmium sulfide electrodes", J. Phys. Chem., 84:3232 (1980).

V. A. Tyagai, A. V. Gorodyskii, R. R. Dogonadze and G. Ya. Kolbasov, "Light emission at Si electrode in IR-region in course of cathodic reduction of hydrogen ion and hydrogen peroxide", Elektrokhimiya., 17:314 (1981).

H. Yoneyama, N. Nishimura and H. Tamura, "Photodeposition of palladium and platinum onto titanium dioxide single crystals", J. Phys. Chem., 85:268 (1981).

Reviews

W. J. Albery and A. W. Foulds, "Photogalvanic Cells", J. Photochem., 10:41 (1979).

A. J. Bard, "Photoelectrochemistry and heterogeneous photocatalysis at semiconductors", J. Photochem., 10:59 (1979).

H. Gerischer, "Light-induced charge separation at the solid/electrolyte interface", in: "Light-Induced Charge Separation in Biology and Chemistry", H. Gerischer and J. J. Katz, eds., Verlag Chemie (1979) p. 61.

H. Gerischer, "Photoassisted interfacial electron transfer", Surface Sci., 101:518 (1980).

H. Gerischer, "The influence of surface orientation and crystal imperfections of photoelectrochemical reactions of semiconductor electrodes", in: "Photoeffects at Semiconductor-electrolyte Interfaces", A. J. Nozik, ed., American Chemical Society (1981) p. 1.

A. Heller, "Conversion of sunlight into electrical power and photoassisted electrolysis of water in photoelectrochemical cells", Accounts Chem Res., 14:154 (1981).

K. Honda, A. Fujishima and T. Watanabe, "Electrode processes of photoexcited species: sensitization of photoelectrochemical processes", in: "Surface Electrochemistry", T. Takamura and A. Kozawa, eds., Japan Scientific Societies Press (1978) p. 141.

J. Manassen, G. Hodes and D. Cahen, "Photoelectrochemical cells", Chem. Technol., 11:112 (1981).

INDEX